Genetic Toxicology
and Cancer Risk
Assessment

Genetic Toxicology and Cancer Risk Assessment

edited by
Wai Nang Choy
*Schering-Plough Research Institute
Lafayette, New Jersey*

RC268.65
G46
2001

MARCEL DEKKER, INC.   NEW YORK · BASEL

ISBN: 0-8247-0294-8

This book is printed on acid-free paper.

Headquarters
Marcel Dekker, Inc.
270 Madison Avenue, New York, NY 10016
tel: 212-696-9000; fax: 212-685-4540

Eastern Hemisphere Distribution
Marcel Dekker AG
Hutgasse 4, Postfach 812, CH-4001 Basel, Switzerland
tel: 41-61-261-8482; fax: 41-61-261-8896

World Wide Web
http://www.dekker.com

The publisher offers discounts on this book when ordered in bulk quantities. For more information, write to Special Sales/Professional Marketing at the headquarters address above.

Copyright © 2001 by Marcel Dekker, Inc. All Rights Reserved.

Neither this book nor any part may be reproduced or transmitted in any form or by any means, electronic or mechanical, including photocopying, microfilming, and recording, or by any information storage and retrieval system, without permission in writing from the publisher.

Current printing (last digit):
10 9 8 7 6 5 4 3 2 1

PRINTED IN THE UNITED STATES OF AMERICA

In memory of my parents,
for their love and guidance

Preface

Genetic toxicology studies have been used to assess mutagenicity of environmental, pharmaceutical, and agricultural chemicals for over 30 years. Most of the activities in this field have been focused on the development of new test methods, standardization of test procedures, generation of mutagen databases, and cross-validation of test results. After many years of continual evaluations and refinements, contemporary routine genetic toxicology tests are reliable tools for the detection of environmental mutagens.

Since genetic damage to humans can cause serious conditions such as birth defects and cancers, identifying mutagens and carcinogens and minimizing their risks to humans are obvious goals for the protection of public health. When exposure is unavoidable, an accurate estimation of human risk as a result of exposure is essential for making regulatory decisions.

Quantitative cancer risk assessment is an intricate process that utilizes knowledge from many different scientific disciplines, including genetic toxicology, rodent cancer bioassays, epidemiology, metabolism, pharmacokinetics, and biostatistics. Because of this diversity, the application of each discipline at different stages of risk assessment is often not apparent to scientists in different areas. For example, the routine genetic toxicology tests may seem peculiar to biostatisticians, and, in contrast, the use of genotoxicity results beyond mutagen identification may be obscure to genetic toxicologists. An objective of this book is to provide a comprehensive overview of the role of genetic toxicology in the various phases of cancer risk assessment and to bridge the gap of communication among scientists participate in this multi-faceted process.

This book has five parts—the introductory chapters (Chapters 1–3) followed by four distinct subdivisions. Chapter 1–3 contains brief reviews of the rationale and practice of genetic toxicology, its performance in predicting carcinogens, and the complications of genotoxic and nongenotoxic mechanisms of carcinogenesis. This part provides the background information for the discussions of controversial issues in the subsequent chapters. The four parts are organized according to the four components of risk assessment: hazard identification (Chapters 4–10), dose–response relationship (Chapter 11), exposure assessment (Chapter 12), and risk characterization (Chapters 13–15).

In recent years, much effort and many advances have been made in the improvement of cancer risk assessment methodologies. The most notable achievements are the incorporation of the physiologically based pharmacokinetic models (PBPKs), and the development of the biologically based two-stage clonal expansion risk assessment model with parameters for cell proliferation kinetics. These new models will undoubtedly improve the reliability of quantitative risk assessment.

It is interesting to note that with recent studies on the mechanisms of carcinogenesis of nongenotoxic carcinogens, the distinction between genotoxic and nongenotoxic carcinogens has become ambiguous. Several nongenotoxic carcinogens, as defined by their negative results in the standard genetic toxicology tests, were found to set off a process that leads to DNA damage after their initial perturbations to the cells. This finding has raised the question as to whether these chemicals should be treated as genotoxic carcinogens for risk analysis. This appears to be an issue of timing of the first genetic alternations in multistage carcinogenesis. Since genetic alternations are observed in all cancer cells, nongenotoxic carcinogenesis may be considered as an additional step prior to the first genetic change that is induced by genotoxic carcinogens. This finding does not appear to have new implications for the current practice of risk assessment—that if the first insult is not genotoxic, the nongenotoxic risk assessment method should apply.

This issue is actually related to the threshold concept of chemical carcinogenesis. The conventional assumption is that nongenotoxic carcinogenesis has a threshold dose and genotoxic carcinogenesis does not. In reality, the dose–response curves of tumor induction are rarely, if ever, linear because each curve is a composite curve of many dose–response curves. They are dose-responses of metabolism, pharmacokinetics and damages at the cellular and molecular levels. It is conceivable that even if the dose–response of DNA damage is linear, the tumor incidence is not. Neoplastic transformation from DNA damage to tumor is controlled by another set of variables. All the processes have to be linear for the dose–response curve of tumor induction to be linear. Therefore, it is difficult to generalize the shape of the dose–response relationship at low doses for all carcinogens based simply genotoxicity of the carcinogen. A recent

Preface

dose–response analysis based on experimental data and consideration of the background tumor incidences showed a J-shape curve for rare tumors and a near-linear curve for tumors with high background incidence. This interpretation provides a new insight for the study of carcinogenic thresholds.

This book is a collection of a large amount of practical information on the current practice of regulatory genetic toxicology and quantitative cancer risk assessment. All the invited authors are eminent scientists in their respective area of expertise, or leaders of risk assessment programs in regulatory agencies. I am very grateful to each of them for their time and effort in sharing their knowledge and wisdom with us in this book. Their contributions are not only informative, but also inspiring. In addition, I wish to thank Ms. Sandra Beberman at Marcel Dekker, Inc., for her suggestion to write this book and her continued support of this project, and Ms. Annie Cok for her patience and excellent editorial skills in the production of this book.

Wai Nang Choy

Contents

Preface		*v*
Contributors		*xi*
1.	Human Cancer Genetics *Jason W. Barlow and David Malkin*	1
2.	Genetic Toxicity Tests for Predicting Carcinogenicity *Errol Zeiger*	29
3.	Genotoxic and Nongenotoxic Mechanisms of Carcinogenesis *Wai Nang Choy*	47
4.	Structure–Activity Relationship (SAR) and Its Role in Cancer Risk Assessment *David Y. Lai and Yin-tak Woo*	73
5.	Regulatory Genetic Toxicology Tests *Wai Nang Choy*	93
6.	Complementary Genetic Toxicology Assays *B. Bhaskar Gollapudi*	115

Contents

7. The Mouse Lymphoma Assay (MLA) Using the Microwell Method — 141
   Masamitsu Honma and Toshio Sofuni

8. The In Vitro Micronucleus Assay — 163
   Marilyn J. Aardema and Micheline Kirsch-Volders

9. ICH Guidances on Genotoxicity and Carcinogenicity: Scientific Background and Regulatory Practice — 187
   Lutz Müller, Peter Kasper and Leonard Schechtman

10. New OECD Genetic Toxicology Guidelines and Interpretation of Results — 223
    Michael C. Cimino

11. Dose–Response Relationships in Chemical Carcinogenesis and Cancer Risk Assessment — 249
    Werner K. Lutz

12. Molecular Epidemiology and Biomarkers — 271
    Jia-Sheng Wang, Jonathan M. Links, John D. Groopman

13. Quantitative Cancer Risk Assessment of Nongenotoxic Carcinogens — 299
    Anna M. Fan and Robert A. Howd

14. Risk Assessment of Genotoxic Carcinogens — 321
    Lauren Zeise

15. Biologically Based Cancer Risk Assessment Models — 355
    Suresh H. Moolgavkar

Index — *373*

Contributors

Marilyn J. Aardema Human and Environmental Safety Division, The Procter & Gamble Company, Cincinnati, Ohio

Jason W. Barlow Department of Pediatrics, Hospital for Sick Children, University of Toronto, Toronto, Ontario, Canada

Wai Nang Choy Safety Evaluation Center, Schering-Plough Research Institute, Lafayette, New Jersey

Michael C. Cimino Risk Assessment Division, Office of Pollution Prevention and Toxics, U.S. Environmental Protection Agency, Washington, D.C.

Anna M. Fan Office of Environmental Health Hazard Assessment, California Environmental Protection Agency, Oakland, California

B. Bhaskar Gollapudi Toxicology & Environmental Research and Consulting, The Dow Chemical Company, Midland, Michigan

John D. Groopman Department of Environmental Health Sciences, School of Hygiene and Public Health, Johns Hopkins University, Baltimore, Maryland

Masamitsu Honma Division of Genetics and Mutagenesis, National Institute of Health Sciences, Tokyo, Japan

Robert A. Howd Office of Environmental Health Hazard Assessment, California Environmental Protection Agency, Oakland, California

Peter Kasper Department of Experimental Pharmacology and Toxicology, Federal Institute for Drugs and Medical Devices, Bonn, Germany

Micheline Kirsch-Volders Laboratorium voor Cellulaire Genetica, Vrije Universiteit Brussel, Brussels, Belgium

David Y. Lai Office of Pollution Prevention and Toxics, U.S. Environmental Protection Agency, Washington, D.C.

Jonathan M. Links Department of Environmental Health Sciences, School of Hygiene and Public Health, Johns Hopkins University, Baltimore, Maryland

Werner K. Lutz Institute of Pharmacology and Toxicology, University of Würzburg, Würzburg, Germany

David Malkin Department of Pediatrics, Hospital for Sick Children, University of Toronto, Toronto, Ontario, Canada

Suresh H. Moolgavkar Department of Biostatistics, Fred Hutchinson Cancer Research Center, Seattle, Washington

Lutz Müller Department of Toxicology/Pathology, Novartis Pharma AG, Basel, Switzerland

Leonard Schechtman National Center for Toxicological Research (NCTR), NCTR Washington Operations, Rockville, Maryland

Toshio Sofuni Life Science Technology Research Center, Olympus Optical Co. Ltd., Tokyo, Japan

Jia-Sheng Wang Department of Environmental Health Sciences, School of Hygiene and Public Health, Johns Hopkins University, Baltimore, Maryland

Yin-tak Woo Office of Pollution Prevention and Toxics, U.S. Environmental Protection Agency, Washington, D.C.

Contributors

Errol Zeiger Environmental Toxicology Program, National Institute of Environmental Health Sciences, Research Triangle Park, North Carolina

Lauren Zeise Office of Environmental Health Hazard Assessment, California Environmental Protection Agency, Oakland, California

1

Human Cancer Genetics

Jason W. Barlow
University of Toronto, Toronto, Ontario, Canada

David Malkin
University of Toronto, Toronto, Ontario, Canada

I. INTRODUCTION

The study of molecular genetics is growing at a remarkable pace. In the past century, science has rediscovered the work of Gregor Mendel, determined the structure of DNA, and learned that DNA is the means of inheritance. This area of biology is tremendously exciting and has the potential to alter our view of the world, as physics did in the first half of the Twentieth century with quantum mechanics and general relativity.

The study of human genetics has numerous applications to human health. Some of the earliest diseases studied had resulted from inborn errors of metabolism caused by loss or mutation of a single gene. An example of such a disease is phenylketonuria (PKU), which is characterized by mutation of the gene that produces phenylalanine hydroxylase. It is hoped that one day, transfer of a normal copy of the gene back into a patient may restore production of functional protein, providing a viable therapeutic option in the treatment of these diseases. Thus far, this technique has proven exceedingly difficult; defenses mounted by the human immune system result in the destruction of the therapeutic vector, severely limiting the in vivo efficiency of gene transfer techniques.

Cancer defines a more complex group of diseases characterized by unregulated proliferation of cells, which have the potential to invade adjacent tissues and metastasize to distant organs. Alterations of genes and genetic pathways that lead to cancer result from complex interactions that involve many environmental factors, including food, ultraviolet radiation (sun exposure), and numerous chemicals, which may interact with the host genome to induce carcinogenic changes in affected cells.

Deoxyribonucleic acid may be altered or damaged in a number of ways, including simple point mutations, deletions, insertions, translocations, inversions, and amplifications (Fig. 1). Often, several alterations must occur in an orderly and sequential manner for the initiation and progression of cancer, as described for colon cancer (1). In this model, a well-defined pattern of frequently observed genetic alterations is observed. Mutation of the adenomatous polyposis coli (*APC*) gene on chromosome 5q results in increased cellular proliferation of colonic epithelial cells. Hypomethylation of DNA, activation of the Kirsten ras (*K-ras*) proto-oncogene, and loss of the deleted-in-colon carcinoma (*DCC*) gene

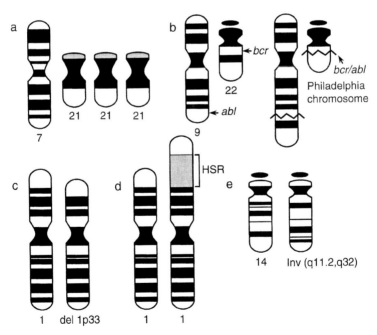

FIG. 1. a, Aneuploidy: monosomy 7 and trisomy 21. **b**, Interchromosomal translocation t(9:22)(qq). **c**, Interstitial deletion (1p33). **d**, Homogeneous staining region (HSR). **e**, Inversion 14(q11.2, q32).

on chromosome 18q follow this. Loss of the *p53* tumor suppressor gene and other less consistent chromosomal alterations are involved in the final steps to malignant transformation and metastasis (2). This multihit model of carcinogenesis invokes much of what we have learned about cancer and is applicable to many types of cancer (3). Thus, cancer results from a stepwise progression of many genetic events, each contributing to genetic instability, uncontrolled growth, or inappropriate survival, the end result being tissue invasion and metastasis.

Fortunately, the human genome is not static in the face of these events. All cells possess the capacity to repair DNA breaks, point mutations, and nucleotide repeats. However, the genes responsible for these processes may become the targets of mutations themselves, and their inactivation or loss also promotes genetic instability. Damage to the genome can occur at three different times in the life span of any given organism: (a) in the germline, in which case every cell produced postmeiosis will carry the same mutation; (b) in a somatic cell so that only the progeny of the affected parent cell will carry the alteration; and (c) at an intermediate point, in which a change in a pluripotent somatic stem cell early in its differentiation can affect derivative tissues but not the whole organism. An example of such alterations occur in leukemia, in which thousands of affected cells are produced from the stem cell, each carrying the same genetic defects.

II. CHROMOSOMAL ABERRATIONS

The earliest days of cancer genetics began with the analysis of metaphase chromosomes by light microscopy for gross chromosomal abnormalities. With the exception of point mutations, most of the alterations listed above could be detected this way. *Aneuploidy* is the loss or gain of chromosome number. In its simplest form, this involves loss or gain of a single chromosome (e.g., trisomy 21 or monosomy 7) (Fig. 1a). More complex chromosome anomalies, such as amplifications, deletions, and translocations have been characterized in greater detail by cytogenetic analysis of the banding patterns of metaphase chromosomes. For example, in a translocation, a new chromosome is formed by the fusion of fragments of two different chromosomes, which may be clearly demonstrated by comparing the banding pattern of the new chromosome with that observed in the contributing normal chromosomes (Fig. 1b).

Translocations have been identified in the majority of hematopoietic malignancies and in many solid tumors. Many of these translocations are virtually diagnostic of a specific malignancy. Translocations juxtapose genes or portions of genes normally remote from each other and result in the production of novel fusion proteins or the deregulation of normal gene expression, which may provide a growth advantage to the cell. The first disease-associated translocation described, the *Philadelphia chromosome*, was associated with malignancy in patients with chronic myelogenous leukemia. It involves fusion of the carboxy terminus of the c-*abl* gene on chromosome 9 with the amino terminus of the *BCR* gene on chromosome 22 (Fig. 1b) (4, 5, 6). Alveolar rhabdomyosarcoma (A-RMS) is characterized at the molecular level by the presence of a complete reciprocal translocation t(2:13)(q35;q14) involving *Pax3*, a homeobox gene involved in early neurogenesis and myogenesis in mice (7), and the *fork head* in rhabdomyosarcoma (*FKHR*) gene, which is a transcription factor (8). The resulting protein consists of an intact Pax3 DNA-binding domain, part of the fork head DNA-binding domain, and the C-terminal FKHR transcription activation domain (8). This tumorigenic hybrid transcription factor activates genes at an inappropriate time and in an unregulated manner. The result is an important early step in the genesis of A-RMS.

Embryonal rhabdomyosarcoma (E-RMS) is defined by deletion of the chromosomal region 11p15.5. The smallest region of overlap among several deletions is 3–5 megabases (9). Although this region encompasses numerous genes involved in regulation of cell growth, including *WT2, GOK, p57^{kip2}*, and the IGFII gene loci (10), the gene(s) within this region involved in the induction of E-RMS have not been identified.

On a finer scale, chromosomal alterations may be detected by fluorescence in situ hybridization (FISH), which incorporates biotin or digoxigenin into DNA probes specific to a gene, a chromosomal region, or a whole chromosome. These probes are hybridized to denatured metaphase chromosomes and then detected using a fluorochrome, such as fluorescein isothiocyanate (FITC) bound to avidin or antibodies against digoxigenin. However, FISH requires prior knowledge of where to look in the genome and access to an appropriate specific probe.

If FISH cannot be used, a different approach, such as comparative genomic hybridization (CGH) (11), may be applicable (Fig. 2). Recent advances in cytogenetic analysis have enhanced our ability to identify gross chromosomal alterations in tumor tissue, particularly gain or loss of genetic material. The CGH technique represents an advance in this technology and allows the determination of DNA copy number along each chromosome relative to another genome. The technique highlights gains (amplification, duplication) or losses (deletion) of genomic regions through the use of two different fluorochromes. The DNA of interest is labeled with one fluorochrome (fluorescein), and normal reference DNA is labeled with another (rhodamine). These pools of DNA are simultaneously hybridized to normal metaphase chromosomes. By measuring the ratio of the two fluorochromes, gains, indicated by an increased ratio, or losses, indicated by a decreased ratio of genetic material, can be detected. A ratio of one indicates two copies of a chromosomal region. Comparative genomic hybridization can detect the gain or loss of one copy of a chromosome and a gain of up to four copies. It has an amplification or deletion detection limit of 5 megabases, meaning that amplifications or deletions smaller than this will not be detected. Additionally, this form of analysis does not detect reciprocal translocations and inversions.

Human Cancer Genetics

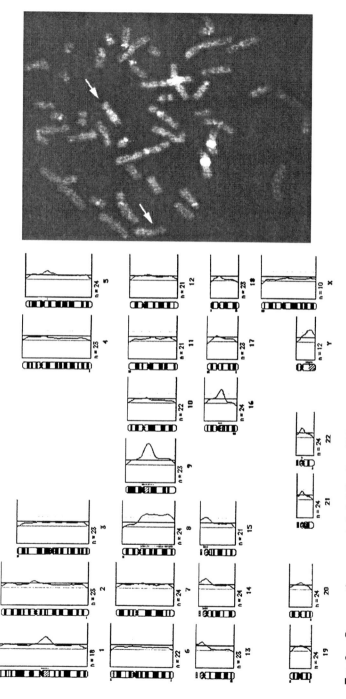

FIG. 2. Comparative genomic hybridization (CGH). Shown is a normal metaphase spread competitively hybridized through use of two differently labeled whole genomic probes in equal amounts. The fluorescence intensity ratio between the two fluorochromes reveals gains and losses along each chromosome. **Right panel:** CGH metaphase spread after CGH analysis using prostate cancer DNA. Arrows show gain of long arm of chromosome 8 determined by stronger FITC signal derived from the trisomy 8 within the tumor DNA. **Left panel:** Composite CGH profile composed of several karyotyped metaphase spreads of prostate cancer sample. Lines to the left of the median indicate chromosomal loss; lines to the right of the median indicate chromosomal gain. Note that the confidence intervals have been removed for simplicity. (Courtesy of Jeremy Squire, Ontario Cancer Institute.)

Techniques such as spectral karyotyping (SKY) (12) and multifluor in situ hybridization (M-FISH) (13) both use the differential display of colored chromosome specific paints. Spectral Karyotyping uses 23 different probes, referred to as paints, that individually recognize all human chromosomes in a metaphase plate (Fig. 3). A computer detects differences in the fluorochrome labeling, and a unique color is assigned to each chromosome. These produce detailed chromosomal maps, which allow the detection of highly complex rearrangements previously undetected by karyotyping.

A. DNA Repair

Cells are constantly subjected to DNA damage caused by exposure to oxidative free radicals, gamma radiation, ultraviolet (UV) light, or endogenous and exogenous chemical agents. These agents induce single- and double-strand DNA breaks, nucleotide base alterations, and inter- and intrastrand cross-links. Additionally, the process of DNA replication itself is not error free and can occasionally result in incorrect base incorporation or addition or deletion of base sequences, particularly in the presence of repeated nucleotide sequences such as GCGCGCGCGCGCGCGC (abbreviated as $(GC)_8$). These repeated dinucleotides are referred to as microsatellites. In the context of these repeats, slippage of DNA polymerase is possible, which allows mistakes to be incorporated into the new strand. Hereditary nonpolyposis colon cancer (HNPCC) is a disease characterized by deficiencies in repair of mismatches brought about by this type of replication error. Such deficiencies cause microsatellite instability, that is, an alteration in the number of repeats. For example, in colon cancers, a repeat of $(G)_8$ in the *Bax* gene is frequently altered to $(G)_7$ or $(G)_9$ (14). This results in a frameshift mutation and loss of Bax function. To improve the fidelity of DNA replication and respond to DNA damage, several repair systems have evolved and are conserved through evolution. These repair processes often involve the interaction of several proteins; these proteins themselves may be targets of inactivation in some disease processes, including cancer. Loss or alteration of a single protein may be sufficient to cripple a particular repair process. Further, because these repair pathways depend on several genes, and because of genetic variation in these genes, repair pathways can confer a predisposition to cancer.

Hypersensitivity to DNA damage characterizes several diseases, including xeroderma pigmentosa (XP), ataxia telangiectasia (AT), Bloom's syndrome (BS), Fanconi's anemia (FA), Cockayne's syndrome (CS), and HNPCC.

Nucleotide excision repair (NER) and mismatch repair are two systems that maintain the fidelity of the genome. In particular, NER is responsible for repairing the damage caused by exogenous mutagens. Many of the genes involved in this process were identified by the study of xeroderma pigmentosa (15–17), perhaps the best characterized of the diseases listed above. Xeroderma pigmentosa is

Karyotype by Spectral Analysis

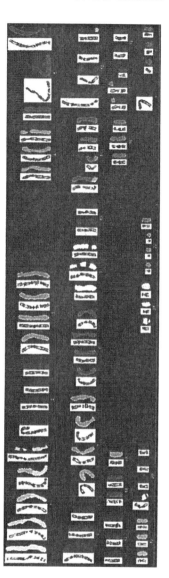
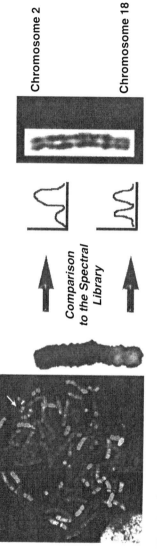

FIG. 3. Spectral karyotyping (SKY). A mixture of DNA probes (labeled with 5 fluorochromes in varied combinations) hybridizes specifically to the metaphase nucleus and identifies each chromosome by its unique color combination. **Bottom left:** Image of hybridized metaphase chromosomes used to assess quality before SKY analysis. **Top:** Karyotype of metaphase spread as a result of spectral analysis, showing each chromosome pairing two images inverted DAPI (**left**) and classified pseudocoloring (**right**). **Bottom right:** A rearranged chromosome identified as an example of SKY analysis. Specifically, a translocation t(2;18)(q10;q10), as determined by comparing the emission spectrum with the spectral library along each chromosome. (Courtesy of Jeremy Squire, Ontario Cancer Institute.)

an autosomal recessive disease characterized by extreme UV sensitivity and predisposition to skin tumors, both of which result from deficiencies in NER. People with XP have an incidence of skin cancer 1000 times higher than the general population. Cell lines taken from these patients can be fused in vitro and then tested for sensitivity to UV radiation. The fused cell lines fall into distinct complementation groups, that is, if two cells with the same defect are fused, they will remain UV sensitive; these cells belong to the same complementation group. If two cell lines from different complementation groups are fused, then one may complement the defect of the other and the fused cell will no longer be UV sensitive. Complementation analysis of XP cell lines demonstrated a total of seven distinct groups (18–20), each of which may represent a defect in a different gene involved in the nucleotide excision repair process. It is not always the case, however, that each complementation group represents a different gene. Four complementation groups were identified in AT and all result from mutations in different regions of the large ATM protein (21).

III. THE CELL CYCLE

In its simplest form, the cell cycle is composed of four phases. Two of these are the major events required for cell division, namely, the synthesis of DNA (S-phase) and mitosis (M-phase). These two events are separated by two gap phases, G_1 and G_2, originally considered relatively uninteresting periods in which the cell grows, metabolizes proteins, and prepares for the S-and M-phases. Ongoing studies of the cell cycle have shown that it is not this simple and several important features have been added (Fig. 4).

In particular, a restriction point (R point) just prior to the G_1/S transition was identified, first in yeast and subsequently in other eukaryotic cells (22). If growth conditions are appropriate, a cell will pass this point and go on to complete the cycle, regardless of subsequent environmental changes (e.g., removal of mitogenic growth factors) (22). Before the restriction point, however, changes in environmental conditions, such as serum deprivation or high cell density, will cause the cell to enter a phase known as *quiescence*, designated G_0, in which DNA synthesis does not occur. Further study expanded this concept to include checkpoints, defined as biochemical pathways involving dependence of one process upon another that are otherwise biochemically unrelated (23). For example, before DNA replication can occur, the signaling processes must indicate that DNA is intact and that the surrounding environment will promote growth. The two major checkpoints occur at the transition points of G_1/S and G_2/M (Fig. 4). The G_1/S checkpoint occurs before DNA synthesis and depends on mechanisms that monitor DNA integrity. Detectable DNA damage will arrest the cell cycle at this checkpoint. The G_2/M checkpoint occurs after DNA duplication and monitors the complete and accurate duplication of the genome, as deter-

Human Cancer Genetics

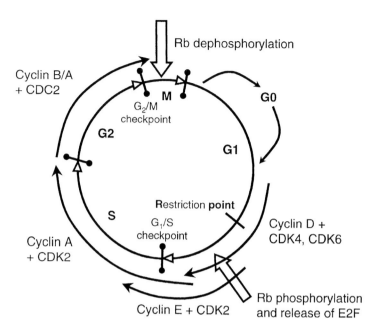

FIG. 4. Cell cycle showing the four phases of the cell cycle (G1, S, G2, M, and G0). Several key transition points are shown the G_1/S and G_2/M checkpoints and the restriction point. The restriction point is a point of no return. Once a cell moves past this point, it will complete the cell cycle regardless of changes in the environment. Also indicated are the points where Rb phosphorylation changes. This is a key determinant of the cell cycle. Activation of cyclin D/CDK4 and cyclin E/CDK2 is key to phosphorylation of Rb, which causes release of E2F—a critical step in the cell cycle.

mined in yeast (24). Less is known about the homologous process involved in the mammalian G_2/M checkpoint (25). The *ATM* gene shares homology with some of the yeast genes involved in this checkpoint, and cells from patients with AT exhibit limited arrest at G_2 in response to DNA damage.

A. Cell Cycle Inhibitors

Inhibitors of cyclin-dependent kinases (CDK) usually mediate the process of cell cycle arrest. There are two major groups of cyclin-dependent kinase inhibitors, the first of which is the $p21^{Cip1}$, $p27^{Kip1}$, and $p57^{Kip2}$ family of structurally related proteins. These are universal inhibitors of cyclin/CDK complexes, capable of inducing cell cycle arrest when overexpressed (26–29). $p21^{Cip1}$ is a component of active cyclin-CDK-PCNA (proliferating cell nuclear antigen) quaternary complexes required for cell cycle progression (30), but increased levels of $p21^{Cip1}$ induce cell cycle arrest (31). Thus, complexes containing one $p21^{Cip1}$ molecule are believed to be active, and those with more than one are thought to be inactive. New evidence suggests, however, that this may not necessarily be the case, in that one molecule of $p21^{Cip1}$ may be sufficient to inhibit the activity of these complexes (32). $p21^{Cip1}$, $p27^{KIP1}$, and $p57^{KIP2}$ inhibit cyclin/CDK complexes by physically binding to CDKs 1 through 6 (28, 29, 33–35). Additionally, these inhibitors preferentially bind cyclin/CDK complexes rather than individual cyclin or CDK proteins (36). $p21^{Cip1}$ is up-regulated by p53 in response to DNA damage, inducing cell cycle arrest. Similarly, induction of $p21^{Cip1}$ by MyoD is required to induce cell cycle arrest, which allows differentiation of muscle cells (37).

The second family of CDK inhibitors, the INK4 (*In*hibitors of CD*K4*) family, consists of $p16^{INK4a}$, $p15^{INK4b}$, and $p18^{INK4c}$. These proteins act more specifically than the universal inhibitors, binding to and inhibiting the function of CDK4 and CDK6. $p16^{INK4a}$ and $p15^{INK4b}$ bind CDK4 or CDK6, dislodging cyclin D1 and inactivating the kinase activity of these complexes, thus preventing phosphorylation of Rb (38, 39). The net result is that Rb remains bound to the E2F transcription factor, and cell cycle arrest ensues (40). Mutant *p16*^{INK4a} alleles produce proteins that have lost their ability to arrest the cell cycle (41). $p19^{ARF}$, a fourth member of this gene family and another product of the *p16* gene has a unique first exon encoded by an *a*lternate *r*eading *f*rame (ARF). $p19^{ARF}$ induces cell cycle arrest through the p53 pathway rather than the Rb pathway. More specifically, $p19^{ARF}$ binds and targets MDM2 for degradation, thus stabilizing p53, which can go on to induce cell cycle arrest through induction of $p21^{Cip1}$ (42, 43).

To understand how these inhibitors act, it is necessary to know more about the details of the cell cycle. Key players in cell cycling are enzyme complexes (holoenzymes), which consist of two major components, a CDK subunit and a cyclin, an activating subunit. CDKs are expressed throughout the cell cycle; cyclins are produced at specific periods during the cell cycle and are rate limiting. The activities of these complexes are regulated by their phosphorylation status and by the expression of CDK inhibitors. A CDK is converted to an inactive form by phosphorylation at three amino acid residues on the CDK, specifically threonine 14, tyrosine 15, and threonine 161. Dephosphorylation of Thr14 and Tyr15 converts the inactive cyclin/CDK complex to its active form, which allows phosphorylation of Rb and other cell cycle proteins. Also, the phosphorylated form of CDK is targeted for ubiquitin-mediated degradation.

Kinases phosphorylate other proteins and alter the activity of the target protein in a positive or negative manner depending on the protein and the site of phosphorylation. A case in point is the retinoblastoma protein, which can be phosphorylated at different points in the cell cycle by cyclin/CDK complexes. Specifically, cyclin D/CDK4 or cyclin D/CDK6 phosphorylates Rb in G_1. Cyclin

E, which reaches peak levels at the G_1/S transition, will, if conditions are right, phosphorylate Rb a second time (44, 45), releasing the transcription factor E2F which binds unphosphorylated Rb (Fig. 4). Once released, E2F is able to bind DNA and promote transcription of genes needed for continuation of the S-phase of the cell cycle.

IV. ONCOGENES

Proto-oncogenes are normal components of the genome necessary for cell function; however, when altered, they become *oncogenes* and promote cancer. An oncogene is defined as a gene whose protein product, under the appropriate conditions, can be involved in processes leading to the transformation of a normal cell to a malignant cell. Proto-oncogenes may become activated oncogenes by one of several mechanisms. The first is through fusion of the oncogene with another gene by translocation or inversion. These events are frequently observed in leukemias, as evidenced by the Philadelphia chromosome, which is a translocation product involving the *abl* oncogene with the *bcr* gene (46) (Fig. 1b). A second mechanism of oncogene activation involves the juxtaposition of the oncogene with immunoglobulin regulatory elements or T-cell receptor genes in B- and T-cell lymphomas, again through chromosome breakage and reunion. Such rearrangement results in aberrant oncogene expression in immune cells that up-regulate lineage-specific genes necessary for their function. A third means of oncogene deregulation involves gene amplification, detected as homogeneously staining regions or double minutes (47) (Fig. 1d). For example, *Myc* is amplified in the human leukemia cell line HL-60 (48). A final means of oncogene activation involves point mutation. One such mutation occurs in *Ras*, a member of a family of genes encoding proteins involved in signal transduction. When activated, these membrane-bound G-proteins transmit signals by phosphorylating other molecules. The most common *Ras* mutation results in a change from glycine to arginine at amino acid 12, which constitutively activates the protein (49) and sends inappropriate growth signals to the nucleus. Table 1 shows a summary of some oncogenes and their associated cancers.

Oncogenes were first described in studies of cancer-causing viruses, which include the DNA and RNA tumor viruses. The RNA tumor viruses, more specifically the retroviruses, have been known to play a role in the genesis of cancer since Harvey Rous demonstrated in 1911 that a transmissible agent could produce sarcomas in chickens. It was later shown that these viruses contained nucleotide sequences not present in nontransforming retroviruses (50). Further, these same sequences were found in the DNA of normal chickens. Thus, the virus had acquired a gene derived from a normal cellular gene. In this case the viral oncogene was designated v-*src* and the cellular proto-oncogene was c-*src*. Several

TABLE 1. A Summary of Oncogenes, Their Chromosome Location, the Normal Function of the Encoded Protein, and Cancers Frequently Associated with Alterations of These Genes

Gene	Location	Function	Associated cancers
H-ras	11p15	GTPase	Many carcinomas and leukemias
N-ras	1p13.2	GTPase	Leukemia and lymphoma
K-ras	12p12.1	GTPase	Colorectal and pancreatic carcinoma, lung cancer
Mdm-2	12q13-q14	p53 degradation	Sarcomas
c-abl	9q34.1	Tyrosine kinase	Chronic myelogenous leukemia, acute lymphocytic leukemia
c-sis	22q12.3-13.1	Platelet-derived growth factor	Gliomas
c-Myc	8q24.12-24.13	Transcription factor	Lymphomas
L-Myc	1p34.3	Transcription factor	Small cell lung cancer
N-Myc	2p24.1	Transcription factor	Neuroblastoma

other retroviruses transform cells with a gene derived from a normal cell by a similar mechanism (51).

It is important to keep in mind that oncogenes are not only associated with viruses but are frequently involved in cancer independent of viruses. Viruses have helped us learn about the causes of cancer; however, it may be frustrating to realize that the root cause of cancer is thought not to occur through an infectious agent but through an alteration of our own genes. These genes have, through one of several mechanisms, been corrupted to accelerate the growth of the transformed cells.

c-Myc (cellular Myc) is a proto-oncogene first described in 1982 as the homologue of v-Myc (viral-Myc) found in the avian myelocytomatosis retrovirus (52). The c-Myc gene encodes Myc1 and Myc2, two highly conserved nuclear phosphoproteins, from two distinct alternative promoters. Myc2, the more common form, is 14 amino acids shorter. Increased expression of c-Myc has been described in numerous cancers, including breast, colon, cervical carcinomas, Burkitt's lymphoma, small-cell lung carcinoma, osteosarcoma, glioblastoma, and myeloid leukemia. In Burkitt's lymphoma, the activating alteration is a translocation between Myc on chromosome 8 and an immunoglobulin gene on chromosome 2, 14, or 22. The role of c-Myc in the genesis of these cancers is not fully understood but is likely related to enhanced cell cycle progression, inhibition of terminal differentiation, and induction of apoptosis. The oncogenic activity of *Myc* depends on association with Max (*Myc-associated* factor X), with which it forms heterodimers involved in transcription. This suggests a mechanism for *Myc*-induced transformation through its effects on transcription of target genes (53). Additionally, *Myc* has a role in gene repression through the binding of Miz-1, a protein that specifically interacts with the HLH domain of *Myc*. In the absence of *Myc*, Miz-1 binds the cyclin D1 promoter and activates transcription, but in the presence of *Myc*, Miz-1 will preferentially bind *Myc*, thereby repressing transcription (54)

N-*Myc* and L-*Myc* are homologous *Myc* family members encoding the proteins N-Myc and L-Myc, which share a common exon structure and some conserved domains with *Myc*. These include *Myc* box 1 (MbI), *Myc* box 2 (MbII), and the basic helix-loop-helix leucine zipper region (BR-HLH-Zip). *Myc* box 2 has been shown to be essential for all known *Myc* functions (55) and both of the *Myc* boxes appear to bind cellular proteins (56, 57).

Myc deregulation, resulting in inappropriate, nonphysiological increase in *Myc* activity (57), can occur through a variety of genetic alterations, including translocation, proviral insertion, gene amplification, and mutation (48, 58–60). Normal c-*Myc* promotes cell cycle progression. Consequently, its expression is tightly regulated and the protein has a half-life of less than 30 minutes (61). Additionally, expression of the protein correlates with the proliferative potential of the cell. Thus, quiescent cells have essentially undetectable expression. Mitogen or serum stimulation results in increased c-*Myc* RNA and protein expression as the cell enters G_1. c-*Myc* is also regulated through phosphorylation; however, with more than a dozen available phosphorylation sites, the ones of biological relevance remain unclear. Threonine 58 and serine 62 located within MbII appear to be involved in *Myc* transactivation; however, their precise roles must be determined. It has been suggested that the potent transforming potential of c-*Myc* may be related to its role in blocking terminal differentiation and preventing cells from exiting the cell cycle, thereby allowing the expansion of a population of cells that may acquire additional transforming mutations (62).

V. SPORADIC VERSUS HEREDITARY GENETIC CHANGES

Cancer is a disease caused by defective or damaged genetic material. Because our genetic material is entirely inherited, it may seem paradoxical that the majority of human cancers are sporadic and affect people "at random." In some kindreds, however, numerous family members develop cancers that may be considered familial, the affected individuals being genetically predisposed to develop malignancies. These familial cancer aggregations are relatively uncommon, but they have been extremely helpful in the identification of specific genes involved in some cancers.

A. Tumor Suppressor Genes

1. Retinoblastoma, the Definitive Tumor Suppressor Gene

The gene associated with susceptibility to retinoblastoma (RB), a rare human cancer, was cloned in 1986 (63). Further investigation determined that the protein encoded by this gene was a DNA-binding nuclear phosphoprotein (64) and that inactivation of the gene required loss or alteration of both alleles. Genes of this type are known as tumor suppressor genes (TSG), which are functionally recessive genes inherited in an autosomal-dominant pattern that act to suppress or regulate cell growth. Inactivation of these genes allows cells to proliferate unchecked, an important step in the progression of a cancer. Additional evidence for the existence of tumor suppressor genes comes from the study of the fusion of normal cells with tumor cells. The hybrid cells were nontumorigenic—a reversion attributed to the reconstitution of lost tumor suppressors (65).

Many cancers are believed to result from the accumulation of as many as 3–7 distinct gene alterations (2, 66), the development of retinoblastoma requires only two mutational events, both to the same gene. It provides one of the most cited examples of the molecular difference between sporadic and hereditary cancers. Forty percent of children with RB have a hereditary form of the disease manifested by earlier age of onset and a higher frequency of bilateral tumors than sporadic RB. This proved to be because these individuals inherit one mutated *Rb* susceptibility gene. The cancer then occurs in a somatic cell when the second allele of the gene is mutated. By contrast, in the sporadic form of the disease, both alleles in the same somatic cell must receive mutational "hits" to generate the tumor (Fig. 5). The rate at which individual hits occur is 2×10^{-7} per year (67). The probability of this event occurring twice in the same gene in the same cell to produce the tumor is much less than the likelihood of a second hit in a cell already carrying the first mutation, as is the case in the hereditary form of the disease. The process by which these mutational events occur forms the basis of Knudson's "two hit theory" (67).

Subsequently, it was found that the *Rb* gene was also disrupted in tumors other than retinoblastoma, including breast cancer and osteosarcoma (68). The *Rb* gene was found to play a vital role in the regulation of cell cycle checkpoints. Inactivation of this function is critical to the development of the majority of human cancers, as described in detail in this book.

Other studies have identified putative tumor suppressor genes (TSGs) that are mutated or deleted in childhood cancers, such as neuroblastoma, and Wilms' tumor. The gene associated with some forms of Wilms' tumors and the associated WAGR (Wilms' tumor, aniridia, genitourinary abnormalities, and mental retardation) syndrome was identified in 1990 (69). This gene, known as Wilms' tumor 1 (*WT1*), is localized to chromosome 11p13, whereas a second Wilms' tumor gene

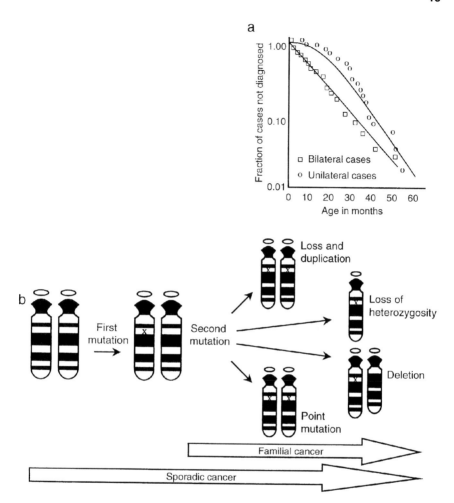

FIG. 5. Knudson's "two hit theory" of tumor suppressor genes (TSG). Both alleles of retinoblastoma must be mutated to inhibit their ability to suppress cell growth. **a**, Demonstrates that when tumors occur bilaterally (familial cancer), the time required for tumor formation is represented by a straight line, indicating that only one hit is required as these people inherit one hit. For unilateral cases, the line is curved, indicating two independent events (hits) must occur. (Courtesy of A. Knudson.) **b**, Inactivation of a TSG requires two steps. In sporadic tumors, these are independent events that occur in the same cell at separate times. The first mutation is typically a somatic point mutation or small deletion. The second mutation can occur by one of several mechanisms, as indicated. For familial cancers, the first hit is an inherited somatic mutation and the second occurs by one of several mechanisms indicated.

(*WT2*) has been identified, localized to chromosome 11p15.5 and is deleted in some tumors (70, 71). Table 2 summarizes some TSGs, indicating their chromosome location, associated cancers, and the familial cancer syndromes that assisted in their identification.

B. DNA Tumor Viruses

Just as RNA tumor viruses helped to elucidate the role of oncogenes, DNA tumor viruses have served to highlight the role of TSGs. Exposure of cells to a DNA virus may result in a productive infection and a subsequent lytic cycle, or it may transform the cell randomly, integrating its own genome into host chromosomes. When cells are transformed, they lose contact inhibition, require less serum, and grow to a high cell density, producing multilayered colonies. When grown in semisolid agar, they exhibit anchorage-independent growth. Further, these transformed cells can produce tumors upon injection into susceptible animals. The mechanism by which several diverse types of DNA viruses exert these effects is consistent.

Simian virus 40 (SV40) is a monkey polyomavirus. There is some evidence that SV40 DNA occurs in human tumors (72), but it is generally believed that this virus does not pose a significant risk to humans. Molecular analysis of the DNA sequence of SV40 virus revealed coding sequences for two proteins, the large T and small t antigens. The large T antigen has been shown to bind the protein product of the *p53* TSG inactivating its function (73, 74). T antigen also binds members of the Rb family including pRb, p107, and p130. In this manner, it disrupts the ability of pRb to bind and sequester the E2F transcription factor. Consequently, E2F is free to promote cell cycle progression. Large T antigen also interacts with p300, a known cofactor of transcription factors such as p53 and pRb, acting as a bridge to connect DNA-binding transcription factors with the basal transcription machinery. However, it is not known if this interaction between T antigen and p300 is independent of the interaction between p53 and T antigen.

Human adenoviruses are responsible for upper respiratory tract infections. Many rodent cells can be transformed with adenoviruses. The transformed cells do not contain a complete adenovirus genome but do consistently include the adenovirus E1A and E1B genes. Conserved regions of E1A bind the pRb, p107, and p130 family of proteins, as well as p300. Simultaneously, the E1B protein binds the N-terminus of p53 and blocks its transactivation function. Because of these properties, adenoviruses have found great utility as vectors for numerous novel approaches to gene transfer therapy in the treatment of various cancers.

A third oncogenic DNA virus is the human papilloma virus (HPV) of which over 60 serological types are known (75). Of these, at least 29 are associated with genital infection-causing warts; viral DNA is found in more than 85%

TABLE 2. A Summary of Tumor Suppressor Genes, Their Chromosome Location, Associated Cancers, and Familial Cancer Syndrome

Gene	Location	Associated cancers	Familial cancer syndrome
APC	5q21-q22	Colorectal carcinoma, hepatoblastoma	Adenomatous polyposis coli, familial polyposis
ATM	11q22-23	Breast carcinoma, leukemia, lymphoma	Ataxia telangiectasia
BRCA1	17q21	Breast and ovarian carcinoma	Hereditary early onset breast and ovarian cancer
BRCA2	13q12.3	Breast cancer	Familial early onset breast cancer
$P16^{INK4A}$/CDKN2A	9p21	Breast, bladder, lung, ovarian, esophageal, pancreatic, renal, head and neck carcinoma, astrocytoma, glioma, osteosarcoma, leukemia, mesothelioma	Malignant melanoma
Rb1	13q14.2	Bladder, prostatic, cervical, lung, retinoblastoma, osteosarcoma	Retinoblastoma
p53	17p13.1	Breast, colorectal, lung, esophageal, stomach, ovarian, prostatic carcinoma, glioma and others	Li-Fraumeni syndrome
NF1	17q11.2	Glioma, sarcoma, leukemia	Neurofibromatosis type 1
VHL	3p25-26	Renal cell cancer	Von Hippel-Lindau disease
WT1	11p13	Wilms' tumor	WAGR syndrome (Wilms' tumor, aniridia, genitourinary abnormalities, mental retardation)
WT2	11p15.5	Wilms' tumor	N/A

TABLE 3. Common DNA Tumor Viruses, the Proteins They Express, and the Proteins Inactivated to Induce Cellular Transformation

Virus	Viral family	Proteins encoded	Proteins functionally inactivated
SV40	Polyomaviridae	Large T antigen	p53, p300 and pRb, p107, p130
Adenovirus	Adenoviridae	E1B and E1A	p53, p300 and pRb
Human papilloma virus	Papovaviridae	E6 and E7	p53 and pRb, p107, p130

of cervical cancers. A subset, including HPV 16, 18, 31, 33, 39, 51, 58, is more frequently associated with malignancy and considered high risk. These types consistently contain and express proteins from the E6 and E7 regions of the viral genome. The E6 protein binds the product of the *p53* tumor suppressor gene, transports it to the cytosol, and targets it for ubiquitin degradation (76). The E7 protein has a nuclear localization signal where it binds hypophosphorylated Rb, preventing Rb from sequestering the E2F transcription factor (77). Consequently, E2F enhances the transcription of genes promoting cell cycle progression. Significantly, there is a consistent difference in the coding sequence of cancer-associated forms of the E7 viral protein (resulting in an aspartic acid in place of a glycine) compared with benign forms. The cancer-associated E7 protein has a greater affinity for Rb (78). Previous studies have indicated that three key pathways must be altered to form a cancer: inactivation of the p53 pathway, inactivation of the Rb pathway, and activation of telomerase (79). Human papilloma virus fulfills two of these three requirements, making the virus a potent risk factor for development of cervical cancer. Table 3 summarizes the proteins produced by the DNA viruses mentioned and the corresponding cellular proteins that are functionally inactivated.

C. p53

p53 is a nuclear phosphoprotein and transcription factor originally identified as a protein associated with SV40 large T antigen (73, 74). The protein is composed of several functional domains including tetramerization, DNA binding, and transactivation domains. Numerous mutations of *p53* have been reported in a wide variety of tumors, and although some are nonsense mutations, most are missense, occurring predominantly in the DNA binding domain (80). Many of these mutations have been shown to alter the function of the protein, specifically its role in G_1/S and G_2/M checkpoint control (81). Typically, one allele of *p53* is mutated and, mutation or deletion, inactivates the second allele (80, 82), although in up to

40% of tumors, the wild-type allele is retained. In addition to point mutations and deletions, p53 protein inactivation can occur when other cellular proteins, such as WT1 or viral oncoproteins including SV40 T antigen, bind p53. Further, regulatory proteins such as mdm-2 target p53 for degradation, eliminating its function. Mutations at different residues of p53 have varying effects, such as loss of tumor suppressor ability, dominant-negative function, and dominant oncogenic functions (83). The role of p53 in cell cycle arrest and apoptosis (programmed cell death), particularly in response to DNA damage, is part of a complex process involving numerous proteins. p53 has been shown to induce cell cycle arrest (84), and it was later determined that this occurred through up-regulation of $p21^{Cip1}$ (26), as described previously.

People born with germline mutations of *p53* develop normally, yet they are very cancer-prone. Statistical modeling has suggested that 50% of carriers of germline *p53* mutations will develop invasive cancer by age 30, compared with 1% of the general population (85). The apparently normal development of people harboring germline *p53* mutations suggests that one or more additional steps are required for tumor development, which may account for the latency (10–30 years) of tumor formation. The Li-Fraumeni syndrome (LFS) is a familial, autosomal dominantly inherited disorder in which carriers develop a wide range of cancers, including breast cancer, sarcomas, brain tumors, adrenocortical carcinoma, and leukemias (86, 87). Approximately 70% of families with LFS harbor germline *p53* mutations (88–93). Because *p53* plays a role in cell cycle arrest and apoptosis, mutations in this gene provide an appealing explanation for the high rate of cancer development in these LFS families. An explanation for the cancers in the remaining 30% of LFS families is lacking. It is possible that current techniques have failed to detect some p53 mutations, such as deletions. Alternatively, there may be mutations in areas of the gene, such as the promoter region, that have not yet been examined, or these cancers may result from alterations of other genes in the p53 regulatory pathway. A final possibility is that the criteria used to define LFS may be incorrect and by redefining the syndrome, germline p53 mutations may be sufficient to fully explain the disease.

Three major groups of *p53*-related proteins may be functionally significant in the genesis of LFS: upstream regulators of p53, downstream effectors of p53, and proteins that functionally interact with p53. Upsteam regulators of p53 include kinases and phosphatases responsible for regulating the phosphorylation state of p53. One such protein is ATM, the product of the ataxia telangiectasia gene, which has been shown to phosphorylate serine 15 of p53 in response to ionizing radiation (94, 95). Phosphorylation at serine 15 increases the stability of the protein and consequently there is an accumulation of p53. Downstream effectors of p53 are the proteins transcriptionally regulated by p53 such as $p21^{Cip1}$ and Bax. These proteins function in G_1 arrest and apoptosis, respectively. Inactivation of these genes could mimic loss of p53. However, to date, mutations of these genes have not been observed. p53 has been shown to interact with numerous proteins. Of these, a functional role has been demonstrated for p300 and MDM2. p300 is a protein that links nuclear transcription factors such as p53 to the transcriptional machinery. p300 mutations could inhibit the transcriptional activity of p53, but again, mutations of p300 have not been observed, and the activity of this protein is not limited to involvement with p53; therefore, its inactivation could have broader implications. MDM2, another protein interacting with p53, acts as a negative regulator by binding p53 and targeting it for ubiquitin-mediated degradation. Amplification of the mdm2 gene has been observed in sarcomas, which are important component tumors of LFS; however, deregulation of this gene is not a heritable means of transmitting LFS.

VI. TELOMERASE

The duration of the finite human lifecycle varies widely, although an upper limit of 120 years has been suggested. It has also been observed that after 40–50 divisions in culture, cells cycle progressively more slowly and eventually stop dividing altogether, a process termed *proliferative senescence*. Cells derived from older individuals undergo fewer divisions than cells from young individuals. A live cell no longer capable of proliferating is said to be senescent, and the limitation on the number of cell divisions is now referred to as the Hayflick limit (96). The mechanism responsible for this phenomenon was unknown for many years, and it seemed remarkable that cells were able to count the number of times they had divided. Explanations eventually emerged.

The ends of chromosomes (telomeres) are important in maintenance of the stability of chromosomes and the DNA within them. A mechanism is required to ensure that the DNA at the ends of chromosomes does not come apart. Additionally, there is what is known as the end replication problem that arises from the mechanism by which DNA is replicated. Synthesis of DNA can only occur in the 5´ to 3´ direction. Thus, DNA synthesis on the leading strand is continuous. Synthesis on the lagging strand is discontinuous and occurs in a series of 5´ to 3´ reactions from Okazaki fragments on the antisense template. However, with a linear genome, a portion of the lagging strand is not always replicated because there is not enough room on the end of the chromosome for another Okazaki fragment; Thus, there is no template from which to initiate DNA polymerization. With each successive nuclear division, the ends of the chromosomes become progressively shorter. To deal with this problem, chromosomes have evolved telomeres, which consist of many copies of a nucleotide repeat on the ends of chromosomes. The precise nature of this sequence varies between species, but it is typically guanine rich. In the case of humans, the repeat is TTAGGG (97), repeated many times at the ends of both strands of the chromosome. The repeat extends 12 to 16 nucleotides beyond the end of the complimentary strand (97), long enough to allow human cells about 50 cell divisions before the telomeres become too short to allow faithful chromosome replication. At this point, such cells stop dividing. It has been suggested that the ends of chromosomes help preserve chromosome stability, which prevents chromosome fusion, breakage, and loss through nucleotide-nucleotide or DNA-protein interactions, or both (98). Additionally, cells have evolved a mechanism, involving an enzyme known as telomerase, that consists of an RNA/protein complex, able to produce and maintain telomeres. The limited capacity of cells to divide may be a result of a linear genome, but a collateral advantage could be that cells that have achieved their allotted 50 divisions are unlikely to proceed to unregulated growth leading to cancer; in many cases, cancerous cells are found to have activated telomerase.

VII. SUMMARY

Cancer defines a group of diseases with a strong genetic basis represented by the culmination of multiple genomic alterations. These alterations can either be inherited or can occur in response to exogenous factors such as radiation or endogenous factors such as superoxide radicals. The alterations manifest as point mutations, deletions, insertions, translocation, inversions, and amplifications. Cytogenetic techniques such as karyotyping, CGH and SKY have enhanced our ability to detect these alterations within cancerous cells. This work has helped to direct research to particular regions of chromosomes and to specific genes. Each cell is equipped with mechanisms to repair these genetic mutations. Sometimes, these mechanisms themselves become targets of inactivation, which may accelerate the rate at which mutations accumulate. The progression to cancer is not always linear, as not all genes respond to mutation in the same manner. Many parts of the genome may be damaged without material effect. What we have learned is that genes involved in particular processes have pronounced effects on cell growth and survival when altered. Hence, these genes are frequently involved in a broad range of cancers. These two classes of genes are oncogenes and tumor suppressor genes. Oncogenes are cellular genes that have become deregulated by genetic alteration. This deregulation results in enhanced growth in the context of a relevant cell type. Several oncogenes were identified in retroviruses associated with cancer. Inactivation of many tumor suppressors promotes unregulated progression of the cell cycle. Equally important to the discovery of TSGs were studies involving DNA tumor viruses. These viruses produce proteins that bind to and inactivate cellular proteins, which promotes transformation. Finally, for a transformed cell to become immortal, it must overcome the limitation of shortened telomeres. This is typically the result of activation of the enzyme telomerase, which maintains the ends of chromosomes. As we continue to learn more about the causes of cancer, it is possible to detect cancer earlier and identify the genes involved. To some extent, this may allow prediction of the severity of a cancer and dictate the best approach to therapy. The future should bring more sophisticated means of detecting genetic alterations. This information would help determine which forms of therapy might be most effective and would lead to an improved clinical outcome.

REFERENCES

1. B Vogelstein, ER Fearon, SE Kern, SR Hamilton, AC Preisinger, Y Nakamura, R White. Allelotype of colorectal carcinomas. Science 244:207–211, 1989.
2. ER Fearon, B Vogelstein. A genetic model of colorectal tumorigenesis. Cell 61:759–767, 1990.
3. B Vogelstein, KW Kinzler. The multistep nature of cancer. Trends Genet 9:138–141, 1993.
4. JD Rowley. Letter: A new consistent chromosomal abnormality in chronic myelogenous leukaemia identified by quinacrine fluorescence and Giemsa staining. Nature 243:290–293, 1973.
5. N Heisterkamp, JR Stephenson, J Groffen, PF Hansen, A de Klein, CR Bartram, G Grosveld. Localization of the c-abl oncogene adjacent to a translocation break point in chronic myelocytic leukaemia. Nature 306:239–242, 1983.
6. J Groffen, JR Stevenson, N Heisterkamp, A de Klein, CR Bartram, G Grosveld. Philadelphia chromosomal breakpoints are clustered within a limited region, bcr, on chromosome 22. Cell 36:93–99, 1984.
7. MD Goulding, G Chalepakis, U Deutsch, JR Erselius, P Gruss. Pax-3, a novel murine DNA binding protein expressed during early neurogenesis. EMBO J 10:1135–1147, 1991.
8. N Galili, RJ Davis, WJ Fredericks, S Mukhopadhyay, FJ Rauscher, BS Emanuel, G Rovera, FG Barr. Fusion of a fork head domain gene to Pax3 in the solid tumour alveolar rhabdomyosarcoma. Nat Genet 5:230–235, 1993.
9. M Visser, C Sijmons, J Bras, RJ Arceci, M Godfried, LJ Valentijn, PA Voute, F Baas. Allelotype of pediatric rhabdomyosarcoma. Oncogene 15:1309–1314, 1997.
10. RJ Hu, MP Lee, TD Connors, LA Johnson, TC Burn, K Su, GM Landes, AP Feinberg. A 2.5Mb transcript map of a tumor-suppressing subchromosomal transferable fragment from 11p15.5 and isolation and sequence analysis of three novel genes. Genomics 46:9–17, 1997.
11. A Kallioniemi, O Kallioniemi, D Sudar, D Rutovitz, JW Gray, F Waldman, D Pinkel. Comparative genomic hybridization for molecular cytogenetic analysis of solid tumors. Science 258:818–821, 1992.
12. E Schrock, S du Manoir, T Veldman, B Schoell, J Wienberg, MA Ferguson-Smith, Y Ning, DH Ledbetter, I Bar-Am, D Soenksen, Y Garini, T Ried. Multicolor spectral karyotyping of human chromosomes. Science 273:494–497, 1996.
13. MR Speicher, S Gwyn Ballard, DC Ward. Karyotyping human chromosomes by combinatorial multi-fluor FISH. Nat Genet 12:368–375, 1996.

14. N Rampino, H Yamamoto, Y Ionov, Y Li, H Sawai, JC Reed, M Perucho. Somatic frameshift mutations in the bax gene in colon cancers of the microsatellite mutator phenotype. Science 275:967–969, 1997.
15. P Sung, V Bailly, C Weber, LH Thompson, L Prakash, S Prakash. Human xeroderma pigmentosum group D gene encodes a DNA helicase. Nature 365:852–855, 1993.
16. AM Sijbers, WL deLaat, RR Ariza, M Bidderstaff, YF Wei, JG Moggs, KC Carter, BK Shell, E Evans, MC de Jong, S Rademakers, J de Rooij, NG Jaspers, JH Hoeijmakers, RD Wood. Xeroderma pigmentosum group F caused by a defect in a structure-specific DNA repair endomuclease. Cell 86:811–822, 1996.
17. JH Hoeijmakers. Human nucleotide excision repair syndromes: Molecular clues to unexpected intricacies. Eur J Cancer 30A:1912–1924, 1994.
18. EA De Weerd-Kasteline, W Keijzer, D Bootsma. Genetic heterogeneity of xeroderma pigmentosum demonstrated by somatic cell hybridization. Nat New Biol 238:80–83, 1972.
19. D Bootsma, EA De Weerd-Kastelein, WJ Kleijer, W Keyzez. Genetic complementation analysis of xeroderma pigmentosum. Basic Life Sci 5B:725–728, 1975.
20. W Vermeulen, M Stefanini, S Giliani, JH Hoeijmakers, D Bootsma. Xeroderma pigmentosum complementation group H falls into complementation group D. Mutant Res 255:201–208, 1991.
21. K Savitsky, A Bar-Shiva, S Gilad, G Rotman, Y Ziv, L Vanagaite, DA Tagle, S Smith, T Uziel, S Sfez, M Ashkenazi, I Pecker, M Frydman, R Harnik, Sankhavaram, R Patanjali, A Simmons, GA Clines, A Sartiel, RA Gatti, L Chessa, O Sanal, MF Lavin, NGJ Jaspers, A Malcolm, R Taylor, CF Arlett, T Miki, SM Weissman, M Lovett, FS Collins, Y Shiloh. A single ataxia telangiectasia gene with a product similar to PI-3 kinase. Science 268:1749–1753, 1995.
22. AB Pardee. G1 events and regulation of cell proliferation. Science 246:603–608, 1989.
23. SJ Elledge. Cell cycle checkpoints: Preventing an identity crisis. Science 274:1664–1672, 1996.
24. TA Weinert, LH Hartwell. The RAD9 gene controls the cell cycle response to DNA damage in Saccharomyces cerevisiae. Science 241:317–322, 1988.
25. LH Hartwell, MB Kastan. Cell cycle control and cancer. Science 266:1821–1828, 1994.
26. WS El-Deiry, T Tokino, VE Velculescu, DB Levy, R Parsons, JM Trent, D Lin, WE Mercer, KW Kinzler, B Vogelstein. WAF1, a potential mediator of p53 tumor suppression. Cell 75:817–825, 1993.
27. W Gu, JW Schneider, G Condorelli, S Kaushal, V Mahdavi, B Nadal-Ginard. Interaction of myogenic factors and the retinoblastoma protein mediates muscle cell commitment and differentiation. Cell 72:309–324, 1993.
28. JW Harper, GR Adami, N Wei, K Keyomarsi, SJ Elledge. The p21 Cdk-interacting protein Cip1 is a potent inhibitor of G1 cyclin-dependent kinases. Cell 75:805–816, 1993.
29. Y Xiong, GJ Hannon, H Zhang, D Casso, R Kobayashi, D Beach. p21 is a universal inhibitor of cyclin kinases. Nature 366:701–704, 1993a.

30. H Zhang, GJ Hannon, D Beach. p21-containing cyclin kinases exist in both active and inactive states. Genes Dev 8:1750–1758, 1994.
31. JW Harper, SJ Elledge, K Keyomarsi, B Dynlacht, LH Tsai, P Zhang, S Dobrowolski, C Bai, L Connell-Crowley, E Swindell, MP Fox, N Wei. Inhibition of cyclin-dependent kinases by p21. Mol Biol Cell 6:387–400, 1995.
32. L Hengst, U Gopfert, HA Lashuel, SI Reed. Complete inhibition of Cdk/cyclin by one molecule of p21Cip1. Genes Dev 12:3882–3888, 1998.
33. Y Xiong, H Zhang, D Beach. Subunit rearrangement of the cyclin-dependent kinases is associated with cellular transformation. Genes Dev 7:1572–1583, 1993.
34. H Zhang, Y Xiong, D Beach. Proliferating cell nuclear antigen and p21 are components of multiple cell cycle kinase complexes. Mol Biol Cell 4:897–906, 1993.
35. V Dulic, WK Kaufmann, SJ Wilson, TD Tlsty, E Lees, JW Harper, SJ Elledge, SI Reed. p53-dependent inhibition of cyclin-dependent kinase activities in human fibroblasts during radiation-induced G1 arrest. Cell 76:1013–1023, 1994.
36. M Hall, S Bates, G Peters. Evidence for different modes of action of cyclin dependent kinase inhibitors: p15 and p16 to kinases, p21 and p27 bind to cyclins. Oncogene 11:1581–1588, 1995.
37. O Halevy, BG Novitch, DB Spicer, SX Skapek, J Rhee, GJ Hannon, D Beach, AB Lassar. Correlation of terminal cell cycle arrest of skeletal muscle with induction of p21 by MyoD. Science 267:1018–1021, 1995.
38. SW Tam, JW Shay, M Pagano. Differential expression and cell cycle regulation of the cyclin-dependent kinase 4 inhibitor p16ink4. Cancer Res 54:5816–5820, 1994.
39. C Sandhu, N Bhattacharya, J Daksis. Transforming growth factor beta stabilizes p15INK4b protein, increases p15INK4b-Cdk4 complexes and inhibits cyclin D1-cdk4 association in human mammary epithelial cells. Mol Cell Biol 17:2458–2467, 1997.
40. J Lukas, D Parry, L Aagaard, DJ Mann, J Bartkova, M Strauss, G Peters, J Bartek. Retinoblastoma-protein-dependent cell-cycle inhibition by the tumour suppressor p16. Nature 375:503–506, 1995.
41. J Koh, GH Enders, BD Dynlacht, E Harlow. Tumour-derived p16 alleles encoding proteins defective in cell-cycle inhibition. Nature 375:506–510, 1995.
42. J Pomerantz, N Schreiber-Agus, NJ Liegeois, A Silverman, L Alland, L Chin, J Potes, K Chen, I Orlow, H Lee, C Cordon-Cardo, RA DePinho. The Ink4a tumor suppressor gene product, p19Arf, interacts with MDM2 and neutralizes MDM2's inhibition of p53. Cell 92:713–723, 1998.
43. T Kamijo, JD Weber, G Zambetti, F Zindy, MF Roussel, CJ Sherr. Functional and physical interactions of the ARF tumor suppressor with p53 and Mdm2. Proc Natl Acad Sci U S A 95:8292–8297, 1998.
44. PW Hinds, S Mittnacht, V Dulic, A Arnold, SI Reed, RA Weinberg. Regulation of retinoblastoma protein functions by ectopic expression of human cyclins. Cell 70:993–1006, 1992.
45. M Hatakeyama, JA Brill, GR Fink, RA Weinberg. Collaboration of G1 cyclins in the functional inactivation of the retinoblastoma protein. Genes Dev 8:1759–1771, 1994.
46. E Shtivelman, B Lifshitz, RP Gale, E Canaani. Fused transcript of abl and bcr genes in chronic myelogenous leukemia. Nature 315:550–554, 1985.

47. NE Kohl, N Kanda, RR Schreck, G Bruns, SA Latt, F Gilbert, FW Alt. Transposition and amplification of oncogene-related sequences in human neuroblastomas. Cell 35:359–367, 1983.
48. S Collins, M Groudine. Amplification of endogenous myc-related DNA sequences in a human myeloid leukaemia cell line. Nature 298:679–681, 1982.
49. E Santos, D Martin-Zanca, EP Reddy, MA Pierotti, G Della Porta, M Barbacid. Malignant activation of a K-ras oncogene in lung carcinoma but not in normal tissue of the same patient. Science 223:661–664, 1984.
50. D Stehelin, HE Varmus, JM Bishop, PK Vogt. DNA related to the transforming gene(s) of avian sarcoma viruses is present in normal avian DNA. Nature 260:170–173, 1976.
51. JM Bishop. Viral oncogenes. Cell 42:23–38, 1986.
52. B Vennstrom, D Sheiness, J Zabielski, JM Bishop. Isolation and characterization of c-myc, a cellular homolog of the oncogene (v-myc) of avian myelocytomatosis virus strain 29. J Virol 42:773–779, 1982.
53. B Amati, MW Brooks, N Levy, TD Littlewood, GI Evan, H Land. Oncogenic activity of the c-Myc protein requires dimerization with Max. Cell 72:233–245, 1993.
54. K Peukert, P Staller, A Schneider, G Carmichael, F Hanel, M Eilers. An alternative pathway for gene regulation by Myc. EMBO J 16:5672–5686, 1997.
55. J Stone, T de Lange, G Ramsay, E Jakobovits, JM Bishop, H Varmus, W Lee. Definition of regions in human cmyc that are involved in transformation and nuclear localization. Mol Cell Biol 7:1697–1709, 1987.
56. D Sakamuro, KJ Elliott, R Wechsler-Reya, GC Prendergast. BIN1 is a novel MYC-interacting protein with features of a tumour suppressor. Nat Genet 14:69–77, 1996.
57. LM Facchini, LZ Penn. The molecular role of Myc in growth and transformation: Recent discoveries lead to new insights. FASEB J 12:633–651, 1998.
58. R Taub, C Moulding, J Battey, W Murphy, T Vasicek, GM Lenoir, P Leder. Activation and somatic mutation of the translocated c-myc gene in Burkitt lymphoma cells. Cell 36:339–348, 1984.
59. M Graham, JM Adams, S Cory. Murine T lymphomas with retroviral inserts in the chromosomal 15 locus for plasmacytoma variant translocations. Nature 314:740–743, 1985.
60. K Bhatia, K Huppi, G Spangler, D Siwarski, R Iyer, I Magrath. Point mutations in the c-Myc transactivation domain are common in Burkitt's lymphoma and mouse plasmacytomas. Nat Genet 5:56–61, 1993.
61. PH Rabbitts, JV Watson, A Lamond, A Forster, MA Stinson, G Evan, W Fischer, E Atherton, R Sheppard, TH Rabbitts. Metabolism of c-myc gene products: c-Myc mRNA and protein expression in the cell cycle. EMBO J 4:2009–2015, 1985.
62. MC Nussenzweig, EV Schmidt, AC Shaw, E Sinn, J Campos-Torres, B Mathey-Prevot, PK Pattengale, P Leder. A human immunoglobulin gene reduces the incidence of lymphomas in c-Myc-bearing transgenic mice. Nature 336:446–450, 1988.
63. SH Friend, R Bernards, S Rogelj, RA Weinberg, JM Rapaport, DM Albert, TP Dryja. A human DNA segment with properties of the gene that predisposes to retinoblastoma and osteosarcoma. Nature 323:643–646, 1986.

64. WH Lee, JY Shew, FD Hong, TW Sery, LA Donoso, LJ Young, R Bookstein, EY Lee. The retinoblastoma susceptibility gene encodes a nuclear phosphoprotein associated with DNA binding activity. Nature 329:642–645, 1987.
65. MJ Anderson, EJ Stanbridge. Tumor suppressor genes studied by cell hybridization and chromosome transfer. FASEB J 7:826–833, 1993.
66. DJB Ashley. The two "hit" and multiple "hit" theories of carcinogenesis. Br J Cancer 23:313–328, 1969.
67. AG Knudson. Mutation and cancer: Statistical study of retinoblastoma. Proc Natl Acad Sci U S A 68:820–823, 1971.
68. SH Friend, JM Horowitz, MR Gerber, XF Wang, E Bogenmann, FP Li, RA Weinberg. Deletions of a DNA sequence in retinoblastomas and mesenchymal tumors: Organization of the sequence and its encoded protein. Proc Natl Acad Sci U S A 84:9059–9063, 1987.
69. KM Call, T Glaser, CY Ito, AJ Buckler, J Pelletier, DA Haber, EA Rose, A Kral, H Yeger, WH Lewis, C Jones, DE Housman. Isolation and characterization of a zinc finger polypeptide gene at the human chromosome 11 Wilms' tumour locus. Cell 60:509–520, 1990.
70. MJ Coppes, CE Campbell, BR Williams. The role of WT1 in Wilms tumorigenesis. FASEB J 7:886–895, 1993.
71. SF Dowdy, CL Fasching, D Araujo, KM Lai, E Livanos, BE Weissman, EJ Stanbridge. Suppression of tumorigenicity in Wilms tumor by the p15.5-p14 region of chromosome 11. Science 254:293–295, 1991.
72. E Pennisi. Monkey virus DNA found in rare human cancers. Science 275:748, 1997.
73. DI Linzer, AJ Levine. Characterization of a 54K dalton cellular SV40 tumor antigen present in SV40-transformed cells and uninfected embryonal carcinoma cells. Cell 17:43–52, 1979.
74. DP Lane, LV Crawford. T antigen is bound to a host protein in SV40-transformed cells. Nature 278:261–263, 1979.
75. EM DeVilliers. Heterogeneity of the human papillomavirus group. J Virol 63:4898–4903, 1989.
76. M Scheffner, BA Werness, JM Huibregtse, AJ Levine, PM Howley. The E6 oncoprotein encoded by human papillomavirus types 16 and 18 promotes the degradation of p53. Cell 63:1129–1136, 1990.
77. N Dyson, PM Howley, K Munger, E Harlow. The human papilloma virus-16 E7 oncoprotein is able to bind to the retinoblastoma gene product. Science 243:934–937, 1989.
78. K Munger, BA Werness, N Dyson, WC Phelps, E Harlow, PM Howley. Complex formation of human papillomavirus E7 proteins with the retinoblastoma tumor suppressor gene product. EMBO J 8:4099–4105, 1989.
79. L Chin, J Pomerantz, RA DePinho. The INK4a/ARF tumor suppressor: One gene—two products—two pathways. Trends Biochem Sci 8:291–296, 1998.
80. JM Nigro, SJ Baker, AC Preisinger, JM Jessup, R Hostetter, K Cleary, SH Bigner, N Davidson, S Baylin, P Devilee, T Glover, FS Collins, A Weston, R Modali, CC Harris, B Vogelstein. Mutations in the p53 gene occur in diverse human tumour types. Nature 342:705–708, 1989.

81. K Goi, M Takagi, S Iwata, D Delia, M Asada, R Donghi, Y Tsunematsu, S Nakazawa, H Yamamoto, J Yokota, K Tamura, Y Saeki, J Utsunomiya, T Takahashi, R Ueda, C Ishioka, M Eguchi, N Kamata, S Mizutani. DNA damage-associated dysregulation of the cell cycle and apoptosis control in cells with germ-line p53 mutation. Cancer Res 57:1895–1902, 1997.
82. SJ Baker, ER Fearon, JM Nigro, SR Hamilton, AC Preisinger, JM Jessup, P van Tuinen, DH Ledbetter, DF Barker, Y Nakamura, R White, B Vogelstein. Chromosome 17 deletions and p53 gene mutations in colorectal carcinomas. Science 244:217–221, 1989.
83. M Harvey, H Vogel, D Morris, A Bradley, A Bernstein, LA Donehower. A mutant p53 transgene accelerates tumor development in heterozygous but not nullizygous p53-deficient mice. Nat Genet 9:305–311, 1995.
84. J Martinez, I Georgoff, J Martinez, AJ Levine. Cellular localization and cell cycle regulation by a temperature-sensitive p53 protein. Genes Dev 5:151–159, 1991.
85. LC Strong, WR Williams. The genetic implications of long-term survival of childhood cancer. A conceptual framework. Am J Pediatr Hematol Oncol 9:99–103, 1987.
86. FP Li, JF Fraumeni Jr. Soft-tissue sarcomas, breast cancer, and other neoplasms: A familial syndrome? Ann Intern Med 71:747–752, 1969.
87. JE Garber, AM Goldstein, AF Kantor, MD Dreyfus, JF Fraumeni Jr. Follow-up study of twenty-four families with Li-Fraumeni syndrome. Cancer Res 51:6094–6097, 1991.
88. D Malkin, FP Li, LC Strong, JF Fraumeni Jr, CE Nelson, DH Kim, J Kassel, MA Gryka, FZ Bischoff, MA Tainsky, SH Friend. Germline p53 mutations in a familial syndrome of breast cancer, sarcomas, and other neoplasms. Science 250:1233–1238, 1990.
89. L Brugieres, M Gardes, C Moutou, A Chompret, V Meresse, A Martin, N Poisson, F Flamant, C Bonaiti-Pellie, J Lemerle, J Feunteun. Screening for germline p53 mutations in children with malignant tumors and family history for cancer. Cancer Res 53:452–455, 1993.
90. JM Birch, AL Hartley, KJ Tricker, J Prosser, A Condie, AM Kelsey, M Harris, PHM Jones, A Binchy, D Crowther, AW Craft, OB Eden, DGR Evans, E Thompson, JR Mann, J Martin, ELD Mitchell, MF Santibanez-Koref. Prevalence and diversity of constitutional mutations in the p53 gene among 21 Li-Fraumeni families. Cancer Res 54:1298–1304, 1994.
91. T Frebourg, N Barbier, Y Yan, JE Garber, M Dreyfus, J Fraumani, FP Li, SH Friend. Germline p53 mutations in 15 families with Li-Fraumeni syndrome. Am J Hum Genet 56:608–615, 1995.
92. JM Varley, G McGown, M Thorncroft, MF Santibanez-Koref, AM Kelsey, KJ Tricker, D Gareth, R Evans, JM Birch. Germline mutations of TP53 in Li-Fraumeni families: An extended study of 39 families. Cancer Res 57:3245–3252, 1997.
93. S Srivastava, Z Zou, K Pirollo, W Blattner, EH Chang. Germline transmission of a mutated p53 gene in a cancer-prone family with Li-Fraumeni syndrome. Nature 348:747–749, 1990.
94. CE Canman, D Lim, KA Cimprich, Y Taya, K Tamai, K Sakaguchi, E Appella, MB Kastan, JD Siliciano. Activation of the ATM kinase by ionizing radiation and phosphorylation of p53. Science 281:1677–1679, 1998.

95. S Banin, L Moyal, S Shieh, Y Taya, CW Anderson, L Chessa, NI Smorodinsky, C Prives, Y Reiss, Y Shiloh, Y Ziv. Enhanced phosphorylation of p53 by ATM in response to DNA damage. Science 281:1674–1677, 1998.
96. L Hayflick. The limited in vitro lifetime of diploid cell strains. Exp Cell Res 37:614–636, 1965.
97. EH Blackburn. Structure and function of telomeres. Nature 350:569–573, 1991.
98. B McClintock. The stability of broken ends of chromosomes in *Zea Mays*. Genetics 41:234, 1941.

2
Genetic Toxicity Tests for Predicting Carcinogenicity

Errol Zeiger
National Institute of Environmental Health Sciences, Research Triangle Park, North Carolina

I. INTRODUCTION

Genetic toxicology is the study of substances that can damage the DNA and chromosomes of cells. This damage is usually measured as mutations, chromosome aberrations, DNA strand breaks, or as DNA adducts or interferences with the repair of damage. A number of in vitro and in vivo test systems have been developed to study the effects of chemicals and radiation on cellular DNA and chromosomes. The most commonly used in vitro tests for routine screening are gene mutation systems in bacteria and gene and chromosome damage systems in cultured rodent cells. In vivo, chromosome damage is typically measured in bone marrow cells of mice or rats.

Two major concerns have led to the use of genetic toxicity tests to identify mutagenic chemicals and characterize their effects. One is the induction in the germ cells of mutations that can either affect reproductive performance or result in genetic disease in future generations. The other concern is based on the role of somatic cell mutations in the initiation and progression of cancer. It can be argued that the induction of mutations in germ cells has a potentially greater impact on the species and on future health concerns. However, the incidence and immediacy of cancer in the population have made somatic cell mutations of greater immediate concern to researchers and regulatory agencies.

Genetic toxicity is not a measure of carcinogenicity, but it is often used as a surrogate for cancer. This is because the tests measure an initiating or intermediate event in tumorigenesis (1), and because of the reported high associations between positive responses in genetic toxicity tests and rodent and human carcinogenicity (2–4). As a result of these considerations, genetic toxicity testing is used routinely as an initial toxicological screen in chemical and drug development.

The purpose of this chapter is not to describe individual tests or protocols for their use. Rather, it is to describe the usefulness of some of the most commonly used tests and the interpretation of their results with respect to the potential animal and human carcinogenicity of the test chemicals.

II. GENETIC TOXICITY TESTING

A. Testing History

Genetic toxicity testing received its greatest impetus from studies by Ames and colleagues (2, 5) and others (3, 4) who reported that approximately 90% of the known carcinogens were mutagenic in the bacterium *Salmonella typhimurium* and that a positive mutagenic response was highly predictive for rodent carcinogenicity (Table 1). These early studies were based on the database of carcinogens that was available before the testing of chemicals in mice and rats using the protocol developed by the U.S. National Cancer Institute (NCI) and later adopted by the U.S. National Toxicology Program (NTP). Most of these early carcinogens were electrophilic alkylating agents that were generally detected as carcinogens after limited treatment regimens that used relatively small numbers of animals. Later compendia that incorporated data from the NCI/NTP cancer tests produced lower concordance values, approximately 60% (Table 1). Many of these carcinogens were structurally different from the earlier carcinogens and fewer were potential electrophiles. In addition, most later carcinogens required long-term feeding at high doses for their activity, and many were judged to be carcinogenic based solely on an increase in an already high spontaneous liver tumor rate in mice.

Subsequent to the initial reports about the effectiveness of the *Salmonella* test as a screen for carcinogens, numerous other genetic test systems were developed to measure gene and chromosomal mutations. Numerous claims were made on the effectiveness of these other test systems for predicting carcinogenicity, although often based on very limited data. Many tests have been developed and proposed for identifying carcinogens (Table 2), but relatively few have been used extensively for a wide range of chemicals and in multiple laboratories. As a result, industry uses and regulatory requirements have restricted themselves to the

TABLE 1. Performance of the *Salmonella* Test for the Identification of Carcinogens*

Dataset	prev.	sens.	spec.	+pred.	conc.
Purchase et al., 1978 (3)	.48	.91	.94	.93	.93
Sugimura et al., 1976 (6)	.70	.93	.74	.89	.87
McCann et al., 1975 (2)	.72	.89	.76	.90	.86
Kier et al., 1986 (7)	.96	.80	.67	.98	.79
Auletta and Ashby, 1988 (8)	.83	.78	.62	.91	.76
Heddle and Bruce, 1977 (9)	.70	.66	.88	.93	.72
Zeiger et al., 1990 (10)	.59	.48	.91	.89	.66
Zeiger, 1998 (11)	.56	.54	.79	.77	.60

*Listed in descending order of concordance.
Abbreviations: *prev.*, prevalence (proportion) of carcinogens among the tested chemicals; *sens.*, sensitivity, the proportion of carcinogens correctly identified by the test; *spec.*, specificity, the proportion of noncarcinogens correctly identified; *+pred.*, positive predictivity, the proportion of carcinogens among chemicals giving a positive test result; *conc.*, concordance, the proportion of correct identifications of carcinogens and noncarcinogens.

tests with the largest databases. This is not to suggest that the other tests would not be as effective or more effective than those currently used, but that there is insufficient experience with them to justify their routine use.

Genetic toxicity tests are used by industry to screen chemicals under development for potential carcinogenicity, prioritize chemicals for additional testing, characterize the spectrum of genetic effects that can be produced by specific chemicals, and aid in the interpretation of toxicity and carcinogenicity test results. A positive genetic toxicity response suggests potential carcinogenicity, and the chemical is either deferred in preference to one that is not genotoxic or its potential carcinogenicity guides its further development and testing. Genetic toxicity testing is often required by government regulatory agencies before a food additive, drug, pesticide, or other industrial chemical can be produced or mar-

TABLE 2. Genetic Toxicity Test Compilations

Compilation	No. of test systems or endpoints*
Hollstein et al., 1979 (12)	119
Green et al., 1980 (13)	72
IARC, 1987 (14)	173
Lohman et al., 1992 (15)	85

*There is incomplete overlap among the different compilations.

keted. Regulatory agencies use the test results to classify chemicals as to their mutagenicity as an indication of their potential carcinogenicity or mammalian germ cell mutagenicity, and to aid in interpreting carcinogenicity test results.

B. Genetic Toxicity Tests

Tests for routine screening of chemicals measure gene mutations and gross chromosome damage, such as breaks or rearrangements. This is based on the assumptions that although most genotoxic chemicals produce both gene mutations and chromosome aberrations, there are chemicals that produce only one of these two types of damage, and that all chemicals that can produce genetic damage are equally able to induce cancer.

The most widely used gene mutation tests are the *Salmonella* (Ames) test and various in vitro mammalian cell tests, usually the mouse lymphoma L5178Y *tk* test, Chinese hamster ovary (CHO) or lung (V79) cell hypoxanthine-guanine phosphoribosyl transferase (HPRT) test, or the CHO-AS52 *gpt* test. Chromosome aberrations are usually measured in vitro using the CHO or V79 cell lines or indirectly in mouse lymphoma L5178Y cells. Most in vivo genetic toxicity testing is based on the measurement of chromosome aberrations or micronuclei in mouse bone marrow cells.

1. *Salmonella* Mammalian Microsome Test

The *Salmonella* test (Ames test; SAL) (16–20) uses bacteria containing defined point mutations in the histidine (*his*) operon, such that the organisms cannot grow and form colonies in the absence of the amino acid histidine. These tester strains of *Salmonella* have been further genetically engineered to increase their sensitivity to mutagens. Mutations in the affected *his* genes can reverse the effects of the original mutation and enable the cells to grow without added histidine. Various tester strains containing different mutations allow the identification of mutagens that act by producing basepair substitution or frameshift mutations.

2. Mammalian Cell Gene Mutation Test

The L5178Y mouse lymphoma cell mutation test (MLA) measures the induction of mutations and chromosome damage in cultured mouse lymphoma cells (21–24). This test measures the ability of the cells to grow and form colonies in the presence of trifluorothymidine (TFT). Normal cells will take up and metabolize TFT to a toxic substance and will not grow or survive in its presence. Cells that have mutations in their thymidine kinase (*tk*) genes that prevent them from metabolizing TFT can survive its presence and form colonies. The *tk* gene can be inactivated by either point mutations or chromosome damage (that is, rearrangements or deletions of the section of chromosome containing the gene). The test yields TFT-resistant mutants that have a bimodal distribution of colony sizes. The large colonies are associated with point mutations and the small colonies are associated with chromosomal damage such as deletions and rearrangements (25–28). As a consequence, the MLA has been proposed as both a gene mutation and chromosome aberration test (23, 29, 30). Point mutation assays in Chinese hamster cells are also used (29, 31).

3. In Vitro Chromosome Aberration Test

Typically, in the in vitro chromosome aberration (ABS) test, CHO or V79 cells are treated with the test chemical for 1–1.5 cell cycles. The cells are transferred, after exposure to hypotonic treatment and fixation, to microscope slides and stained, and their chromosomes are examined microscopically in their first metaphase after treatment for the presence of breaks or rearrangements (32–37). The data are evaluated as the percentage of cells containing aberrations and sometimes as aberrations per cell. Compared with cells containing point mutations where the damage is measured in subsequent generations of cells, most cells containing chromosome damage will not survive to complete mitosis or will produce nonviable daughter cells.

4. In Vivo Rodent Chromosome Aberration and Micronucleus Tests

In vivo chromosome damage is typically measured in bone marrow cells because this is a rapidly dividing population and is therefore more susceptible to chemically induced chromosome damage, and the cells are more easily recovered than from other tissues. In the tests for chromosome aberrations or micronuclei, the animals are treated with either a single dose or a series of doses, and the femoral bone marrow is removed. The time of removal after chemical treatment depends on the treatment regimen and the endpoint to be examined. The bone marrow cells are spread on a microscope slide, stained, and scored for the presence of chromosome aberrations or micronuclei.

The most widely used in vivo test at this time is the erythrocyte micronucleus (MN) test (32, 38–41). Micronuclei appear in daughter cells after damage in the parent cells. Chromosome fragments that result from breaks may not be incorporated into the main nucleus of the daughter cell after mitosis. A nuclear membrane will form around the fragment which will be visible as a separate, smaller (micro) nucleus that is present in addition to the cell nucleus. Micronuclei can also form from a complete chromosome when there is damage to the cell's spindle apparatus or to the chromosome itself. In such a situation, the MN would also contain the chromosome's centromere and kinetochore, which could be detected using specific DNA probes and antibodies, respectively.

Micronuclei are scored in polychromatic erythrocytes (PCE) in the bone marrow of rats or mice. Alternatively, in mice, MN can be scored in circulating erythrocytes because unlike the rat and human spleens, the mouse spleen does not remove MN-containing erythrocytes from the blood. Scoring normochromatic erythrocytes (NCE) in peripheral blood for MN permits the effect of long-term treatment on the accumulation of MN-erythrocytes in peripheral blood to be examined (42).

5. Metabolic Activation In Vitro

Many chemicals are not toxic or mutagenic by themselves but must be metabolized to mutagenic forms. Bacteria and most cultured mammalian cells cannot perform most of the metabolic conversions found in mammals and humans because they do not contain the necessary drug metabolizing enzyme systems. This limitation has been partially overcome by the development of exogenous metabolic activation systems that can be added to the test procedure. These systems usually consist of homogenates of liver fractions (S-9) of rodents, usually rats, that had been pretreated with substances to enhance (induce) the levels of the preferred metabolic enzymes. NADP and other enzyme cofactors are added to the S-9 to enable its activity in vitro (17, 43–46). A complete in vitro testing protocol includes tests with and without S-9 fractions.

III. THE EFFECTIVENESS OF GENETIC TOXICITY TESTS FOR IDENTIFYING CARCINOGENS AND NONCARCINOGENS

The use of in vitro and in vivo genetic toxicity tests for identifying carcinogens and noncarcinogens are based on several premises: (a) tests for both gene mutations and chromosome aberrations are needed; (b) mammalian cell tests are more relevant than microbial or other nonmammalian tests, and; (c) in vivo mammalian tests are more relevant than in vitro tests. These premises guide most genetic toxicity testing requirements, whether from government regulatory agencies or industrial chemical or drug development scientists. The premises have recently been re-examined using the databases of chemicals tested by the U.S.-NTP for rodent carcinogenicity and genetic toxicity.

The U.S.-NTP databases of carcinogenicity and genetic toxicity test results were used for the evaluation. These studies are characterized by defined, standardized test protocols and evaluation criteria and the availability of the peer-reviewed test data and summary conclusions (47–49). Other, unpublished genetic toxicity test results are available but were not used in the evaluations. NTP rodent carcinogenicity test results are available on more than 400 chemicals (48, 50), the majority of which have been tested for genetic toxicity in at least one in vitro test. All chemicals were tested for mutagenicity in *Salmonella*. Fewer chemicals were tested in vitro in the MLA or chromosome aberration tests or in vivo for MN in rodent erythrocytes.

Genetic Toxicity Tests

Carcinogenicity was tested in male and female B6C3F1 mice and Fischer 344 rats. A positive response (increased tumor incidence) in a single sex of a single species was sufficient to label the chemical a carcinogen for the purposes of this evaluation. For a chemical to be considered a noncarcinogen, it must have been adequately tested and have produced negative results in all four sex/species combinations. For the purposes of these evaluations, equivocal genetic and carcinogenicity responses were considered to be negative.

For evaluating the different genetic toxicity tests for their ability to identify carcinogens and noncarcinogens and for comparing the tests, the most useful performance characteristics are the sensitivity, specificity, positive predictivity, and accuracy. The sensitivity of the test is the proportion of carcinogens tested that gave positive results, and the specificity is the proportion of noncarcinogens that gave negative results. Positive predictivity is the proportion of carcinogens among chemicals that gave positive responses (the "true positives"); the concordance is the proportion of correct identifications of carcinogens and noncarcinogens. The performance characteristics from the NTP databases for three in vitro tests and the mouse bone marrow MN test are in Table 3 (11).

The following premises were then examined:

1. Tests for both gene mutations and chromosome aberrations are needed. The Salmonella test identified .54 (111/205) of the rodent carcinogens, with a positive predictivity for cancer of .77; the in vitro ABS test identified .52 (71/136) with a positive predictivity of .73 (11) (Table 3). Test results were available for 218 chemicals tested in both (Table 4). When both tests were positive, the predictivity was .90. However, the predictivity of SAL alone for these selected chemicals was the same as that of the combination (.90), whereas the predictivity of ABS alone was .73. Among the chemicals that produced

TABLE 3. Performance of In Vitro Short-Term Tests in the NTP Database for Predicting Rodent Carcinogenicity

Test	Chems	Prev.	Sens.	Spec.	+Pred.	Conc.
SAL	363	.56	.54	.79	.77	.65
ABS	218	.62	.52	.68	.73	.58
MLA	191	.59	.74	.32	.61	.57
MN	83	.60	.28	.82	.70	.49

Abbreviations: *prev.*, prevalence (proportion) of carcinogens among the tested chemicals; *sens.*, sensitivity, the proportion of carcinogens correctly identified by the test; *spec.*, specificity, the proportion of noncarcinogens correctly identified; *+pred.*, positive predictivity, the proportion of carcinogens among chemicals giving a positive test result; *conc.*, concordance, the proportion of correct identifications of carcinogens and noncarcinogens.

TABLE 4. Predictivity of the Combination of SAL and ABS Tests for Carcinogenicity

SAL	ABS	Carcinogen	Noncarcinogen	Positive predictivity
+	+	44	5	.90
+	−	18	2	.90
−	+	27	21	.56
−	−	47	54	.47

chromosome aberrations but not mutations in SAL, the predictivity of the positive ABS response was .56. This demonstrates that the combination of SAL and ABS are no more effective for prospectively identifying carcinogens than SAL alone.

2. Mammalian cell tests are more relevant than microbial, or other nonmammalian, tests. The sensitivity and positive predictivities of the SAL, ABS, and MLA tests were compared. The MLA test correctly identified .74 of the carcinogens, with a positive predictivity of .61 (84/113) (Table 3). Among the 188 chemicals tested in SAL and MLA, .46 (51/111) of the carcinogens were correctly identified by SAL. When both MLA and SAL were positive, the predictivity was .86. Among the chemicals tested in both the SAL and MLA tests, SAL had a positive predictivity of .85 (51/60) compared with .61 (83/135) for MLA (Table 5). Although MLA identified more carcinogens than did SAL, it also incorrectly identified a higher proportion of noncarcinogens.

There was agreement in test results on 92/149 chemicals tested in both ABS and MLA (Table 6). Where there was disagreement, it was more likely that MLA was positive and ABS negative. When both tests were positive, the sensitivity was .69 and the predictivity for cancer was .77; however, when both were neg-

TABLE 5. Predictivity of the Combination of SAL and MLA Tests for Carcinogenicity

SAL	MLA	Carcinogen	Noncarcinogen	Positive predictivity
+	+	48	8	.86
+	−	3	1	.75
−	+	35	44	.44
−	−	25	24	.51

Genetic Toxicity Tests

TABLE 6. Predictivity of the Combination of ABS and MLA Tests for Carcinogenicity

ABS	MLA	Carcinogen	Noncarcinogen	Positive predictivity
+	+	44	13	.77
+	−	3	5	.38
−	+	24	25	.49
−	−	20	15	.57

ative, the specificity was .54 and the predictivity for noncarcinogenicity was .43. This can be compared with the sensitivity of .47 and positive predictivity of .72 for ABS alone and a sensitivity of .67 and positive predictivity of .64 for MLA alone, among these 149 chemicals (Table 6).

If the MLA test is detecting both point mutations and chromosome aberrations, it would be expected to be more sensitive than either the gene mutation or chromosome aberration tests. The test is more sensitive. It detected .75 (83/111) of all carcinogens also tested in SAL and .75 (68/91) of all carcinogens also tested in ABS. This can be compared with sensitivities for these same chemicals of .44 (48/111) and .52 (47/91) for SAL and ABS, respectively. However, this higher sensitivity was counterbalanced by higher false positive rates—.38 (52/135) and .36 (38/106) versus .15 (9/60) and .27 (18/65)—when compared with SAL and ABS, respectively (Tables 5, 6).

There were 48 chemicals positive in the ABS test and negative in SAL. Of these, 56% were carcinogens (Table 4). Similarly, there were 79 chemicals positive in the MLA test and negative in SAL. Of these, 44% were carcinogens (Table 5). It can be concluded that these SAL negative chemicals predominantly produce chromosome aberrations. However, these additional positive responses had no effective predictive value for carcinogenicity because of the high proportion of noncarcinogens that were also identified as positive.

3. In vivo mammalian tests are more relevant than in vitro tests. The ability of a chemical to cause mutations or chromosome damage in cells in vivo is obviously of more concern than the same effects in bacteria or cultured mammalian cells. Similarly, it is assumed that there is less likelihood of obtaining false positive results in vivo because there are none of the artifacts associated with in vitro testing. There are 70 NTP chemicals with published ABS and bone marrow MN test results: 23 were clastogenic only in vitro and 6 were clastogenic only in vivo. Among the in vitro-only clastogens, 15 (65%) were carcinogens, whereas 3 (50%) of the in vivo-only clastogens were carcinogens. The positive predictivi-

TABLE 7. Performance of the Combination of the ABS and Bone Marrow MN Tests

ABS	MN	Carcinogen	Noncarcinogen	Positive predictivity
+	+	8	3	.73
+	−	15	8	.65
−	+	3	3	.50
−	−	16	14	.53

ties of both tests were equivalent—.68 and .65 for the ABS and MN tests, respectively—which shows that both tests also had equivalent false positive rates. As anticipated, the in vivo MN test was less sensitive (.26; 11/42) than the corresponding in vitro ABS test (.55; 23/42) (Table 7).

When SAL and MN were evaluated, the combination was no more effective for predicting carcinogenicity than was SAL alone, and MN had a higher proportion of false positive results than did SAL (Table 8).

Many regulatory testing guidelines require the SAL, ABS, and MN tests. When this combination was examined for the 70 NTP chemicals with published MN results, the following relationships were seen (Table 9). The results in this table can be compared with the predictivities of the individual tests: .88, .68, and .65 for SAL, ABS, and MN, respectively. Of the 19 chemicals that were negative in SAL and positive in the ABS or MN tests, six were carcinogens and 11 were not. These results, and those described above (Tables 4, 8) suggest that the induction of structural chromosome damage in vitro or in vivo, although indicative of genetic activity and useful as a biomarker, is not as good a predictor of carcinogenesis as are point mutations.

TABLE 8. Performance of the Combination of the SAL and Bone Marrow MN Tests

SAL	MN	Carcinogen	Noncarcinogen	Positive predictivity
+	+	7	1	.88
+	−	14	2	.88
−	+	4	5	.44
−	−	17	20	.46

TABLE 9. Performance of the Combination of SAL, ABS, and Bone Marrow MN Tests

SAL	ABS	MN	Carcinogen	Noncarcinogen	Positive predictivity
+	+	+	6	1	.86
+	+	−	11	2	.85
+	−	+	1	0	—
+	−	−	3	0	1.00
−	+	+	2	2	.50
−	+	−	4	6	.40
−	−	+	2	3	.40
−	−	−	13	14	.48
			42	28	

IV. DISCUSSION AND CONCLUSIONS

These analyses show that the premises supporting the use of these genetic toxicity tests are not fully supported by the data. The use of in vitro mammalian cell tests, whether for measuring gene mutations or ABS, in addition to Salmonella, does not result in a predictivity greater than that of Salmonella alone. It introduces a higher proportion of false positive results. The higher number of positive responses obtained by combining tests does not enhance the predictive value because the increased number of true positives is counterbalanced by an equally large increase in the number of false positives. Similarly, the in vivo MN test results did not improve the predictivity for carcinogens and introduced a number of false positives. The concordance values presented here are based on noncongeneric groups of chemicals. It is anticipated that some chemical classes would show higher concordances (e.g., [51]); however, there are insufficient numbers of carcinogens and noncarcinogens in each chemical class for such a determination.

Among the individual tests, the highest concordance between genetic toxicity and rodent cancer was .65 (for SAL) (Table 3); the concordance between in vivo MN induction and carcinogenicity was .49. The low concordance for the MN test may, in part, be related to the anticipated low sensitivity of the test. Although the in vitro and in vivo genetic toxicity tests are not 100% predictive of cancer, neither are the rodent carcinogenicity tests 100% predictive of rodent carcinogens. As can be seen in Table 10, when tested by the NTP using the same batch of chemical, in parallel in the same laboratory in similar protocols, the concordance between the rat and mouse was 74%. This is slightly higher than, but comparable to, the concordances between the in vitro tests and (overall) rodent carcinogenicity.

TABLE 10. Concordance of Carcinogenicity in Rats and Mice

Carcinogenicity			
Rats	Mice	1987*	1999†
+	+	67	111
+	−	32	47
−	+	36	54
−	−	131	180
Concordance		74% (198/266)	74% (291/392)

*Source: Ref. 52
†NTP, January 1999

The contribution of carcinogenic and mutagenic potency to predictivity was also evaluated. There are two ways to approach the question of potency of response. The most common approach is to define potency as a function of the dose and the magnitude of the response, usually measured as a slope or as an effective dose (for example, TD50, or lowest effective dose) (53–55). Another approach is to define potency as a function of the breadth of biological effects (that is, number of sites or species affected) that can be induced by the chemical (51, 56, 57).

Recent evaluations have used a number of different measures of potency of the responses, including slopes, lowest effective doses, and fold-increases over the control. Regardless of the measure used, there is minimal correlation between the quantitative response in Salmonella or in vitro chromosome aberration tests and qualitative or quantitative results in carcinogenicity tests (58, 59), which suggests that the use of quantitative genetic toxicity information to predict carcinogenicity is of no practical value.

However, the broader the range of carcinogenic effects induced, the more likely the chemical will be identified by its mutagenicity (that is, the higher the sensitivity of the genetic toxicity tests). For example, .69 of the chemicals tumorigenic in both rats and mice were mutagenic in SAL, whereas .40 of the single-species carcinogens were mutagenic (Table 11). This can be compared to SAL's sensitivity of .54 for all NTP rodent carcinogens (Table 3).

There are more differences between mutagenic (in SAL) and nonmutagenic carcinogens then their responses in genetic toxicity tests. Chemicals that are *Salmonella* mutagens tend to produce tumors in both rats and mice, whereas the nonmutagens are primarily single-species carcinogens. Similarly, the mutagenic chemicals tend to induce tumors at multiple sites, and there appear to be tissue-specific sites for the mutagenic and nonmutagenic carcinogens; however, the data are open to interpretation (60, 61).

TABLE 11. Rodent Tumor Response and *Salmonella* Mutagenicity

	Carcinogenicity	#	% SAL+
2 species;	Common site	46	78
	Not common site	31	55
	Total 2 species.	77	69
Single species;	Multiple sites	20	40
	Single site	62	40

Adapted from Ref. 56.

The proportion (prevalence) of carcinogens in the database strongly influences the performance characteristics evaluated. For example, if a test has a low sensitivity and high specificity, it will perform well when the proportion of carcinogens in the database is low. Alternatively, if the test is highly sensitive with low specificity, it will perform well when the proportion of carcinogens is high and poorly when the proportion is low.

V. SUMMARY

Analysis of the NTP's genetic toxicity database of noncongeneric chemicals shows that the most effective predictor of rodent carcinogenicity is the *Salmonella* test. Positive results in other tests, in the absence of a SAL positive, had no predictive value. Similarly, negative results in SAL or the other tests do not indicate noncarcinogenicity. The nonmutagenic carcinogens may be of less concern because the mutagenic carcinogens may act by mechanisms that are relevant to more than a single strain or species and tend to be multispecies carcinogens. These multispecies carcinogens are more likely to be human carcinogens than those that act in rodents by a single, species-specific mechanism.

REFERENCES

1. ER Fearon, B Vogelstein. A genetic model for colorectal tumorigenesis. Cell 61:759–767, 1990.
2. J McCann, E Choi, E Yamasaki, BN Ames. Detection of carcinogens in the *Salmonella*/microsome test: Assay of 300 chemicals. Proc Natl Acad Sci USA 72:5135–5139, 1975.
3. IFH Purchase, E Longstaff, J Ashby, JA Styles, D Anderson, PA Lefevre, FR Westwood. An evaluation of 6 short-term tests for detecting organic chemical carcinogens. Br J Cancer 37:873–959, 1978.

4. T Sugimura, S Sato, M Nagao, T Yahagi, T Matsushima, Y Seino, M Takeuchi, T Kawachi. Overlapping of carcinogens and mutagens. In: PN Magee, S Takayama, T Sugimura, T Matsushima, eds. Fundamentals of Cancer Prevention. Baltimore: University Park Press, 1976, pp 191–215.
5. BN Ames, WE Durston, E Yamasaki, FD Lee. Carcinogens are mutagens: A simple test system for combining liver homogenates for activation and bacteria for detection. Proc Natl Acad Sci U S A 70:2281–2285, 1973.
6. T Sugimura, T Yahagi, M Nagao, M Takeuchi, T Kawachi, K Hara, E Yamasaki, T Matsushima, Y Hashimoto, M Okada. Validity of mutagenicity tests using microbes as a rapid screening method for environmental carcinogens. In: R Montesano, H Bartsch, L Tomatis, eds. Screening Tests in Chemical Carcinogenesis, vol 12. Lyon: IARC Scientific Publications, 1976, pp 81–104.
7. LD Kier, DJ Brusick, AE Auletta, ES Von Halle, VF Simmon, MM Brown, VC Dunkel, J McCann, K Mortelmans, MJ Prival, TK Rao, VA Ray. The *Salmonella typhimurium*/mammalian microsome mutagenicity assay. A report of the U.S. Environmental Protection Agency Gene-Tox Program. Mutat Res 168:67–238, 1986.
8. A Auletta, J Ashby. Workshop on the relationship between short-term test information and carcinogenicity. Williamsburg, Virginia, January 20–23, 1987. Environ Mol Mutagen 11:135–145, 1988.
9. JA Heddle, WR Bruce. Comparison of tests for mutagenicity or carcinogenicity using assays for sperm abnormalities, formation of micronuclei, and mutations in *Salmonella*. In: HH Hiatt, JD Watson, JA Winsten, eds. Origins of Human Cancer. Book C. Human Risk Assessment. Cold Spring Harbor: Cold Spring Harbor Laboratory, 1977, pp 1549–1557.
10. E Zeiger, JK Haseman, MD Shelby, BH Margolin, RW Tennant. Evaluation of four in vitro genetic toxicity tests for predicting rodent carcinogenicity: Confirmation of earlier results with 41 additional chemicals. Environ Mol Mutagen 16 (suppl. 18): 1–14, 1990.
11. E Zeiger. Identification of rodent carcinogens and noncarcinogens using genetic toxicity tests: Premises, promises, and performance. Regul Toxicol Pharmacol 28:85–95, 1998.
12. M Hollstein, J McCann, F Angelosanto, W Nichols. Short-term tests for carcinogens and mutagens. Mutat Res 65:133–226, 1979.
13. S Green, A Auletta. Editorial introduction to the reports of the Gene-Tox program. An evaluation of bioassays in genetic toxicology. Mutat Res 76:165–168, 1980.
14. IARC. Genetic and Related Effects: An Updating of Selected IARC Monographs from Volumes 1 to 42, Supplement 6. Lyon: International Agency for Research on Cancer, 1987.
15. PHM Lohman, ML Mendelsohn, DH Moore II, MD Waters, DJ Brusick, J Ashby, WJA Lohman. A method for comparing and combining short-term genotoxicity test data: The basic system. Mutat Res 266:7–25, 1992.
16. BN Ames, J McCann, E Yamasaki. Methods for detecting carcinogens and mutagens with the *Salmonella*/mammalian microsome mutagenicity test. Mutat Res 31: 347–364, 1975.
17. D Maron, BN Ames. Revised methods for the *Salmonella* mutagenicity test. Mutat Res 113:173–212, 1983.

18. LD Claxton, J Allen, A Auletta, K Mortelmans, E Nestmann, E Zeiger. Guide for the *Salmonella typhimurium*/mammalian microsome tests for bacterial mutagenicity. Mutat Res 189:83–91, 1987.
19. D Gatehouse, S Haworth, T Cebula, E Gocke, L Kier, T Matsushima, C Melcion, T Nohmi, T Ohta, S Venitt, E Zeiger. Recommendations for the performance of bacterial mutation assays. Mutat Res 312:217–233, 1994.
20. OECD. Guideline for the Testing of Chemicals: No. 471, Bacterial Reverse Mutation Test: Organization for Economic Cooperation and Development, Paris, 1997.
21. D Clive, R McCuen, JFS Spector, C Piper, KH Mavournin. Specific gene mutations in L5178Y cells in culture. A report of the U.S. Environmental Protection Agency Gene-Tox Program. Mutat Res 115:225–251, 1983.
22. WJ Caspary, YJ Lee, S Poulton, BC Myhr, AD Mitchell, CJ Rudd. Evaluation of the L1578Y mouse lymphoma cell mutagenesis assay: Quality-control guidelines and response categories. Environ Mol Mutagen 12 (suppl 13):19–36, 1988.
23. AD Mitchell, AE Auletta, D Clive, PE Kirby, MM Moore, BC Myhr. The L5178Y/tk$^{+/-}$ mouse lymphoma specific gene and chromosomal mutation assay. A phase III report of the U.S. Environmental Protection Agency Gene-Tox program. Mutat Res 394:177–303, 1997.
24. OECD. Guideline for the Testing of Chemicals: No. 476, In Vitro Mammalian Cell Gene Mutation Test: Organization for Economic Cooperation and Development, Paris, 1997.
25. J Hozier, J Sawyer, D Clive, MM Moore. Chromosome 11 aberrations in small colony L5178Y TK$^{-/-}$ mutants early in their clonal history. Mutat Res 147:237–242, 1985.
26. J Hozier, J Sawyer, D Clive, M Moore. Cytogenetic distinction between TK$^+$ and TK$^-$ chromosomes in the L5178YTK$^{+/-}$ 3.7.2C cell line. Mutat Res 105:451–456, 1982.
27. WF Blazak, BE Stewart, I Galperin, KL Allen, CJ Rudd, AD Mitchell, WJ Caspary. Chromosome analysis of trifluorothymidine-resistant L5178Y mouse lymphoma cell colonies. Environ Mutagen 8:229–240, 1986.
28. ML Applegate, MM Moore, CB Broder, A Burrell, J Juhn, KL Kaswek, P-F Lin, A Wadhams, JC Hozier. Molecular dissection of mutations at the heterozygous thymidine kinase locus in mouse lymphoma cells. Proc Natl Acad Sci USA 87:51–55, 1990.
29. AE Auletta, KL Dearfield, MC Cimino. Mutagenicity test schemes and guidelines: U.S. EPA Office of Pollution Prevention and Toxics and Office of Pesticide Programs. Environ Mol Mutagen 21:38–45, 1993.
30. RD Combes, H Stopper, WJ Caspary. The use of L5178Y mouse lymphoma cells to assess the mutagenic, clastogenic and aneugenic properties of chemicals. Mutagenesis 10:403–408, 1995.
31. AW Hsie, DA Casciano, DB Couch, DF Krahn, JP O'Neill, BL Whitfield. The use of Chinese hamster ovary cells to quantify specific locus mutation and to determine mutagenicity of chemicals. Mutat Res 86:193–214, 1981.
32. RJ Preston, W Au, MA Bender, JG Brewen, AV Carrano, JA Heddle, AF McFee, S Wolff, JS Wassom. Mammalian in vivo and in vitro cytogenetic assays: A report of the U.S. EPA's Gene-Tox Program. Mutat Res 87:143–188, 1981.

33. OECD. Guideline for the Testing of Chemicals: No. 473, In Vitro Mammalian Chromosomal Aberration Test: Organization for Economic Cooperation and Development, Paris, 1997.
34. T Sofuni, A Matsuoka, M Sawada, M Ishidate Jr., E Zeiger, MD Shelby. A comparison of chromosome aberration induction in the CHO and CHL cell systems by 25 compounds. Mutat Res 241:175–213, 1990.
35. DJ Kirkland. Chromosomal aberration tests in vitro: Problems with protocol design and interpretation of results. Mutagenesis 7:95–106, 1992.
36. SM Galloway, AD Bloom, M Resnick, BH Margolin, F Nakamura, P Archer, E Zeiger. Development of a standard protocol for in vitro cytogenetic testing with Chinese hamster ovary cells. Comparison of results for 22 compounds in two laboratories. Environ Mutagen 7:1–51, 1985.
37. BH Margolin, MA Resnick, JY Rimpo, P Archer, SM Galloway, AD Bloom, E Zeiger. Statistical analyses for in vitro cytogenetic assays using Chinese hamster ovary cells. Environ Mutagen 8:183–204, 1986.
38. OECD. Guideline for the Testing of Chemicals: No. 474, Mammalian Erythrocyte Micronucleus Test: Organization for Economic Cooperation and Development, Paris, 1997.
39. MD Shelby, GL Erexson, GJ Hook, RR Tice. Evaluation of a three-exposure mouse bone marrow micronucleus protocol: Results with 49 chemicals. Environ Mol Mutagen 21:160–179, 1993.
40. RR Tice, GL Erexson, MD Shelby. The induction of micronucleated polychromatic erythrocytes in mice using single and multiple treatments. Mutat Res 234:187–194, 1990.
41. T Morita, N Asano, T Awogi, YF Sasaki, S-I Sato, H Shimada, S Sutou, T Suzuki, A Wakata, T Sofuni, M Hayashi. Evaluation of the rodent micronucleus assay in the screening of IARC carcinogens (groups 1, 2A and 2B). The summary report of the 6th collaborative study by CSWGMT/JEMS.MMS. Mutat Res 389:3–122, 1997.
42. JT MacGregor, CM Wehr, PR Henika, MD Shelby. The in vivo erythrocyte micronucleus test: Measurement at steady state increases assay efficiency and permits integration with toxicity studies. Fundam Appl Toxicol 14:513–522, 1990.
43. AS Wright. The role of metabolism in chemical mutagenesis and chemical carcinogenesis. Mutat Res 75:215–241, 1980.
44. E Zeiger, RS Chhabra, BH Margolin. The effects of the hepatic S9 fraction from Aroclor 1254 treated rats on the mutagenicity of benzo(a)pyrene and 2-aminoanthracene in the *Salmonella*/microsome assay. Mutat Res 64:378–389, 1979.
45. E Zeiger. The *Salmonella* mutagenicity assay for identification of presumptive carcinogens. In: HA Milman, EK Weisburger, eds. Handbook of Carcinogen Testing. New Jersey: Noyes Publications, 1985, pp 83–99.
46. MJ Prival, VD Mitchell. Analysis of a method for testing azo dyes for mutagenic activity in *Salmonella typhimurium* in the presence of flavin mononucleotide and hamster liver S9. Mutat Res 97:103–116, 1982.
47. RS Chhabra, JE Huff, BS Schwetz, J Selkirk. An overview of prechronic and chronic toxicity/carcinogenicity experimental study designs and criteria used by the National Toxicology Program. Environ Health Perspect 86:313–321, 1990.

48. JK Selkirk, SM Soward. Compendium of abstracts from long-term cancer studies reported by the National Toxicology Program from 1976 to 1992. Environ Health Perspect 101 (suppl 1):3–281, 1993.
49. E Zeiger. Genotoxicity database. In: LS Gold, E Zeiger, eds. Handbook of Carcinogenic Potency and Genotoxicity Databases, Boca Raton: CRC Press, 1997, pp 687–729.
50. NTP web site. http://ntp-server.niehs.nih.gov/htdoc/pub.html
51. E Zeiger. Carcinogenicity of mutagens: Predictive capability of the *Salmonella* mutagenesis assay for rodent carcinogenicity. Cancer Res 47:1287–1296, 1987.
52. JK Haseman, JE Huff, E Zeiger, EE McConnell. Comparative results of 327 chemical carcinogenicity studies. Environ Health Perspect 74:229–235, 1987.
53. LS Gold, L Bernstein, J Kaldor, G Backman, D Hoel. An empirical comparison of methods used to estimate carcinogenic potency in long-term animal bioassays: lifetable vs summary incidence data. Fundam Appl Toxicol 6:263–269.
54. LS Gold, E Zeiger, eds. Handbook of Carcinogenic Potency and Genotoxicity Databases. Boca Raton: CRC Press, 1997.
55. R Peto, MC Pike, L Bernstein, LS Gold, BN Ames. The TD50: A proposed general convention for the numerical description of the carcinogenic potency of chemicals in chronic-exposure animal experiments. Environ Health Perspect 58:1–8, 1984.
56. RW Tennant. Stratification of rodent carcinogenicity bioassay results to reflect relative human hazard. Mutat Res 286:111–118, 1993.
57. E Dybing, T Sanner, H Roelfzema, D Kroese, RW Tennant. T25: A simplified carcinogenic potency index: Description of the system and study of correlations between carcinogenic potency and species/site specificity and mutagenicity. Pharmacol Toxicol 80:272–279, 1997.
58. BA Fetterman, BS Kim, BH Margolin, JS Schildcrout, MG Smith, SM Wagner, E Zeiger. Predicting rodent carcinogenicity from mutagenic potency measured in the Ames *Salmonella* assay. Environ Mol Mutagen 29:312–322, 1997.
59. JS Schildcrout, BH Margolin, E Zeiger. Predicting rodent carcinogenicity using potency measures of the in vitro sister chromatid exchange and chromosome aberration assays. Environ Mol Mutagen 33:59–64, 1999.
60. J Ashby, RW Tennant. Chemical structure, *Salmonella* mutagenicity and extent of carcinogenicity as indicators of genotoxic carcinogenesis among 222 chemicals tested in rodents by the U.S. NCI/NTP. Mutat Res 204:17–115, 1988.
61. LS Gold, TH Slone, BR Stern, L Bernstein. Comparison of target organs of carcinogenicity for mutagenic and non-mutagenic chemicals. Mutat Res 286:75–100, 1993.

3

Genotoxic and Nongenotoxic Mechanisms of Carcinogenesis

Wai Nang Choy
Schering-Plough Research Institute, Lafayette, New Jersey

I. INTRODUCTION

The mechanism of carcinogenesis has always been a fascinating topic of discussion in cancer biology, although our knowledge of it is still incomplete. Numerous mechanisms have been proposed and are often inspired by new findings in cancer research. In the past century, hypotheses on the etiology of human cancer have shifted many times, from chemical to viral, back to chemical, and in a few cases, chemical and viral. The mechanisms proposed are mutations of cancer genes, activation of oncogenes, inactivation of tumor suppressor genes, enhanced cell proliferation and with various combinations. These processes can occur sequentially or concurrently. The progression of carcinogenesis was initially divided into initiation and promotion and later multistage. The number of stages is difficult to estimate experimentally, but results of cancer epidemiological studies suggested there could be three to seven stages, based on the age-dependent increase of cancer incidence in humans (1). Although carcinogenesis is a multifaceted process, the resultant cancer cells are always genetically altered with multiple mutations and chromosome alternations. Human cancer is thus firmly considered as a "genetic disease" (2).

For chemical carcinogenesis, chemicals are customarily classified as genotoxic or nongenotoxic carcinogens based on the mutagenicity of the chemical. The initial perturbation of a genotoxic carcinogen to the cell is as genetic changes that result in mutations. Nongenotoxic carcinogens, on the other hand, induce epigenetic changes that result in abnormal gene expression.

The terms genotoxic and nongenotoxic carcinogenesis and risk assessment often have different meanings to scientists in different disciplines. In this discussion, genotoxic and nongenotoxic carcinogens are referred to the initial perturbation, or the "first stage," of carcinogenesis. Changes after the initial perturbation, genetic or epigenetic, do not affect the classification of genotoxicity of the chemical. Risk assessment is referred to as risk characterization, a process for quantitative estimation of cancer risks in humans, based on the genotoxicity and the potency of the carcinogens.

II. GENETIC CHANGES IN CANCER CELLS

Based on the somatic cell mutation theory of carcinogenesis proposed in the 1920s (3), gene mutation has long been considered as the primary mechanism of chemical carcinogenesis. The most convincing evidence of this theory comes from studies of inheritable cancers (4, 5) and various tumors that are known to associate with specific genes in humans (2). These findings confirmed the presence of human cancer genes and that they are chromosomally located. Mutations of these genes could result in neoplasm, although allelic mutations alone are not considered sufficient to acquire malignancy (6). Human cancer genes are often tumor suppressor genes, proto-oncogenes, and DNA repair genes. There are about 15 tumor suppressor genes, and more than 100 proto-oncogenes (7).

All human and rodent tumor cells contain mutations and chromosome aberrations (8, 9), with multiple, and sometimes specific, DNA sequence changes (10, 11). Studies of neoplastic transformation in cultured cells (12, 13) and in precancerous stages of human tumor development (14) indicated that the first genetic change always happens at an early or intermediate stage of carcinogenesis.

III. GENETIC AND EPIGENETIC MECHANISMS OF CARCINOGENESIS

In practice, the genotoxicity of a carcinogen is determined by a battery of genetic toxicology tests. A battery typically consists of three to four mutagenicity tests considered reliable for the detection of mutagens. They are a bacterial mutagenicity test (such as the Ames test), an in vitro cytogenetic assay in mammalian cells (chromosome aberration), an in vivo cytogenetic assay in rodents (such as micronucleus, chromosome aberration), and sometimes a mammalian cell gene mutation assay (such as mouse lymphoma, Chinese hamster ovary cells) (15).

Mechanisms of Carcinogenesis

The test procedures and specific requirement of these tests are well established and have been extensively described in two recent international guidelines (16–18). If positive or equivocal findings are observed, additional tests may need to be conducted (such as unscheduled DNA synthesis, DNA adduct), and final decisions are often made by weight of evidence and sound scientific judgment. Based on the results of these tests, a carcinogen is defined as genotoxic or nongenotoxic.

Because all cancer cells are genetically altered, the genetic and epigenetic mechanisms of chemical carcinogenesis can only be distinguished at a stage prior to the first genetic change. For genotoxic chemicals, mutation is assumed to occur at the first chemical-DNA interaction, which is the first stage. For nongenotoxic chemicals, mutation occurs after the first stage, depending on the mode of action of the carcinogen (Fig. 1).

A. The Genotoxic Mechanism

Because most carcinogens are mutagens, regulatory agencies often consider genotoxic mechanism as the default mechanism, unless there is sufficient evidence to indicate otherwise. Correlation studies of mutagenicity and carcinogenicity of chemicals in rodent bioassays showed a positive correlation ranging from ~70% to > 90%, depending on the database (19–21). The genotoxic mechanism undoubtedly accounts for most cancers.

In carcinogenesis, mutation may result in a gain or loss of function, depending on the role of the affected gene in cancer development. In general, activation (usually mutation) of oncogenes is considered a gain of function, for the new phenotype is the appearance of a new function. Oncogene activation often involves point mutations, chromosome translocations, and sometimes gene amplifications (22). In contrast, mutations in the tumor suppressor genes are loss of function, for which the mutated genes can no longer suppress tumor growth (7). Mutations in suppressor genes are often point mutations or deletions. Because the two copies of the tumor suppressor genes have to be inactivated for tumor expression, the term "loss of heterozygocity" (LOH) is used to denote the loss of the second copy of the allelic gene in a heterozygous genotype with a pre-existing mutation.

Mutated cells presumably acquire selective advantages, such as rapid growth rate, growth in adverse environment, and genome instability to permit further genetic alterations (23). Accumulation of genetic changes over time results in tumor cells.

The common types of genotoxic damage are described below.

1. DNA Sequence Changes

A genetic change is defined as a change in the DNA sequence. A change in a single nucleotide is a point mutation. The two common types of point mutations

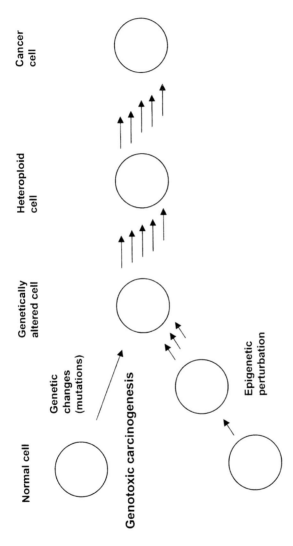

Fig. 1. Multistage carcinogenesis. The major difference in genotoxic and nongenotoxic carcinogenesis is at the first stage, in which genotoxic carcinogens induce mutations and nongenotoxic carcinogens create a susceptible state for genetic changes. Genetic changes occur at subsequent stages for both mechanisms.

Mechanisms of Carcinogenesis 51

are base-pair substitution and frameshift mutations. Base-pair substitution can be induced by a variety of chemicals including nucleoside analogs (such as bromodeoxyuridine, aminopurine) which result in mismatch of base pairs (transition or transversion) after DNA replication. It can also be induced by misrepair of DNA damage (excision or post-replicational repair) or imbalance of the nucleotide pool (such as excessive thymidine). Frameshift mutations involve an insertion or deletion of nucleotides, which results in a change in the reading frame of the coding sequence. This type of mutation can be caused by intercalating agents (such as 9-aminoacridine, ICR-191) or by deletions (such as depurination, illegitimate recombination).

Changes in multiple nucleotides, depending on their severity, can be detected as chromosome aberrations by cytogenetic techniques. These changes involve different magnitudes of deletion, translocation, insertion, inversion, or amplification. Deletion is caused by DNA breaks, translocation, or illegitimate recombinations. Gene amplification is caused by illegitimate repetitive recombinations. It also generates a homogeneous staining region (HSR) in the chromosome and double-minute chromosomes containing the amplified DNA sequence. Changes at specific regions of the chromosome can be identified by chromosome banding (G-band, Q-band) or, if the DNA sequence and DNA probes are available, by fluorescence in situ hybridization (FISH) (24).

The genetic consequence of these changes depends on the mutation site (intron, exon) and the function of the mutated gene (structural, regulatory). These changes result in altered gene transcription, mRNA processing, or translation that causes the absence of or defective gene products.

Studies on the interactions between chemical and DNA reveal several common mechanisms of mutagenesis. Two mechanisms often discussed are DNA adducts and free radical damage.

a. DNA Adducts. The term DNA adduct has been loosely ascribed to different types of DNA modifications. The discussion below is confined to the alkylated DNA adducts and macromolecular (bulky) DNA adducts. The DNA modifications by oxidative damage, sometimes also referred to as DNA adducts, are discussed in a separate section.

Certain classes of chemicals, notably alkyl sulfates, N-nitrosamines, N-nitrosoureas, and nitrogen mustards can alkylate DNA to form DNA adducts (25). The alkyl sulfates and nitrogen mustards react with DNA to form N^7-alkylguanine and N^3-alkyladenine, and the N-nitroso compounds react with oxygen to generate O^6-alkylguanine and O^4-alkylthymidine (26). These alkylated adducts, if not repaired timely or correctly (27), are carcinogenic. Unrepaired adducts often undergo spontaneous depurination (28), and if misrepaired (29), will result in base-pair substitutions. Another well-known mutagenic adduct is N^2-3-ethenoguanine, which is induced by chloroethylene oxide, the metabolite of vinyl chloride (30). These alkylated DNA adducts can cause DNA configuration changes that interfere with DNA replication and transcription (31, 32).

A few chemicals are known to induce macromolecular DNA adducts. They are the polycyclic aromatic hydrocarbons (PAH) (such as benzo(a)pyrene) (33), mycotoxins (such as aflatoxin B) (34), chemotherapeutic agents (such as cisplatinum, 8-methoxypsoralen) (35, 36) and environmental chemicals (such as styrene oxide) (37). These bulky adducts, as expected, change DNA configuration and cause DNA damage. In fact, bulky adducts are often detected in tumor tissues and are used as biomarkers to study molecular dosimetry for exposure assessment (38, 39).

Recent studies revealed a class of bulky DNA adducts known as I-compounds. These are covalent adducts generated by endogenous reactive intermediates of oxygen metabolism. The I-compounds are classified as type I and type II. The type I compounds are species-specific adducts and their formation is affected by environmental factors. An increase of type I compound was associated with a decrease of preneoplastic lesions, which has led to the speculation that they may be protective for carcinogenesis. The type II compounds, in contrast, are generated by DNA damages or DNA cross-links. They are found in tumor tissues and are associated with tumor development (40).

b. Free Radical DNA Damage. Free radicals are reactive atoms with unpaired electrons. Radicals of the oxygen and nitrogen species are known to oxidize DNA and induce damage. Oxidative damage is associated with inflammation, neurodegenerative diseases, aging, and cancer (41–43). Acute and large amount of oxidative damage are cytotoxic to the cells and chronic low level of damage may interfere with cell cycle control (44).

The common reactive oxygen species are superoxide (O_2^-), hydroxyl radical ($OH^{\bullet}$), peroxyl radical ($RO^{\bullet}_2$), alkoxyl radical ($RO^{\bullet}$), ozone (O_3), singlet oxygen ($^1O_2$), hydrogen peroxide (H_2O_2), and hypochlodrite (HOCl). The reactive nitrogen species are nitric oxide (NO_2) and peroxynitrite ($ONOO^-$). The reactive oxidative radicals are generated by normal chemical metabolism, and they can sometimes function as mediators of cell division and signal transduction (45–47). Chemicals such as peroxisome proliferators (48) and asbestos fibers are known to generate reactive oxidants (49)

As expected, oxidative DNA damage, if not repaired correctly (50), results in gene mutations, DNA breaks, and cross-links (51, 52). A common DNA modification involves the reaction with the hydroxyl radical ($OH^{\bullet}$). Hydroxylation of DNA forms 8-hydroxydeoxyguanosine and 8-hydroxyguanine (53), but adducts of the other nucleotides were also reported (54). In addition, the $OH^{\bullet}$ radical can also oxidize or hydrolyze the deoxyribose moiety of DNA, generating apurinic or apyrimidinic sites susceptible to mutations or DNA breaks (52).

B. The Nongenotoxic Mechanisms

The nongenotoxic mechanisms of carcinogenesis do not involve DNA sequence changes at an early stage. Nongenotoxic carcinogens are defined by their absence of mutagenicity in genetic toxicology test (16–18). The nongenotoxic mechanisms are also referred to as "secondary mechanisms" (55), with the implication that they are alternate mechanisms. Unlike the genotoxic mechanism that is based simply on mutations, the nongenotoxic mechanisms are much more diverse, and in some cases, as yet speculative (56). In fact, most nongenotoxic carcinogenesis are species-, sex-, or tissue-specific, and they are often observed after high dose or prolonged exposures (57, 58). These mechanisms have been extensively reviewed (55, 59). A brief summary of the major mechanisms is included below.

In the absence of mutation, phenotypic changes can only be achieved by changes of gene expression. The most well-studied mechanism of gene regulation is methylation of DNA. DNA methylation often occurs at cytosine and is mediated by methyltransferase to form 5-methylcytosine. Hypermethylation suppresses gene expression, and conversely, hypomethylation enhances expression. Unregulated hypomethylation thus has been proposed to be a mechanism for nongenotoxic carcinogenesis (60, 61). However, hypermethylation has also been observed in carcinogenesis (62). This inconsistency may be explained by the functions of the genes that are involved, whether they have to be activated or silenced in neoplasm, such as the tumor suppressor and the senescence genes (63). Incidentally, methylation can be reversed by an inhibitor of methyltransferase, 5-aza 2´-deoxycytidine, which is commonly used in experimental studies. Recent studies showed that although methylation of DNA has always been considered a nongenotoxic process, hydrolytic deamination of 5-methylcytosine can induce mismatch point mutations and that the DNA methylation sites are often mutation hot spots in human tumors (64).

Several common nongenotoxic mechanisms of carcinogenesis are discussed below.

1. Cell Proliferation

The rationale of cell proliferation-induced cancer is based on the concepts of spontaneous mutation rate and clonal expansion of pre-existing mutants, two different processes for the generation of mutants. In the first process, the number of mutations is a function of the mutation rate of the gene (mutation/cell/generation) and the number of cell divisions. Theoretically, with the accumulation of random mutations over many cell divisions, a mutation in the cancer gene would initiate carcinogenesis (65). The second process is clonal expansion of pre-existing mutants in the cancer gene upon cell divisions. This process is analogous to the classic initiation-promotion model of carcinogenesis in which an initiated cell is promoted by cell proliferation. Enhanced cell proliferation in both cases enhanced the generation of cancer cells and shortened the time for carcinogenesis. The process may appear genetic, but the first stage is stimulation of cell proliferation, not direct damage to the DNA.

Two types of cell proliferation are induced by different mechanisms. They are regenerative (or compensatory) proliferation after toxic damage to the tissue and growth factor-induced cell proliferation. Examples of these two mechanisms are described below.

a. Regenerative Cell Proliferation. Male rat renal tumor: most rodent kidney tumors are induced by genotoxic chemicals. A specific type of tumor, the renal tubule cell tumor in the male rat, has been proposed to be induced by regenerative cell proliferation as a result of chronic kidney damage (66). Unlike in other species, or even in female rats, the male rats produce a large amount of protein called α_{2u}-globulin in the liver. Several kidney carcinogens in male rats such as d-limonene, tetrachloroethylene) are capable of binding to α_{2u}-globulin. These bindings are believed to interfere with the degradation of α_{2u}-globulin, which results in the accumulation of α_{2u}-globulin in the lysosomes of renal proximal tubule cells. It is also possible, however, that undegraded α_{2u}-globulin acts as a vector that carries the carcinogens to the kidney (67). Regardless, prolonged accumulation of α_{2u}-globulin or α_{2u}-globulin-chemical complexes in the kidney appears as hyaline droplets, which are toxic to the cells. Regenerative cell proliferation is stimulated which resulted in atypical tubule hyperplasia, leading to the development of renal tubule tumors (68).

Male rat urinary bladder tumor: male rats are also unique in their susceptibility to urinary bladder tumor induced by high doses of sodium salts of moderate or strong organic acids (e.g., sodium ascorbate, saccharin). These chemicals are known to form calcium phosphate precipitates in the urine at high protein concentration and the amount increases at pH above 6.5 (69–71). The precipitates are microcrystals, amorphous precipitates, or calculi that irritate the surface epithelium of the bladder. The male rats are sensitive because of their relatively high concentration of protein, such as α_{2u}-globulin, in the urine as compared with female rats and other rodent species. Prolonged irritation of the bladder cells causes cell death, which stimulates regenerative cell proliferation and may result in bladder tumors (72).

Rodent forestomach tumor: although most forestomach tumors are induced by genotoxic chemicals, some nongenotoxic carcinogens (such as butylated hydroxyanisole, chlorothalonil) are known to induce forestomach tumors. The mechanism proposed is chemical irritation of the forestomach epithelium followed by cell death and regenerative proliferation (73,74). Early cytotoxic changes were reportedly reversible, which indicated the epigenetic nature of the mechanism. Morphological changes are epithelial damage, hyperplasia, hyperkeratosis, dysplasia, papilloma, and eventually squamous cell carcinoma (75).

Rodent liver tumor: most halogenated hydrocarbons are rodent liver carcinogens. Several nongenotoxic agents, including chloroform, induced liver tumors in mice. A recent rodent bioassay in chloroform showed liver tumors in female mice at high doses. This observation led to the postulation that liver tumor induced by chloroform was caused by regenerative cell proliferation after toxic damage of liver cells (76). Cell death at high doses probably was caused by phosgene, a metabolite of chloroform generated by cytochrome P450 enzymes (77). Again, regenerative cell proliferation was proposed to be the cause of liver tumor induction.

b. Growth Factor-Induced Cell Proliferation. Hormones. Rodent thyroid follicular cell tumor: in contrast to human thyroid tumors that are mostly induced by ionizing radiation, thyroid follicular cell tumors in rodents are induced by genotoxic or nongenotoxic carcinogens. Several nongenotoxic chemicals (such as perchlorate, thioureas, lithium, lupiditine) are known to perturb the balance of the thyroid stimulating hormone (TSH) in rodents. These chemicals are referred to as goitrogens and their modes of action are different. They may deplete iodine accumulation by inhibiting iodine trapping in the thyroid or by blocking binding of iodine and coupling of iodothyronine to form thyroxine and triiodothyronine. Alternately, they may inhibit thyroid hormone secretion by proteolysis or enhance metabolism of thyroxine by inducing liver metabolic enzymes (78). Disruption of thyroxine function activates the pituitary-thyroid feedback mechanism, which increases the production of TSH. An increase of TSH in the thyroid, if sustained, stimulates the proliferation of thyroid follicular cells through a TSH receptor-mediated signal transduction process. Excess cell proliferation may result in thyroid tumors (79, 80).

Estrogen-induced tumor: estrogen and its metabolites are known to induce mammary tumors in rodents and humans, pituitary tumors in rats, and kidney tumors in hamsters. Estrogen is metabolized by cytochrome P450 enzymes mostly in the mammary gland and uterus, which forms hydroxylated metabolites. Chemicals that alter estrogen metabolism (such as benzo(a)pyrene, dimethylbenz(a)anthracene) are known to induce DNA synthesis, decrease apoptosis, and enhance cell proliferation (81). At least two mechanisms have been proposed, and they are not mutually exclusive. One proposal is that estrogen or its metabolites directly activates the estrogen receptors to stimulate cell proliferation. The other proposal is that the 4-hydroxylated catechol metabolites of estrogen can undergo reduction-oxidation to generate free radicals. The free radicals then react with DNA to generate 8-hydroxylguanine and other DNA modifications, resulting in genetic damage (82–85). Several estrogen-like compounds (such as diethylstilbestrol, bisphenol A, nonylphenol, polychlorinated biphenyls) are also known to be carcinogenic, but their mechanisms have not been well studied (86, 87).

Organic chemicals. Several nongenotoxic polychlorinated aromatic hydrocarbons can induce cell proliferations through a common aryl hydrocarbon (Ah) receptor, an example of which is 2,3,78-tetrachlorodibenzo-p-dioxin (TCDD). TCDD is known to bind to the Ah receptor and initiates a signal transduction pathway that triggers cell proliferation. Enhanced cell proliferation is proposed to be the mechanism TCDD carcinogenesis (88).

2. Peroxisome Proliferation

Peroxisomes are membrane-bound organelles that contain oxidation enzymes. They are catalase, hydrogen peroxide-generating oxidases and a fatty acid β-oxidation enzyme system (89, 90). Several classes of chemicals (such as hypolipidemic drugs, phthalate esters, and phenoxy acid herbicides) are known to stimulate the proliferation of peroxisomes, which results in increases in the number and sometimes the size (volume density) of peroxisomes. The peroxisomes in rodent livers are especially susceptible to proliferation stimulation. Peroxisome proliferation in rodent livers resulted in liver cell proliferation, liver hyperplasia, hypertrophy, and neoplasm.

Concomitant to peroxisome proliferation, enzyme activities for fatty acid β-oxidation and P450 enzymes (CYP4A subfamily) in the liver are also increased. Catalase activity, however, is increased but not proportionally (89). This imbalanced increase produces high levels of hydrogen peroxide, which may cause oxidative damage to the DNA in the liver (48).

The stimulation of peroxisome proliferation by chemicals is mediated by peroxisome proliferator-activated receptors (PPAR) (91, 92). The PPAR belong to a steroid hormone receptor superfamily and are present in mice, rats, frogs, and humans. At least three subtypes of PPAR (alpha, beta, and gamma) were identified (93). Peroxisome proliferator-activated receptor alpha was found to mediate the stimulation of peroxisome proliferators in mice, and the total PPAR level increases in response to peroxisome proliferation in rats and mice (94, 95). Activation of PPAR induces the transcription of the liver genes for fatty acid metabolism (94–96). Rats and mice are more susceptible to peroxisome proliferators than hamsters, guinea pigs, and monkeys.

It is important to note that the amount of PPAR alpha in humans is much less than in rodents (97, 98). Studies in humans treated with hypolipidemic drugs (rodent peroxisome proliferators) and in cultured human hepatocytes treated with peroxisome proliferators did not show peroxisome proliferation (92, 99, 100). The role of peroxisome proliferation in human carcinogenesis requires further evaluation. At present, none of the known human liver carcinogens (IARC group 1 carcinogens) is a peroxisome proliferator (101, 102). Rodent peroxisome proliferators do not appear to pose a cancer risk to humans.

3. Intercellular Communication

A defect in the intercellular communication has been proposed to be an epigenetic mechanism of carcinogenesis (103–106). Intercellular communication is mediated through gap junctions, which are membrane structures that allow the transfer of small molecules between cells. The molecules can be nutrition or signal transduction molecules. A gap junction protein, connexin, was identified as an essential protein for cell communication. Mislocation of connexin in the membrane was observed in tumor cells (104). The expression of the connexin gene was also down regulated in tumor cells, and inactivation was caused by hypermethylation (107). The connexin gene appears to act as a tumor suppressor gene (that is, functional in normal cells), for its inactivation resulted in carcinogenesis (103, 107).

4. Genomic Imprinting

Genomic imprinting is an epigenetic modification of a parental allele of a gene in the gamete or zygote, resulting in differential expression of the paternal and maternal alleles in the offspring. This unequal expression is a departure from Mendelian genetics, but it is frequently observed in mammalian genetics. Alternations of the imprinted genes have been observed in many human cancer cells (such as Wilms' tumor, hepatoblastoma, rhabdomyosarcoma, Ewing's sarcoma) (108–110). This change is referred to as loss of imprinting (LOI). A possible consequence of LOI, as related to carcinogenesis, is the activation of an inactivated cancer gene or the inactivation of a tumor suppressor gene. The mechanism of imprinting is DNA methylation, because abnormal methylation patterns are observed in LOI regions. Moreover, studies with 5-aza 2´-deoxycytidine, an inhibitor of methylation, and in methyltransferase-deficient mice indicated that methylation is responsible for imprinting (109, 111). The most carefully studied gene for imprinting is the insulin-like growth factor 2 gene (IGF2). This gene is located on human chromosome 11p15.5, for which LOI was observed in a variety of human cancers, including kidney, bladder, cervical, prostate, testicular, and ovarian (112). In addition to human medical conditions, environmental agents have also been reported to induce LOI (113).

IV. HUMAN VERSUS RODENT CARCINOGENESIS

Of the more than 400 rodent chemical carcinogens in the IARC carcinogen list (group 2), only about 50 human chemical carcinogens (group 1) have been identified so far (101). This large difference may be explained by the lack of sufficient epidemiological studies of rodent carcinogens, which are required for their evaluation as human carcinogens. However, it is also possible that there are inherent genetic and biochemical differences between humans' and rodents' response to chemical carcinogenesis. Evidence to support this proposition is discussed below.

A very significant finding, but seldom discussed, is the unique resistance of human cells to chemical carcinogenesis in culture, as demonstrated in in vitro neoplastic transformation studies. Neoplastic transformation in vitro means the immortalization of a primary cell culture to become established cell lines and to acquire neoplastic phenotypes, including the ability to form a tumor in immunodeficient mice. In these studies, rodent cells have repeatedly demonstrated their ability to transform spontaneously, even without chemical treatments, and transformation can easily be enhanced by chemicals. In contract, normal diploid human cells have never been observed to transform spontaneously, and chemical transformation has been extremely difficult, even by very potent carcinogens. In the past 40 years, only a few highly experienced laboratories have reported successful chemical transformation of diploid human cells in culture (114–119). In many cases, transformation was only achievable under "presensitized" conditions at the S-phase of the cell cycle (120).

These differences between rodent and human cell transformation have been extensively discussed as related to the critical phases of the lifespan of primary cells in culture (121). Briefly, all primary cells in culture are diploid and are capable of replication for a definite number of times characteristic of the cell type before senescence. Diploid human fibroblasts, for example, usually divide about 50 times before entering senescence and cell death (122). Primary rodent cells also enter senescence after an initial period of cell division but, strikingly different from human cells, a small number of rodent cells invariably can escape senescence and rejuvenate as a distinct population of immortal cells (that is, established cell lines). The most prominent change accompanying the escape of senescence is the change of chromosome from diploid to heteroploid (121, 123). This heteroploid conversion presumably creates genetic instability, which favors further genetic changes required for neoplastic transformation.

Diploid human cells in culture cannot spontaneously escape senescence to form heteroploid and established cell lines, and senescence is always terminal. Much speculation has been focused on the intrinsic genetic stability and efficient DNA repair in human cells, but experimental evidence is lacking.

The same sequence of events was observed in chemical transformation of human cells. Chemically transformed human cells have been shown to escape senescence and acquire structural and numerical chromosome changes (124). In addition, increases of telomerase activity associated with immortality (125), and loss of function of tumor suppressor genes (p53, RB-1, p16INK4) (125–129) were also reported.

It is evident that human cells are much more resistant to chemical transformation in culture. This characteristic may imply a protective mechanism from chemical carcinogens in humans. In contrast, susceptibility of human cells to chemical mutagenesis is similar to that of rodent cells. This observation indicated that the resistance of transformation in human cells appears to occur after the initial mutation stage which also confirmed the observation that two allelic mutations are not sufficient to induce human cancers (6). Without much knowledge on the mechanism of this interspecies difference, the application of this finding as related to human risk assessment remains uncertain.

V. GENOTOXICITY AND QUANTITATIVE RISK ASSESSMENT

Current methodologies used for cancer risk characterization depend on the genotoxicity of the carcinogen. Genotoxicity, as mentioned before, is determined by the test results of a battery of genetic toxicology tests (16–18). If test results are equivocal, the weight of evidence will be evaluated. For regulatory risk assessment, the default assumption is always genotoxic because a conservative position is considered prudent for the protection of public health.

With increasing knowledge of the mechanisms of genotoxic and nongenotoxic carcinogenesis, the distinction between these two groups of carcinogens is less clear. Many nongenotoxic carcinogens are known to indirectly induce mutations, and examples of which are oxidative DNA damage by peroxisome proliferators and formation of DNA adducts by estrogen. For the purpose of risk characterization, genotoxicity of a carcinogen should be defined by the "first stage" of carcinogenesis, even though genetic changes may occur in subsequent stages.

For genotoxic chemicals, the commonly used cancer risk assessment models are constructed based on two assumptions: multistage carcinogenesis and the absence of a carcinogenic threshold. The rationale for these assumptions are described below.

A. Multistage Carcinogenesis

Mulistage carcinogenesis has been repeatedly demonstrated in a variety of cancer bioassays. The most notable examples are the inheritable childhood cancers (4, 130) and human colorectal cancer (14), the mouse skin initiation-promotion carcinogenesis (131), the rat liver carcinogenesis (132), and the in vitro Syrian hamster embryo cell transformation assays (12, 133). The numbers of stages described in these studies varied, depending on the number of biomarkers monitored. Also, different biomarkers were used in these studies so that direct comparison of the number of stages is difficult, if at all possible. The number of stages is expected to increase with the availability of additional biomarkers and the improvement of technologies for their detection. Nevertheless, epidemiological analysis of human cancer incidences, as related to age-dependent increases, reported an estimate of three to seven rate-limiting stages (1). For quantitative risk assessment models, however, the number of stages is not critical because the first stage determines the cancer potency of the carcinogen.

B. Threshold

The concept that genotoxic carcinogenesis does not have a threshold dose below which cancer is not induced is a controversial issue. The assumption that a single molecular damage (one-hit) is sufficient to initiate carcinogenesis and produce a linear dose-response curve has been repeatedly challenged. Direct proof of low-dose linearity is difficult, but several studies, by using DNA adduct as the target molecule dose, have demonstrated linear dose-response at the low-dose region (39, 134). It can always be argued, however, that these doses are not low enough to demonstrate linearity at very low doses.

Indeed, linear dose-response should not be argued over the commonly observed sigmoid dose-response curves of tumor induction based on the applied doses in animal bioassays. These dose-response curves are, in fact, composite curves of several dose-response curves that resulted in tumor induction. Specifically, they are the dose-response curves of chemical absorption, distribution, metabolism, excretion, cellular transport, and DNA damage. All of these dose-response curves have to be linear for the composite curve to be linear, which is a rare event, if ever happening. The absence of a linear dose-response for tumor induction cannot rule out a linear dose-response of tumor induction based on DNA damage.

The concept of a threshold for nongenotoxic carcinogens, in contrast, is widely acceptable, based on the assumption that multiple targets (receptors) or many cells (regenerative proliferation, multihit) are required for a nongenotoxic response. Chemical binding to a single receptor or induction of a single cell proliferation (which is unlikely) is considered insufficient for the initiation of carcinogenesis. Experimental evidence, however, is lacking.

VI. RISK CHARACTERIZATION

Risk characterization is a quantitative estimate of the carcinogenicity of a carcinogen. For genotoxic carcinogens, this estimate is derived from the "cancer potency" of the carcinogen, which is the slope of the dose-response curve of tumor induction at low doses (Fig. 2). For nongenotoxic carcinogens, the estimate is often the no effect level (NOEL) of carcinogen response modified by a safety factor.

For genotoxic carcinogens, several cancer risk assessment models have been developed for the calculation of the cancer potency. They are statistical models designed to determine the shape and the slope of the dose-response curve of tumor induction. In most situations in which only rodent cancer bioassays are available, the models extrapolate the slope of the dose-response curve from the high-dose to the low-dose region, which is a typical situation of human exposure. A variety of risk assessment models have been proposed (135–138), but the most commonly used model by regulatory agencies is the linearized multistage model. This model was constructed based on multistage carcinogenesis and the absence

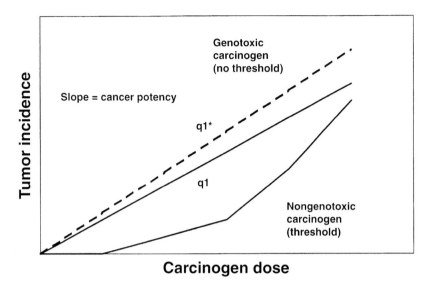

FIG. 2. Theoretical dose-response curves of tumor induction by genotoxic and nongenotoxic carcinogens at the low dose region. Dose-response of genotoxic carcinogens are assumed to be linear without a threshold dose. The cancer potency is the slope of the dose-response curve, designed as (q1) in the linearized multistage risk assessment model. The upper 95% confidence (q1*) is often used for regulatory risk analysis. Nongenotoxic carcinogens are assumed to have a threshold dose.

of a carcinogenic threshold (139–141). This is a default model used by regulatory agencies in the absence of additional information to warrant the inclusion of additional parameters for analysis. Several more sophisticated models are available, the most widely acclaimed being the Moolgavkar-Knudson model (130, 142). This model includes considerations of two allelic gene mutations, cell proliferation rate and cell death rates—a marked improvement over the previous models. However, similar to other biologically or physiologically based models (143–145), experimental values of many parameters required for these models are not available. By using mostly default values, an improvement over the linearized multistage model is questionable. This limitation hampers the application of these newer models for regulatory risk assessment of most carcinogens.

VII. SUMMARY

Various genotoxic and nongenotoxic mechanisms of chemical carcinogenesis are briefly discussed and an overall summary is presented in Table 1. Genotoxicity of

TABLE 1. An Overall Comparison of Genotoxic and Nongenotoxic Carcinogens

	Genotoxic carcinogens	Nongenotoxic carcinogens
Results of genetic toxicology tests	Positive	Negative
Multistage carcinogenesis First stage	Genetic changes Gene mutation Chromosome aberration	Epigenetic changes Cell proliferation Regeneration Growth factor stimulation Peroxisome proliferation Intercellular communication Genomic imprinting
Subsequent stages	Genetic or epigenetic changes	Genetic or epigenetic changes
Dose-response at low doses (assumptions)	No carcinogenic threshold Linear dose-response curve	Carcinogenic threshold Sigmoid dose-response curve
Risk characterization (common methods)	Linearized multistage model Biologically based model	Safety factor method

a carcinogen is defined by the results of a battery of genetic toxicology tests. Genotoxic carcinogens induce direct DNA damage and nongenotoxic carcinogens induce epigenetic changes. The genotoxic mechanisms are different types of mutations, including DNA modifications and DNA adducts. The nongenotoxic mechanisms, however, are more diverse. The most common mechanisms are enhanced cell proliferation induced by cytotoxicity or chemical stimulations, peroxisome proliferation, changes in intercellular communications, and loss of imprinting.

With recent knowledge that several nongenotoxic carcinogens are indirectly responsible for inducing DNA damage, the distinction between genotoxic and nongenotoxic carcinogens is less apparent. Such a distinction is only meaningful at the first stage of carcinogenesis.

The genotoxicity of the carcinogen at the first stage of carcinogenesis has great implications for regulatory quantitative risk assessment. The dose-response of genotoxic carcinogens is considered not to have a threshold, and the dose-response curve at the low dose region is linear. The dose-response of nongenotoxic carcinogens, however, is assumed to have a threshold and the dose-response curve is often sigmoid.

For risk characterization, the linearized multistage model is routinely used by regulatory agencies for the estimation of cancer potency of genotoxic carcinogens. For nongenotoxic carcinogens, the safety factor method is used. Recent advances in biologically based and physiologically based risk assessment modeling will undoubtedly improve the reliability of risk analysis, but the lack of experimental data for new parameters of these models limits their use at this time.

An unexpected but very important observation is that human cells in culture are extremely resistant to chemical carcinogenesis as compared with rodent cells. This difference was not observed in chemical mutagenesis of human cells. This finding raised the possibility that the mechanism of neoplastic transformation in humans, especially at stages beyond the initial mutational changes, is different from that in rodents. Further investigation should be conducted to understand this phenomenon.

REFERENCES

1. DG Miller. On the nature of susceptibility to cancer. The presidential address. Cancer 46:1307–1318, 1980.
2. B Volgelstein, KW Kinzler. The Genetic Basis of Human Cancer. New York: McGraw-Hill, 1998.
3. TH Boveri. The Origin of Malignant Tumors. Baltimore: Williams and Wilkins, 1929.
4. AG Knudson, HW Hethcote, BW Brown. Mutation and childhood cancer. A probabilistic model for the incidence of retinoblastoma. Proc Natl Acad Sci U S A 72:5116–5120, 1975.

5. AG Knudson. Genetics of human cancer. Annu Rev Genet 20:231–251, 1986.
6. AG Knudson. Hereditary predisposition to cancer. Ann NY Acad Sci 833:58–67, 1997
7. ER Fearson. Tumor Suppressor Genes. In: B. Vogelstein, KW Kinzler eds. The Genetic Basis of Human Cancer. New York: McGraw-Hill, 1998, pp 229–236.
8. F Mitelman, S Heim. Chromosome abnormalities in cancer. Cancer Detect Prev 14:527–537, 1990.
9. AA Sandberg. Chromosome abnormalities in human cancer and leukemia. Mutat Res 247:231–249, 1991.
10. RC Sills, GA Boorman, JE Neal, HL Hong, TR Devereux. Mutations in ras genes in experimental tumors of rodents. IARC Science Publication 146, 1999, pp 55–86.
11. T Hernandez-Boussard, P Rodriguez-Tome, R Montesano, P Hainaut. IARC p53 mutation database: A relational database to compile and analyze p53 mutations in human tumors and cell lines. Human Mutat 14:1–8, 1999.
12. JC Barrett, PO Ts'o. Evidence for the progressive nature of neoplastic transformation in vitro. Proc Natl Acad Sci U S A 75:3761–3765, 1978.
13. JC Barrett. Mechanisms of multistep carcinogenesis and carcinogenic risk assessment. Environ Health Perspect 100:9–20, 1993.
14. KW Kinzler, B Vogelstein, Colorectal Tumors. In: B Vogelstein, KW Kinsler, eds. The Genetic Basis of Human Cancer. New York: McGraw-Hill, 1998, pp 565–587,
15. WN Choy. Genetic Toxicology Testing. In: AM Fan, LW Chang, eds. Toxicology and Risk Assessment. New York: Marcel Dekker, 1996, pp 153–169.
16. ICH. International Conference on Harmonisation: Guidance on Specific Aspects of Regulatory Genotoxicity Tests for Pharmaceuticals; Availability. Federal Register 61:18198–18202, 1996.
17. ICH. International Conference on Harmonisation: Guidance on Genotoxicity: A Standard Battery for Genotoxicity Testing of Pharmaceuticals; Availability; Notice. Federal Register 62:62472–62475, 1997.
18. OECD. Organisation for Economic Co-Operation and Development. OECD Guidelines for the Testing of Chemicals. Section 4. Health Effects. OECD, Paris, France, 2000.
19. RW Tennant, JW Spalding, S Stasiewicz, WD Caspary, JM Mason, MA Resnick. Comparative evaluation of genetic toxicity patterns of carcinogens and noncarcinogens: Strategies for predictive use of short-term assays. Environ Health Perspect 75:87–95, 1987.
20. E Zeiger, JK Haseman, MD Shelby, BH Margolin, RW Tennant. Evaluation of four in vitro genetic toxicity tests for predicting rodent carcinogenicity: Confirmation of earlier results with 41 additional chemicals. Environ Mol Mutagen 1:1–14, 1990.
21. LS Gold, TH Slone, BR Stern, L Bernstein. Comparison of target organs of carcinogenicity for mutagenic and non-mutagenic chemicals. Mutat Res 286:75–100, 1993.
22. M Park. Oncogenes. In: B Vogelstein, KW Kinzler, eds. The Genetic Basis of Human Cancer. New York: McGraw-Hill, 1998, pp 205–228.
23. IP Tominson, MR Novelli, WF Bodmer. The mutation rat and cancer. Proc Natl Acad Sci U S A 93:14800–14803, 1996.

24. AK Raap. Advances in fluorescence in situ hybridization. Mutat Res 400:287–298, 1998.
25. CP Wild, R Montesano. Detection of alkylated DNA adducts in human tissues. In: JD Groopman, PL Skipper, eds. Molecular Dosimetry and Human Cancer. Boca Raton, FL: CRC Press: 1991, pp 263–276.
26. B Singer, D Grunberger. Molecular biology of mutagens and carcinogens. New York: Plenum Press, 1983.
27. M Bouziane, F Miao, N Ye, G Holmquist, G Chyzak, TR O'Connor. Repair of DNA alkylation damage. Acta Biochim Pol 45:191–202, 1998.
28. LA Loeb, BD Preston. Mutagenesis by apurinic/apyrimidine sites. Annu Rev Genet 20:201–230, 1986.
29. B Singer. O-Alkyl pyrimidines in mutagenesis and carcinogenesis: Occurrence and significance. Cancer Res 46:4879–4885, 1986.
30. B Singer, SJ Spengler, F Chavez, JT Kusmierrk. The vinyl chloride-derived nucleotide, N2, 3-ethenoguanosine, is a highly efficient mutagen in transcription. Carcinogenesis 8:745–747, 1987.
31. MW Kalnik, M Kouchakdjian, BF Li, PF Swann, DJ Patel. Base pair mismatches and carcinogen-modified bases in DNA: An NMR study of G•T and G•O^{4me}T pairing in dodecanucleotide duplexes. Biochemistry 27:108, 1988.
32. KK Richardson, FC Richardson, RM Crosby, JA Swenberg, TR Skopek. DNA base changes and alkylation following in vivo exposure of Escherichia coli to N-methyl-N-nitrosourea or N-ethyl-N-nitrosourea. Proc Natl Acad Sci U S A 84:344, 1987.
33. KM Shamsuddin, NT Sinopoli, K Hemminki, RR Boesch, CC Harris. Detection of benzo(a)pyrene-DNA adducts in human white blood cells. Cancer Res 45:66–68, 1985.
34. RG Croy, JM Essigmann, VN Reinhold, GN Wogan. Identification of the principal aflatoxin B1-DNA adduct formed in vivo in rat liver. Proc Natl Acad Sci U S A 75:1745–1749, 1978.
35. E Reed, E, S Yuspa, LA Zwelling, RF Ozols, MC Poirier. Quantitation of cis-diaminedichloroplatinum (II) (cisplatin)-DNA-intrastrand adducts in testicular and ovarian cancer patients receiving cisplatin chemotherapy. J Clin Invest 77:545–550, 1986.
36. E Ben-hur, PS Song. The photochemistry and photobiology of furocoumarines (psoralens). Adv Radiat Biol 11:131–171, 1984.
37. F Liu, SM Rappaport, K Pongracz, WJ Bodell. Detection of styrene oxide-DNA adducts in lymphocytes of a worker exposed to styrene. In: H Bartsch, K Hemminki, IK O'Neill, eds. Methods for Detecting DNA Damaging Agents in Humans: Applications in cancer epidemiology and prevention. IARC Scientific Publication No. 89, 1999, pp 159–174.
38. JD Groopman, PL Skipper. Molecular Dosimetry and Human Cancer. Boca Raton: CRC Press, 1991.
39. WN Choy. A review of the dose-response induction of DNA adducts by aflatoxin B1 and its implications to quantitative cancer risk assessment. Mutat Res 296:181–198, 1993.
40. K Randerath, E Randerath, GD Zhou, D Li. Bulky endogenous DNA modifications (I-compounds)—possible structural origins and functional implications. Mutat Res 424:183–194, 1999.

41. G Wiseman, B Halliwell. Damage to DNA by reactive oxygen and nitrogen species: Role in inflammatory disease and progression to cancer. Biochem J 313:17–29, 1966
42. BN Ames. Dietary carcinogens and anticarcinogens. Oxygen radicals and degenerative diseases. Science 221:1256–1264, 1983.
43. B Halliwell. Oxygen and nitrogen are pro-carcinogens. Damage to DNA by reactive oxygen, chlorine and nitrogen species: Measurement, mechanisms and the effects of nutrition. Mutat Res 443:37–52, 1999.
44. JE Klaunig, Y Xu, JS Isenberg, S Bachowski, KL Kolaja, J Jiang, DE Stevenson, EF Walborg Jr. The role of oxidative stress in chemical carcinogenesis. Environ Health Perspect 106:289–295, 1998.
45. C Rice-Evans, R Burdon. Free radical-lipid interactions and their pathological consequences. Prog Lipid Res 32:71–110, 1993.
46. I Fridovich. Oxygen toxicity: A radical explanation. J Exp Biol 201:1203–1209, 1998.
47. AE Aust, JF Eveleigh. Mechanisms of DNA oxidation. Proc Soc Exp Biol Med 222:246–252, 1999.
48. JK Reddy, MS Rao. Peroxisome proliferation and hepatocarcinogenesis. In: H Vainio, PN Magee, DB McGregor, AJ McMichael, eds. Mechanism of Carcinogenesis in Risk Identification. IARC Scientific Publications No. 116, 1992, pp 225–235.
49. VL Kinnula. Oxidant and antioxidant mechanisms of lung disease caused by asbestos fibers. Eur Respir J 14:706–716, 1999.
50. GW Teebor, RJ Boorstein, J Cadet. The repairability of oxidative free radical mediated damage to DNA: A review. Int J Radiat Biol 54:131–150, 1988.
51. LH Breimer. Molecular mechanisms of oxygen radical carcinogenesis and mutagenesis: The role of DNA base damage. Mol Carcinog 3:188–197, 1990.
52. M. Dizdaroglu. Oxidative damage to DNA in mammalian chromatin. Mutat Res 275:331–342, 1992.
53. HJ Helbock, KB Beckman, BN Ames. 8-Hydroxydeoxyguanosine and 8-hydroxyguanine as biomarkers of oxidative DNA damage. Methods Enzymol 300:156–166, 1999.
54. J Cadet, T Delatour, T Douki, D Gasparutto, JP Pouget, JL Ravanat, S Sauvaigo. Hydroxyl radicals and DNA base damage. Mutat Res 424:9–21, 1999.
55. JS Macdonald, HE Scribner. The maximum tolerated dose and secondary mechanisms of carcinogenesis. In: KT Kitchin, ed. Carcinogenicity, Testing, Predicting, and Interpreting Chemical Effects. New York: Marcel Dekker, 1999, pp 125–144.
56. MC MacLeod. A possible role in chemical carcinogenesis for epigenetic, heritable changes in gene expression. Mol Carcinog 15:241–250, 1996.
57. P Grasso, M Sharratt, AJ Cohen. Role of persistent, non-genotoxic tissue damage in rodent cancer and relevance to humans. Annu Rev Pharmacol Toxicol 31:253–287, 1991.
58. E Dybing, T Sanner. Species differences in chemical carcinogenesis of the thyroid gland, kidney and urinary bladder. In: CC Capen, E Dybing, JM Rice, JM, JD Wilbourn, eds. Species Differences in Thyroid, Kidney, and Urinary Bladder Carcinogenesis. IARC Scientific Publication No. 147, 1999, pp 15–32.

59. RL Melnick, MC Kohn, CJ Portier. Implications for risk assessment of suggested nongenotoxic mechanisms of chemical carcinogenesis. Environ Health Perspect 104:123–134, 1996.
60. JI Goodman, JL Count. Hypomethylation of DNA: A possible nongenotoxic mechanism underlying the role of cell proliferation in carcinogenesis. Environ Health Perspect 101:169–172, 1993.
61. JL Counts, JI Goodman. Hypomethylation of DNA: A nongenotoxic mechanism involved in tumor promotion. Toxicol Lett 82:663–672, 1995.
62. YW Lee, CB Klein, B Kargacin, K Salnikow, J Kitahara, K Dowjat, A Zhitkovich, NT Christie, M Costa. Carcinogenic nickel silences gene expression by chromatin condensation and DNA methylation: A new model for epigenetic carcinogenesis. Mol Cell Biol 15:2547–2557, 1995.
63. M Costa. Model for the epigenetic mechanism of action of nongenotoxic carcinogenesis. Am J Clin Nutr 61:666S–669S, 1995.
64. C Schmutte, PA Jones. Involvement of DNA methylation in human carcinogenesis. Biol Chem 379:377–388, 1998.
65. KC Smith. Spontaneous mutagenesis: Experimental, genetic and other factors. Mutat Res 277:139–162, 1992.
66. GC Hard. Mechanisms of chemically induced renal carcinogenesis in the laboratory rodents. Toxicol Pathol 26:104–112, 1998.
67. RL Melnick, MC Kohn MC. Possible mechanisms of induction of renal tubule cell neoplasms in rats associated with alpha 2u-globulin: Role of protein accumulation versus ligand delivery to the kidney. In: CC Capen, E Dybing, JM Rice, JM, JD Wilbourn, eds. Species Differences in Thyroid, Kidney, and Urinary Bladder Carcinogenesis. IARC Scientific Publication No. 147, 1999. pp 119–137.
68. JA Swenberg, LD Lehman-McKeeman. α_2-Urinary globulin associated nephropathy as a mechanism of renal tubule cell carcinogenesis in male rats. In: CC Capen, E Dybing, JM Rice, JM, JD Wilbourn, eds. Species Differences in Thyroid, Kidney, and Urinary Bladder Carcinogenesis. IARC Scientific Publication No. 147, 1999, pp 95–118.
69. SM Cohen. Urinary bladder carcinogenesis. Toxicol Pathol 26:121–127, 1998.
70. SM Cohen, A Anderson, LM de Oliveira, LL Arnold. Tumorigenicity of sodium ascorbate in male rats. Cancer Res 58:2557–2561, 1998.
71. S Fukushima, T Murai. Calculi, precipitates and microcrystalluria associated with irritation and cell proliferation as a mechanism of urinary bladder carcinogenesis in rats and mice. In: CC Capen, E Dybing, JM Rice, JM, JD Wilbourn, eds. Species Differences in Thyroid, Kidney, and Urinary Bladder Carcinogenesis. IARC Scientific Publication No. 147. pp. 159–174. 1999.
72. SM Cohen. Calcium phosphate-containing urinary precipitate in rat urinary bladder. In: CC Capen, E Dybing, JM Rice, JM, JD Wilbourn, eds. Species Differences in Thyroid, Kidney, and Urinary Bladder Carcinogenesis. IARC Scientific Publication No. 147, 1999, pp 175–189.
73. J Whysner, GM Williams. Butylated hydroxyanisole mechanistic data and risk assessment: Conditional species-specific cytotoxicity, enhanced cell proliferation and tumor promotion. Pharmacol Ther 71:137–151, 1996.

74. CF Wilkinson, JC Killeen. A mechanistic interpretation of the oncogenicity of chlorothalonil in rodents and an assessment of human relevance. Regul Toxicol Pharmacol 24:69–84, 1996.
75. R Kroes, PW Wester. Forestomach carcinogenesis: Possible mechanisms of action. Food Chem Toxicol 24:1083–1089, 1986.
76. BE Butterworth, MS Bogdanffy. A comprehensive approach for integration of toxicity and cancer risk assessment. Regul Toxicol Pharmacol 29:23–36, 1999.
77. LR Pohl, JL Martin, AM Taburet, JW George. Oxidative bioactivation of haloforms into hepatotoxins. In: MJ Coon, AH Conney, RW Estabrook, HV Gelboin, JR Gillette, PJ O'Brein, eds. Microsome Drug Oxidations, and Chemical Carcinogenesis, vol II. New York: Academic Press, 1980, pp 881–884.
78. CC Capen. Mechanistic data and risk assessment of selected toxic endpoints of the thyroid gland. Toxicol Pathol 25:39–48, 1999.
79. M McClain. Mechanistic considerations for the relevance of animal data on thyroid neoplasia to human risk assessment. Mutat Res 333:131–142, 1995
80. RM McClain, JM Rice JM. A mechanistic relationship between thyroid follicular cell tumours and hepatocellular neoplasms in rodents In: CC Capen, E Dybing, JM Rice, JM, JD Wilbourn, eds. Species Differences in Thyroid, Kidney, and Urinary Bladder Carcinogenesis. IARC Scientific Publication No. 147, 1999, pp 61–68.
81. NT Telang, M Katdare, HL Bradlow HL, MP Osborne. Estradiol metabolism: An endocrine biomarker for modulation of human mammary carcinogenesis. Environ Health Perspect 3:559–564, 1997
82. JG Liehr. Dual role of oestrogens as hormones and pro-carcinogens: Tumour initiation by metabolic activation of oestrogens. Eur J Cancer Prev 6:3–10, 1997.
83. JG Liehr. Hormone-associated cancer: Mechanistic similarities between human breast cancer and estrogen-induced kidney carcinogenesis in hamsters. Environ Health Perspect 105:565–569, 1997.
84. BT Zhu, AH Conney. Functional role of estrogen metabolism in target cells: Review and perspectives. Carcinogenesis 19:1–27, 1998.
85. D Roy, JG Liehr. Estrogen, DNA damage and mutations. Mutat Res 424:107–115, 1999.
86. DL Davis, NT Telang, MP Osborne, HL Bradlow. Medical hypothesis: Bifunctional genetic-hormonal pathways to breast cancer. Environ Health Perspect 3:571–576, 1997.
87. D Roy, M Palangat, CW Chen, RD Thomas, J Colerangle, A Atkinson, ZJ Yan. Biochemical and molecular changes at the cellular level in response to exposure to environmental estrogen-like chemicals. J Toxicol Environ Health 50:1–29, 1997.
88. JP Whitlock. Mechanistic aspects of dioxin action. Chem Res Toxicol 6:754–763, 1993.
89. AJ Cohen, P Grasso. Review of the hepatic response to hypolipidaemic drugs in rodents and assessment of its toxicological significance to man. Food Cosmet Toxicol 19:585–605, 1981.
90. JK Reddy, ND Lalwani. Carcinogenesis by hepatic peroxisome proliferators: Evaluation of the risk of hypolipidemic drugs and industrial plasticizers to human. CRC Crit Review Toxicol 12:1–53, 1983.

91. I Issemann, S Green. Activation of a member of the steroid hormone receptor superfamily by peroxisome proliferators. Nature 347:645–650, 1990.
92. PR Holden, JD Tugwood. Peroxisome proliferator-activated receptor alpha: Role in rodent liver cancer and species differences. J Mol Endocrinol 22:1–8, 1999.
93. JP Vanden Heuvel. Peroxisome proliferator-activated receptors (PPARS) and carcinogenesis. Toxicol Sci 47:1–8, 1999.
94. JK Reddy, GP Mannaerts. Peroxisomal lipid metabolism. Annu Rev Nutr 14:343–370, 1994.
95. BG Lake. Role of oxidative stress and enhanced cell replication in the hepatocarcinogenicity of peroxisome proliferators. In: G Gibson, B Lake, eds. Peroxisomes: Biology and Importance in Toxicology and Medicine. London: Taylor and Francis, 1993, pp 595–618.
96. B Desvergne, W Wahli. PPAR: A key nuclear factor in nutrient/gene interaction? In: P Baeuerle, eds. Progress in Gene Expression, vol 1. Inducible Gene Expression. Boston, Birkhauser, pp. 142-176, 1994.
97. JM Peters, RC Cattley, FJ Gonzalez. Role of PPAR alpha in the mechanism of action of nongenotoxic carcinogen and peroxisome proliferator Wy-14,643. Carcinogenesis 18:2029–2033, 1997.
98. FJ Gopnzalez, JM Peters, RC Cattley. Mechanism of action of the nongenotoxic peroxisome proliferator: Role of the peroxisome proliferator-activator receptor alpha. J Natl Cancer Inst 90:1702–1709, 1998.
99. J Ashby, A Brady, CR Elcombe, BM Elliott, J Ishmael, J Odum, JD Tugwood, S Kettle, IFH Purchase. Mechanistically-based human hazard assessment of peroxisome proliferator-induced hepatocarcinogenesis. Human Exp Toxicol 13:1–117, 1994.
100. P Bentley, I Calder, C Elcombe, P Grasso, D Stringer, HJ Wiegand, HJ. Hepatic peroxisome proliferation in rodents and its significance for humans. Food Chem Toxicol 31:857–907, 1993.
101. IARC. International Agency for Research on Cancer. Overall Evaluation of Carcinogenicity to Humans. Updated: 11/98. Lyon, France, 1998.
102. WN Choy. Human liver carcinogenesis. In: KT Kitchin, ed. Carcinogenicity, Testing, Predicting, and Interpreting Chemical Effects. New York: Marcel Dekker, pp. 411–438, 1999.
103. V Krutovskikh, H Yamasaki H. The role of gap junction intercellular communication (GJIC) disorders in experimental and human carcinogenesis. Histol Histopathol 12:761–768, 1997.
104. H Yamasaki. Role of disrupted gap junctional intercellular communication in detection and characterization of carcinogens. Mutat Res 365:91–105, 1996.
105. H Yamasaki, V Krutovskikh, M Mesnil, T Tanaka, ML Zaidan-Dagli, Y Omori. Rose of connexin (gap junction) gene in cell growth control and carcinogenesis. C R Acad Sci III 322:11–159, 1999.
106. RG Stern, BN Milestone, RA Gatenby. Carcinogenesis and the plasma membrane. Med Hypothesis 52:367–372, 1999.
107. H Yamasaki, Y Omori, ML Zaidan-Dagli, N Mironv, M Mesnil, V Krutvskikh. Genetic and epigenetic changes of intercellular communication genes during multistage carcinogenesis. Cancer Detect Prev 23:273–279, 1999.

108. S Rainier, LA Johnson, CJ Dobry, AJ Ping, PE Grundy, AP Feinberg. Relaxation of imprinted genes in human cancer. Nature 362:747–749, 1993.
109. AP Feinberg. Genomic Imprinting and Cancer. In: B Vogelstein, KW Kinzler, eds. The Genetic Basis of Human Cancer. NY: MacGraw-Hill 1997, pp 95–107.
110. JG Falls, DJ Pulford, AA Wylie, RL Jirtle. Genomic imprinting: Implications for human disease. Am J Pathol 154:635–647, 1999.
111. B Tycko B. DNA methylation in genomic imprinting. Mutat Res 386:103–105, 1997.
112. K Yun. Genomic imprinting and carcinogenesis. Histol Histopathol 13:42, 1998.
113. DJ Pulford, JG Falls, JK Killian, RL Jirtle. Polymorphisms, genomic imprinting and cancer susceptibility. Mutat Res 436:59–67, 1999.
114. T Kakunaga. Neoplastic transformation of human diploid fibroblast cells by chemical carcinogens. Proc Natl Acad Sci U S A 75:1334–1338, 1978.
115. JW Greiner, CH Evans, JA DiPaolo. Carcinogen-induced anchorage-independent growth and in vivo lethality of human MRC-5 cells. Carcinogenesis 2:359–362, 1981.
116. GE Milo, JW Oldham, R Zimmerman, GG Hatch, SA Weisbrode. Characterization of human cells transformed by chemical and physical carcinogens in vitro. In Vitro 17:719–729, 1981.
117. RJ Zimmerman, JB Little. Characterization of a quantitative assay for the in vitro transformation of normal human diploid fibroblasts to anchorage independence by chemical carcinogenesis. Cancer Res 43:2176–2182, 1983.
118. RJ Zimmerman, JB Little. Characterization of human diploid fibroblasts transformed in vitro by chemical carcinogens. Cancer Res 43:2183–2189, 1983.
119. S Nesnow, G Milo, P Kurian, R Sangaiah, A Gold. Induction of anchorage-independent growth in human diploid fibroblasts by the cyclopenta-polycyclic aromatic hydrocarbon, benz(l)aceanthrylene. Mutat Res 244:221–225, 1990
120. GE Milo, JA DiPaolo. Presensitization of human cells with extrinsic signals to induced chemical carcinogenesis. Int J Cancer 26:805–812, 1980.
121. JW Littlefield. Variation, Senescence, and Neoplasia in Cultured Somatic Cells. Harvard University Press, 1976.
122. L Hayflick. The limited in vitro lifetime of human diploid cell strains. Exp Cell Res 37:614–636, 1965.
123. JW Littlefield, WN Choy, J Epstein. Research on cellular senescence, present and future directions. In: JW Shay, ed. Senescence. Plenum, 1977, pp. 111–120.
124. NC Popescu, SC Amsbaugh, G Milo, JA DiPaolo. Chromosome alterations associated with in vitro exposure of human fibroblasts to chemical or physical carcinogens. Cancer Res 46:4720–4725, 1986.
125. NJ Whitaker, TM Bryan, P Bonnefin, AC Chang, EA Musgrove, AW Braithwaite, RR Reddel. Involvement of RB-1, p53, p16INK4 and telomerase in immortalization of human cells. Oncogene 11:971–976, 1995.
126. JW Shay, WE Wright, H Werbin. Defining the molecular mechanisms of human cell immortalization. Biochim Biophys Acta 1072:1–7, 1991.
127. M Namba, M Iijima, T Kondo, I Jahan, K Mihara. Immortalization of normal human cells is a multistep process and a rate limiting step of neoplastic transformation of the cells. Hum Cell 6:253–259, 1993.

128. M Namba, K Mihara, K Fushimi. Immortalization of human cells and its mechanisms. Crit Rev Oncog 7:19–31, 1996.
129. K Fushimi, M Iijima, C Gao, T Kondo, T Tsuji, T Hashimoto, K Mihara, M Namba. Transformation of normal human fibroblasts into immortalized cells with the mutant p53 gene and X-ray. Int J Cancer 70:135–140, 1997.
130. S Moolgavkar, D Venzon. Two-event models for carcinogenesis: Incidence curves for childhood and adult tumors. Math Biosci 47:55–77, 1979.
131. TJ Slaga, IV Budunova, IB Gimenez-Conti, CM Aldaz. The mouse skin carcinogenesis model. J Investig Dermatol Symp Proc 1:151–156, 1996.
132. HC Pitot, YP Dragan, J Teeguarden, S Hsia, H Compbell. Quantitation of multistage carcinogenesis in rat liver. Toxicol Pathol 24:119–128, 1996.
133. RJ Isfort, RA LeBoeuf. The Syrian hamster embryo (SHE) cell transformation system: A biologically relevant in vitro model-with carcinogen predicting capabilities-of in vivo multistage neoplastic transformation. Crit Rev Oncog 6:251–260, 1995.
134. P Buss, M Caviezel, WK Lutz. Linear dose-response relationship for DNA adducts in rat liver from chronic exposure to aflatoxin B1. Carcinogenesis 11:2133–2135, 1990.
135. CDHS, California Department of Health Services. Guidelines for Chemical Carcinogen Risk Assessment and Their Scientific Rationale. Health and Welfare Agency, State of California, 1985.
136. A Kopp-Schneider. Carcinogenesis models for risk assessment. Stat Methods Med Res 6:317–340, 1997.
137. NP Page, DV Singh, W Farland, JI Goodman, RB Conolly, ME Anderson, HJ Clewell, CB Frederick, H Yamassaki, G Lucier. Implementation of EPA revised cancer assessment guidelines: Incorporation of mechanistic and pharmacokinetic data. Fundam Appl Toxicol 37:16–36, 1997.
138. CD Sherman, CJ Portier. Stochastic simulation of multistage model of carcinogenesis. Math Biosci 134:35–50, 1996.
139. KS Crump, HA Guess, LL Deal. Confidence intervals and test of hypotheses concerning dose-response relations inferred from animal carcinogenicity data. Biometrics 33:436–451, 1977.
140. KS Crump. An improved procedure for low-dose carcinogenic risk assessment from animal data. J Environ Pathol Toxicol 52:675–684, 1981.
141. EL Anderson and US Environmental Protection Agency Carcinogen Assessment Group. Quantitative approaches in use to assess cancer risk. Risk Anal 3:277–295, 1983.
142. S Moolgavkar, A Knudson. Mutation and cancer: A model for human carcinogenesis. J Natl Cancer Inst 66:1037–1052, 1981.
143. KT Bogen. Cell proliferation kinetics and multistage cancer risk models. J Natl Cancer Inst 81:267–277, 1989.
144. YP Dragan, J Hully, K Baker, R Crow, MJ Mass, HC Pitot. Comparison of experimental and theoretical parameters of the Moolgavkar-Venzon-Knudson incidence function for the stages of initiation and promotion in rat hepatocarcinogenesis. Toxicology 1102:161–175, 1995.
145. DW Gaylor, Q Zheng. Risk assessment of nongenotoxic carcinogens based upon cell proliferation/death rates in rodents. Risk Anal 16:221–225, 1996.

4

Structure–Activity Relationship (SAR) and Its Roles in Cancer Risk Assessment

David Y. Lai and Yin-tak Woo
U.S. Environmental Protection Agency, Washington, D.C.

I. INTRODUCTION

It is a general belief that a correlation exists between the molecular structures of chemicals and their biological activities. The study of such a correlation comprises the field of structure-activity relationship (SAR) analysis. Structure-activity relationship analysis has long served as an essential tool in the research and development of industrial, agricultural, and pharmaceutical chemicals for beneficial application purposes. To protect public health from exposure to the large number of existing chemicals for which adequate toxicological data do not exist, SAR analysis is often the first line of approach in hazard evaluation and risk assessment. Structure-activity relationship analysis has also been used by both industry and government agencies for assessing and predicting potential health hazards of chemicals early in the research and premanufacture stage. Structure-activity relationship consideration plays an increasingly important role

---

The views expressed in this chapter are solely those of the authors and do not necessarily reflect those of the U.S. Environmental Protection Agency. Mention of trade names or commercial products does not constitute endoresement or recommendation for use by the Agency.

in prioritizing chemicals for testing, designing test strategies, identifying research needs, and providing mechanistic insight to support risk assessment and regulatory decision-making.

Various SAR approaches and model systems exist for predicting the potential toxicity of chemicals (1). This chapter focuses on the basic principles and concepts of mechanism-based SAR analysis of the carcinogenicity potential of chemicals and the roles of SAR in cancer risk assessment. Some computer-assisted models/systems and the strengths and weaknesses of different approaches for predicting carcinogenicity and mutagenicity of chemicals are also discussed.

II. BASIC PRINCIPLES AND CONCEPTS OF MECHANISM-BASED SAR IN PREDICTING CARCINOGENICITY OF CHEMICALS

Basically, mechanism-based SAR analysis involves comparison of the chemical in question with structurally related compounds with known carcinogenic activity, identification of structural moiety(ies) or fragment(s) that may contribute to carcinogenic activity through a perceived or postulated mechanism, and evaluating the modifying role of the rest of the molecule to which the structural moiety/fragment is attached. Since the pioneer work of the Millers (2), which conceptualized electrophiles as the ultimate carcinogen to interact with DNA to initiate carcinogenesis, considerable knowledge of the mechanisms of chemical carcinogenesis has accrued. For many carcinogen classes, the molecular basis of carcinogenic activity is now known in considerable detail, and the concept of electrophiles provides the most probable rationale for their carcinogenic action. The knowledge of molecular mechanisms and other factors that affect carcinogenesis by various types of chemical carcinogens has established the basis for identifying suspect carcinogens by SAR analysis. The identification of structural features of electrophiles and their precursors has been a centerpiece for SAR prediction of genotoxic carcinogens that constitute the majority of known human carcinogens (3).

Except for some direct-acting carcinogens, most chemical carcinogens require metabolic activation to electrophilic species that bind covalently with nucleophilic sites in DNA to initiate carcinogenesis. The metabolic activation of chemical carcinogens by various metabolic systems has been reviewed (4). Table 1 lists examples of electrophilic species that are possible or probable ultimate carcinogens of a variety of chemical carcinogens. Some of the commonly encountered electrophiles or electrophilic intermediates in these chemical carcinogens include: carbonium ions (alkyl-, aryl-, benzylic), nitrenium and aziridinium ions, epoxides and oxonium ions, episulfonium ions, aldehydes, polarized double bonds (α,β-unsaturated carbonyls or carboxylates), peroxides, free radicals, acylating intermediates, quinone and quinoid intermediates. The structural

Structure-Activity Relationship

TABLE 1. Electrophilic Species of Some Chemical Carcinogens

Electrophilic species	Structures	Examples of carcinogens
Alkylcarbonium ions	$R-CH_2^{\oplus}$	Diethylnitrosamine; dimethylsulfate; 1,2-dimethylhydrazine
Arylcarbonium ions	[benzene ring with $\oplus$]	Benzidine; 1-(4-methoxyphenyl)-1,3-dimethyltriazene
Allylic carbonium ions	$R-\overset{½\oplus}{CH}=CH=\overset{½\oplus}{CH_2}$	Allyl methanesulfonate; lasiocarpine; safrole
Benzylic carbonium ions	[phenyl]$-CH_2^{\oplus}$	Benzyl chloride; 7,12-dimethylbenz(a)anthracene
Formaldehyde	$^{\oplus}CH_2-O^{\ominus}$	Formaldehyde; hexamethylphosphoramide
α,β-Unsaturated carbonyls or carboxylates	$-\underset{\oplus}{C}-C=\overset{O^{\ominus}}{C}-$	Acrylates; arecoline; cyclophosphamide; diallate; ptaquiloside
Acylating moieties	$R-\overset{O}{\underset{\|}{C}}^{\oplus}$	Benzoyl chloride; dimethylcarbamyl chloride
Carbon-centered free radicals or radical cations	$R\cdot \quad Ar-\overset{\cdot}{C}H_2$ $\cdot Ar^{\oplus}-CH_3$	Carbon tetrachloride; 7,12-dimethylbenz(a)anthracene
Carbene	$:C\diagup\diagdown$	Carbon tetrachloride; safrole
Aziridinium ions	$-\overset{\oplus}{N}\diagup^{CH_2}_{CH_2}$	Aziridine; cyclophosphamide; nitrogen mustard
Iminium ions	$\diagdown\overset{\oplus}{N}=CH_2$	Hexamethylmelamine; bis-(morpholino)-methane; hycanthone

(Continued)

TABLE 1. (*Continued*)

Electrophilic species	Structures	Examples of carcinogens
Arylnitrenium ions	$Ar-\overset{\oplus}{N}\diagdown^R$	2-Acetylaminofluuorene; benzidine; N,N-dimethyl-4-aminoazobenzene
Nitrogen-centered free radicals or radicals cations	$Ar-\overset{\cdot}{N}H \quad Ar-\overset{\cdot\oplus}{N}H_2$	Benzidine; 2-naphthylamine
Epoxides	$\overset{\diagdown}{C}-\overset{\diagup}{C} \atop \diagdown_O\diagup$	Aflatoxin B_1; capsaicin; ethylene oxide; vinyl chloride
Oxonium ions (α-haloethers)	$-\overset{\oplus}{C}=CH_2 \leftrightarrow -O-\overset{\oplus}{C}H_2$	Bis(chloromethyl)ether
Peroxy free radicals	$-ROO\cdot$	Di-*t*-butyl peroxide
Episulfonium ions	$-S\overset{\oplus}{\diagup}\overset{CH_2}{\underset{CH_2}{|}}$	1,2-dichloroethane; sulfur mustard
Sulfonium ions (α-halothioesters)	$-\overset{\oplus}{S}=CH_2 \leftrightarrow -S-\overset{\oplus}{C}H_2$	Dichloromethane
Semiquinone and its quinoneimine and diamine/diimine analogs		1,4-Benoquinone; adriamycin; benzidine; 2-naphthylamine; phenacetin
Quinonemethide		Daunomycin; quercetin

moieties with electrophilic potential from chemicals of a large carcinogen database have been identified and constructed into a model structure ("supermolecule") by Ashby and coworkers (5–7). This model structure provides "structural alerts" that represent potential or actual electrophilic centers of carcinogenic and mutagenic chemicals and can be used as a coarse screening tool for predicting potential carcinogens and mutagens.

Structure-Activity Relationship

A review of all structural classes of chemical carcinogens and SAR analysis of the effects of chemical reactivity, molecular geometry, and metabolism on carcinogenicity have revealed some general structural criteria of potential mutagenic and carcinogenic chemicals that are electrophiles or that may generate electrophiles after metabolic transformation (8–12):

1. Polycyclic structures with three to six aromatic rings that mimic the angular ring distribution of carcinogenic polycyclic aromatic hydrocarbon (PAHs). The likelihood of carcinogenicity increases if the molecule is asymmetric, contains bay-region benzo ring, blocked L-region, and free *peri* position adjacent to the bay-region benzo ring;

2. The presence of an amino, dimethylamino, nitroso, or nitro group directly linked to a conjugated double-bonded system, especially if the amino or amine-generating group is at the terminal end of the longest conjugated double-bond system of the molecule;

3. N-Nitroso, hydrazo, aliphatic azo, or aliphatic azoxy structures, 1-aryl-3,3-dialkyltriazenes, and 1,1-diaryl-2-acetylenic carbamates;

4. The presence of a sterically strained ring (e.g., epoxide, aziridine, γ-lactones and δ-sultones) in any type of structure. The likelihood of carcinogenicity may increase if the compound contains two or more of these reactive ring structures;

5. Low-molecular-weight carbamates, thiocarbamates, and thiourea derivatives;

6. The presence of a haloalkyl (particularly 1,2-dihalo), haloalkenyl (both vinylic and allylic), α-haloether, α-haloalkanol, or α-halocarbonyl grouping;

7. Low-molecular-weight aliphatic structures containing conjugated double bonds or isolated double bonds situated at the terminal end of an aliphatic chain;

8. Low-molecular-weight aldehydes. The likelihood of carcinogenicity may increase if the compound contains more than one aldehyde group or if the aldehyde group is terminal or α to a double bond;

9. Benzylic, allylic, or pyrrolic esters if the acyloxy moiety is a good leaving group;

10. Any structural type of alkylating, arylating, or acylating agent or larger molecular assembles incorporating such agents as chemically reactive moieties.

The chemical structures of some representative carcinogens that possess these structural criteria are shown in Figure 1. Although other chemicals that possess these structural criteria have the potential to produce electrophilic reactants, they may not be carcinogenic or mutagenic because the rest of the molecules can play an important role in modifying the reactivity and stability of the electrophilic reactants. Specific chemicals with structural criteria for potential carcinogens or mutagens must be further evaluated for possible influence of other molecular features on their potential to interact with DNA. Some of the molecu-

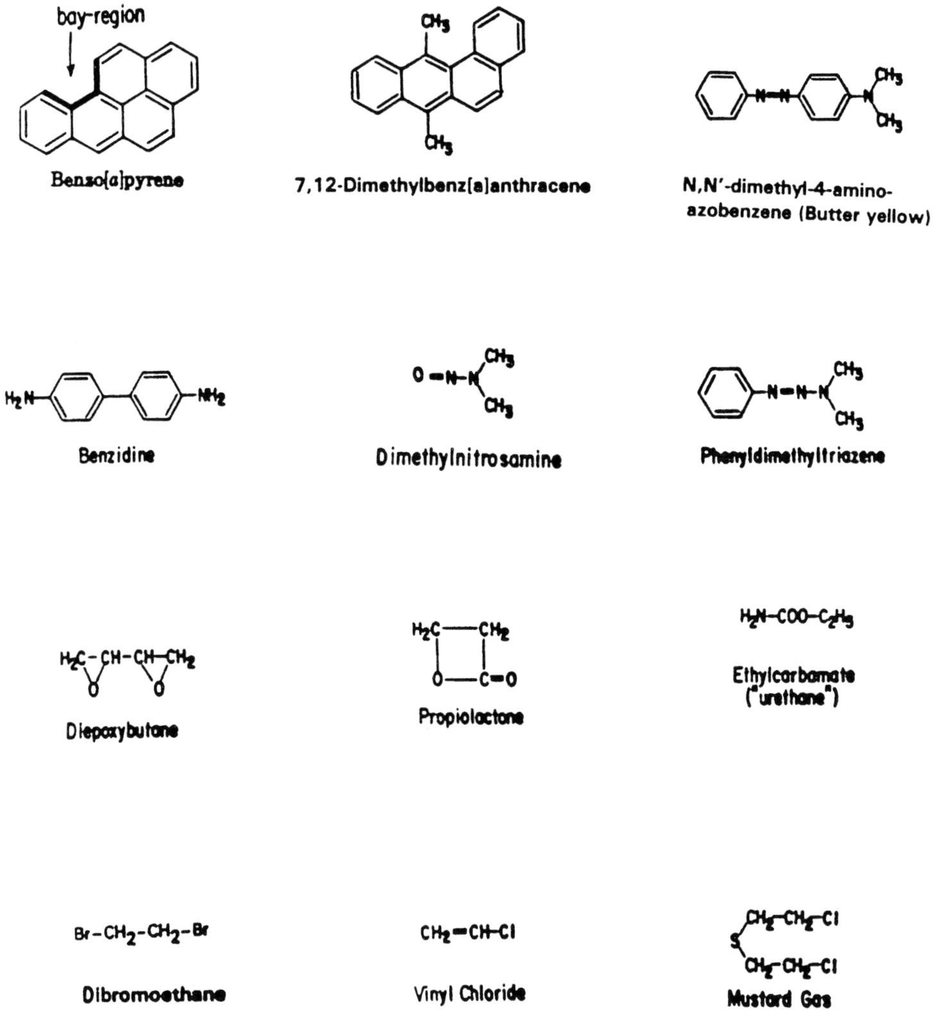

FIG. 1. Chemical structures of some genotoxic carcinogens.

lar parameters that can have significant effects on the carcinogenic potential of electrophilic compounds include:

1. *Molecular size and shape.* In general, compounds with a high molecular weight (more than 1,000) have little chance of being absorbed in significant amounts and, thus, will have less concern for carcinogenicity. Molecular size and geometry of a chemical can also affect its chance to be metabolically activated/detoxicified and its ability to reach target tissues and target macromolecules. Many potent carcinogens have a molecular size and shape favorable for enzyme activation, DNA interaction, or receptor binding. For instance, potent carcinogens such as 2,3,7,8-tetrachloro-dibenzo-*p*-dioxin (TCDD), aflatoxin B_1, benz[a]pyrene, 2-acetylaminofluorene (2-AAF), and other carcinogenic PAHs and aromatic amines are all planar molecules. It has been calculated that most potent carcinogenic PAHs have a planar structure with four to six rings (of which not more than four may be linearly connected) and a molecular size (as measured by incumbrance area of planar molecule) between about 100 to 135 $Å^2$. Virtually all PAHs that are highly elongated or highly symmetrical tend to be inactive (8) (Table 2a). Studies on the 2-, 3-, and 4-isomers of aminobiphenyl have shown good correlations among the mutagenic or carcinogenic activities, the capability of N-hydroxylation (first metabolic activation step), and the planarity of the molecules. Being planar, 4-aminobiphenyl can be readily N-hydroxylated and is a potent carcinogen and mutagen. The presence of the amino group at the 2-position of aminobiphenyl results in marked loss of planarity; being nonplanar, 2-aminobiphenyl does not form any N-hydroxylation intermediates and is noncarcinogenic and nonmutagenic. 3-Aminobiphenyl is somewhat planar and is a weak carcinogen and mutagen (13).

2. *Substituent effects (electronic or steric).* Substitutions on the molecule may exert electronic or steric effects on the reactivity and stability of many ultimate carcinogenic metabolites. For instance, there is evidence that substitution of a chloro, methyl, or methoxy group *ortho* to the amino group of aniline, phenylenediamine, and methylene-*bis*-aniline often enhances carcinogenicity and mutagenicity. The effect has been suggested to be caused by formation of DNA adducts of a favorable conformation for mutagenicity and carcinogenicity (14). However, the effect of *ortho* substitution is related to the size of the subsituents, the larger the subsituents, the less potent the carcinogenic or mutagenic activity. If both *ortho* positions of an amino group of aromatic amines are substituted by large subsituents (e.g., large alkyl groups), an additional decrease in activity occurs. This is because large subsituents at the *ortho* position provide steric hindrance around the amino group, thus inhibiting N-hydroxylation or N-acetylation, the necessary steps required for metabolic activation to the ultimate carcinogenic and mutagenic species (15). Ring substitution with methyl group(s) at detoxification sites (e.g., L-region) of PAHs (e.g., 7, 12-dimethylbenz[a]anthracene) tends to increase carcinogenic activity, whereas ring

TABLE 2. Molecular Parameters That Can Affect the Carcinogenicity of Chemicals

Carcinogenic	Noncarcinogenic
(a) Molecular size and shape	
Benzo[a]pyrene	Pentacene
(b) Substituent effect	
4,4′-Diaminostilbene	4,4′-Diamino-2,2′-stilbene-disulfonic acid, disodium salt
(c) Molecular flexibility and polyfunctionality of reactive groups	
Diepoxyhexane	3,4-Epoxyethylcyclohexane

substitution with bulky substituents invariably decreases activity, particularly if near or at proelectrophilic regions such as the bay region (8, 16).

The introduction of hydrophilic groups (e.g., sulfonyl, carboxyl) into an otherwise carcinogenic compound usually mitigates, and sometimes even totally abolishes, its carcinogenic activity, as compounds that are highly hydrophilic are poorly absorbed, and when absorbed, are readily excreted (17). 4,4′-Diamino-2,2′-stilbenedisulfonic acid was noncarcinogenic in rats and mice, probably because of the substituent effect of the sulfonyc acid group on both the water solubility and the planarity of the molecule (Table 2b).

3. *Molecular flexibility, polyfunctionality, and spacing/distance between reactive groups.* The importance of molecular flexibility can be illustrated by the findings that epoxides on rigid cycloaliphatic rings tend to be considerably less active than epoxides on more flexible noncyclic aliphatic chains. Diepoxides are

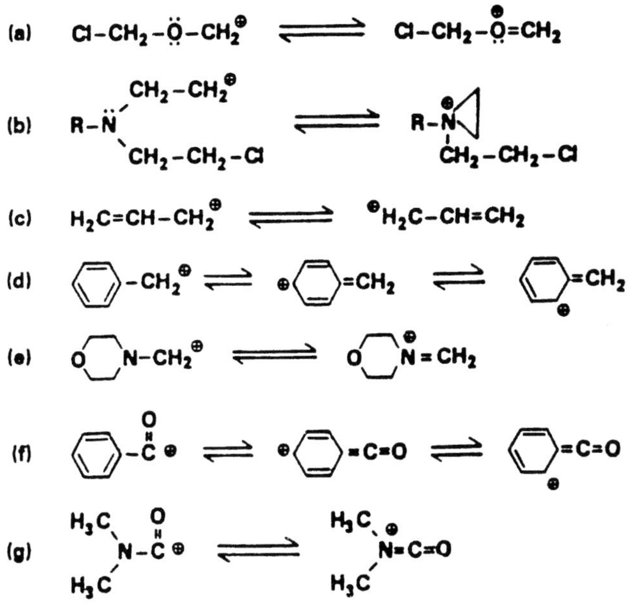

Structure-Activity Relationship

(a) $Cl-CH_2-\ddot{O}-CH_2^{\oplus} \rightleftharpoons Cl-CH_2-\overset{\oplus}{O}=CH_2$

(b) $R-\ddot{N}\begin{smallmatrix}CH_2-CH_2^{\oplus}\\CH_2-CH_2-Cl\end{smallmatrix} \rightleftharpoons R-\overset{\oplus}{N}\begin{smallmatrix}\\\\CH_2-CH_2-Cl\end{smallmatrix}$

(c) $H_2C=CH-CH_2^{\oplus} \rightleftharpoons {}^{\oplus}H_2C-CH=CH_2$

(d) $\text{Ph}-CH_2^{\oplus} \rightleftharpoons {}^{\oplus}\text{C}_6H_5=CH_2 \rightleftharpoons \text{C}_6H_5=CH_2$ (with ⊕ on ring)

(e) $O\underset{}{\overset{}{\bigcirc}}N-CH_2^{\oplus} \rightleftharpoons O\underset{}{\overset{}{\bigcirc}}\overset{\oplus}{N}=CH_2$

(f) $\text{Ph}-\overset{O}{\underset{\parallel}{C}}{\oplus} \rightleftharpoons {}^{\oplus}\text{C}_6H_5=C=O \rightleftharpoons \text{C}_6H_5=C=O$ (with ⊕ on ring)

(g) $\begin{smallmatrix}H_3C\\ \\H_3C\end{smallmatrix}N-\overset{O}{\underset{\parallel}{C}}{\oplus} \rightleftharpoons \begin{smallmatrix}H_3C\\ \\H_3C\end{smallmatrix}\overset{\oplus}{N}=C=O$

FIG. 2. Resonance stabilization of reactive intermediate from (a) bis-(chloromethyl)ether, (b) aliphatic nitrogen mustard, (c) allyl chloride, (d) benzyl chloride, (e) bis-(morpholino)methane, (f) benzoyl chloride, and (g) dimethylcarbamyl chloride.

usually more carcinogenic than monoepoxides, particularly if the two epoxy groups are favorably apart to impart cross-linking activity (18) (Table 2c).

4. *Resonance stabilization of the electrophilic metabolites.* For compounds that require metabolic activation, resonance stabilization of the electrophilic metabolites is important for potent carcinogens, because all electrophiles are reactive and may be readily hydrolyzed or detoxified by cellular protective nucleophiles, such as glutathione. Resonance stabilization gives reactive intermediates a longer lifetime to allow them to have a better chance of remaining reactive during transport from the site of activation to target macromolecules. Structural features that may provide resonance stabilization include conjugated double bonds, aryl moiety (especially those capable of providing long resonance pathway), ring positions that allow several resonance forms, and structures that allow cyclization of reactive intermediate (Fig. 2).

For more details on the state of the art of predicting suspect carcinogens based on SAR analysis, the reader is referred to the following publications (19–21).

III. COMPUTER-ASSISTED SAR PREDICTIVE MODELS AND SYSTEMS

The traditional mechanism-based SAR prediction of potential carcinogens attempts to consider all current knowledge of the chemical and biological processes in tumorigenesis. Yet, there are considerable knowledge gaps in understanding the complexity of the interaction between every chemical structure within a biological system. Predictions of carcinogenic or mutagenic activity often have to be based on limited knowledge and uncertainty of effects of uncommon structural features on the metabolism, toxicokinetics, and postulated mechanisms of the chemicals. Furthermore, this approach, which relies heavily on individuals with cumulative knowledge of and experience in the chemical and biological mechanisms in carcinogenesis, is difficult for nonexperts to put into routine practice. Recent advances in computer technology coupled with increasing economic and societal pressures to reduce animal testing have motivated researchers to develop comprehensive models for screening and predicting carcinogenic or mutagenic activity of the large number of untested chemicals that have diverse structures and, possibly, different modes of action. Carcinogenicity and mutagenicity databases contain a wealth of information encoded in their chemical structures and corresponding biological activities. Analysis of large databases by well-designed computer models has the potential of discerning subtle structural modifiers to biological activity and, thus, the potential for improving current predictive capability. Computer models, such as "expert systems" ("artificial intelligence" systems) can also codify and encapsulate human expertise, reduce or eliminate error and inconsistency, and help nonexperts arrive at expert judgement regarding carcinogenic potential of chemicals.

Using the concept of quantitative structure-activity relationships (QSAR) developed by Hansch (22, 23), a number of commercially available computer-assisted predictive models for toxicologic assessment have been developed. In general, the development of computer-assisted QSAR models involves the graphical entry and storage of structures, generation of molecular structure descriptors, analysis of the descriptors, development of quantitative relationships between descriptors, and biological responses using multivariate statistical or pattern recognition methods. Structural descriptors used for model development include physicochemical, topologic, geometrical, electronic, and other quantum mechanical properties of the molecules.

Two of the most well-known computer-assisted models for carcinogenicity and mutagenicity prediction are the MultiCASE/CASE (Computer Automated Structure Evaluator) (24–26) and the TOPKAT (Toxicity Prediction by Komputer Assisted Technology) (27, 28). Both of these computerized predictive models are based on pattern recognition analysis of learning sets that are carcinogenicity databases of chemicals with noncongeneric structures. They differ from each other primarily in the way substructural fragments are identified and incorporated in the systems.

The CASE/MultiCASE model is completely automated in the generation and inclusion of its molecular substructural fragment descriptors from the learning database. In TOPKAT, the substructural fragments are chosen from a preselected library of more than 3,000 potentially significant fragments compiled on the basis of general organic chemistry functionality and mechanistic considerations. In addition to using substructural fragments as molecular descriptors, TOPKAT and MultiCASE also use molecular topologic indices, molecular shape indices, electronic properties, and partial atomic charges of chemical structures as modifiers from the learning databases for their development. When a new chemical is evaluated, the model will search its structure for the existence of a discriminating feature or a biophore. If a discriminating feature or a biophore is found, the model will then search for the presence of potential modulators to arrive at a projected value for its potency. The statistically based, correlative approach used by these computer-assisted predictive models has the advantage of minimizing human expert input and bias. The knowledge limit and predictive range, determined totally by the information contained within the learning databases, however, are still relatively small for statistical analyses and prediction of the large number of untested chemicals with diverse structures and modes of action. Furthermore, as these models were derived from existing carcinogenicity databases, uncertainties and inaccuracies associated with rodent carcinogenicity studies and with interpretation and assignment of chemicals of carcinogenicity "calls" inherent in such databases are all reflected in the SAR prediction models. A good illustration of the former can be provided by the report of poor performance of the standard MultiCASE and other programs to predict the carcinogenicity potential of pharmaceuticals due to poor coverage for drug molecules and inadequate representation of the molecular diversity of drugs; to correct this deficiency, a new, enhanced FDA-OTR/MCASE software program has been developed with the inclusion of more than 1,000 compounds from the FDA-OTR Carcinogenicity Database to the MultiCASE learning database (29).

Correlative SAR models also have the deficiency of rationalizing molecular mechanisms related to carcinogenicity. Hence, a number of other computer-assisted SAR predictive models also incorporate some mechanistic considerations and expert judgement, as well as using statistical association, in their development. Examples include the k_e carcinogenicity model of Benigni (30, 31) and the COMPACT (Computer-Optimized Molecular Parametric Analysis of Chemical Toxicity) model of Parke (32, 33). The k_e carcinogenicity model measures the k_e (electrophilicity) of a chemical, and the prediction is based on the relevance of the k_e to carcinogenicity. The COMPACT model has been developed on the basis of specific physicochemical conditions (e.g., degree of planarity of the molecule) for the metabolic activity among a particular subfamily of cytochrome P-450 oxidases as a predictor of potential carcinogenic activity. It is a molecular orbital method that predicts whether a chemical may be metabolized by a specific cytochrome P-450 oxidase, that is, CYP1, CYP2E, or CYP3, to form a reactive intermediate that leads to carcinogenicity. One major weakness of these models is that they are tied to a single mechanistic hypothesis and are not modeled for discovering new mechanistic insight that may lead to greater predictive capability.

Rather than using a statistically based, correlative approach, the OncoLogic Cancer Expert System has been developed based on knowledge rules that represent the formalized, codified, and organized SAR knowledge of human experts (34–36). The input of this rule-based computerized expert system includes the chemical structure as well as all available chemical, biological, and mechanistic information (e.g., physicochemical properties, chemical stability, route of exposure, bioactivation and detoxicification, genotoxicity, and other supportive data) critical to the evaluation of carcinogenic potential. In contrast to other SAR predictive models that are not based on mechanistic considerations or use only single mechanistic assumption, the OncoLogic Cancer Expert System uses different sets of knowledge rules specific for different chemical classes or subclasses to account for their different modes of action. This approach maximizes the accuracy of predicting not only organic compounds of diverse structures or substructures, but also other types of chemical substances such as fibers, metals, and polymers. The DEREK (Deductive Estimation of Risk from Existing Knowledge) is another knowledge-based expert system developed for qualitative prediction of carcinogenicity of chemicals. The DEREK system makes its predictions based primarily on a small set of rules, each of which describes the relationship of a structural feature ("toxicophore") to carcinogenicity (37, 38). Because knowledge rules are based on the expertise and intuition of a few individuals, rule-based expert systems are more biased in their perception of SAR knowledge than statistically based, correlative models. Like the traditional mechanism-based SAR approach, current knowledge gaps in mechanism and mode of action also limit the range of applicability of these computerized expert systems. The carcinogenicity of only chemical congeners within chemical classes rich in mechanistic data can be predicted with a high level of confidence.

There are a number of attractive features for computer-assisted SAR models, but difficulties and problems exist involving model development, validation, confirmation, and acceptance issues (39). Nonetheless, computer-assisted predictive models can function as valuable tools for chemical screening and hazard identification if they are applied with adequate oversight and scrutiny. As part of the Predictive-Toxicology Evaluation Project of the U.S. National Institute of Environmental Health Sciences (NIEHS), the state of the art of predicting rodent carcinogenesis and the strengths and weaknesses of different predictive methods have been evaluated (40, 41). The results and conclusions of the first round of prospective predictions made by several different methods on a set of 44 previously untested, noncongeneric chemical substances are: (a) despite the different approaches, there was good agreement among the predictions made for some chemicals that produced unambiguous bioassay results; (b) SAR models that are based primarily on chemical structure and that ignore biological attributes did not perform as accurately as models that use biological information; and (c) models such as human experts that use more extensive and varied information performed better than models based on one or two attributes to represent chemical carcinogenesis.

The strengths and weaknesses of various SAR approaches for carcinogenicity prediction by computer-assisted models have also been reviewed (42).

IV. ROLES OF SAR IN CANCER RISK ASSESSMENT

In the absence of adequate epidemiologic data and based on the findings that virtually all known human carcinogens are also rodent carcinogens, data from lifetime exposure animal bioassays are primarily used for identifying chemical substances that may be associated with carcinogenic effects in humans. However, the high cost of testing, the large number of animals used in 2-year rodent bioassays, and the huge number of chemicals for which carcinogenicity data do not exist have made it impractical and inhibitory to test each of the thousands of existing chemicals for carcinogenicity. Meanwhile, government health regulatory agencies are under increasing mandate to screen large lists of existing chemicals in commerce and the environment for carcinogenic and other toxicologic effects. Structure-activity relationship is serving an increasingly important role in the risk assessment process as the first line approach in hazard identification, prioritizing chemicals for testing, designing test strategies, identifying research needs, and providing mechanistic insight to support risk assessment and regulatory decision-making. The general approach for assessment of potential human carcinogens is depicted in Figure 3. Some examples that illustrate the roles of SAR in the cancer risk assessment process of existing and premanufactured chemicals under various regulatory statutes of the U.S. Environmental Protection Agency (EPA) are presented below.

1. To obtain data necessary for the EPA to assess and manage the risk of hazardous air pollutants (HAPs) regulated under the Clean Air Act, SAR was used as one of the criteria to select seven of the 21 HAPs for testing of carcinogenicity (43).

2. Currently, EPA is using SAR in the prioritization process of selecting representative chemicals from a list of 215 disinfection by-products (DBP) detected in drinking water for potential future carcinogenicity testing (44).

Structure-activity relationships
(SAR)

⇓

Short-term genotoxicity assays

⇓

Chronic animal bioassays

⇓

Epidemiology

FIG. 3. General approach for assessment of potential human carcinogens.

3. A research program was implemented by several U.S. government agencies to generate an integrated body of mechanistic data from studies on a selected group of prototype dyes derived from benzidine and benzidine congeners (45). The objective of the research program was to apply mechanism-based SAR to predict the carcinogenicity of hundreds of benzidine- and benzidine-congener-derived dyes. The results of this program have led to the development of a significant new use rule (SNUR) by EPA under the Toxic Substances Control Act (TSCA), which would require persons to notify EPA at least 90 days before commencing the manufacture, import, or processing of any benzidine-derived dyes (46). A similar SNUR is being developed for azo dyes containing the benzidine congeners o-toluidine and o-dianisidine in the molecules.

4. Although the tumor-promoting activity of organic peroxides is well documented, the carcinogenicity of many of these free radical-forming chemicals has not been adequately tested. Structure-activity relationship and data from a battery of mechanism-based short-term tests specific for peroxides are supported by EPA to be used for predicting the tumor-promoting or carcinogenic activities of organic peroxides (47).

5. On the basis of structural analogy to 2,4,6-trichlorophenol (a known carcinogen) and evidence for formation of semiquinone free radicals and DNA binding activity, 2,4,6-tribromophenol, an existing chemical in a hazardous waste site, was added to the list of hazardous wastes and the list of hazardous constituents regulated under the Resource Conservation and Recovery Act (RCRA) of EPA (48).

6. Since the enactment of the TSCA in 1977, mechanism-based SAR has been used by EPA's Structure Activity Team in the initial hazard assessment of premanufacturing notification (PMN) chemicals with little or no toxicity data (49, 50). Under the PMN program of the TSCA, EPA is required to evaluate each

Structure-Activity Relationship

of the approximately 2,000 PMN submissions each year from companies intending to manufacture or import new chemical substances for any unreasonable risk they may pose to human or the environment. Structure-activity relationship analysis of a PMN chemical substance may result in a regulatory decision of either dropping the case from further consideration or triggering testing requirements to obtain data for further assessment while regulating chemical exposure. To further support a regulatory decision based on carcinogenicity, quantitative cancer risk of the PMN chemical is also estimated through use of exposure data and surrogate dose-response data of close structural analogues, if available.

7. To help the industry design and develop safer chemicals, the recently constructed "Pollution Prevention (P2) Framework" of the Office of Pollution and Prevention and Toxics (OPPT)/EPA provides a number of EPA-developed, SAR-based computerized models to assess the potential hazards and risks of chemicals (51). The "OncoLogic" Expert System (34) is one of the P2 models for estimating the carcinogenicity potential of chemical substances.

In addition to data from animal carcinogenicity studies, SAR is also an important element in the weight of evidence approach for hazard identification of potential human carcinogens (52, 53). Structure-activity relationship can be used to strengthen or weaken carcinogenicity concern. For instance, information on reactive metabolites and potential mode of action consistent with that of a chemical class in question would enhance the confidence of conclusions made from animal bioassays. Mechanism-based SAR can also play a crucial role in modifying the weight of evidence when there are conflicting tumor data and gaps from animal bioassays.

In recent years, a number of national and international programs have arisen to evaluate alternative test methods to replace, reduce, and refine the use of animals for hazard identification and risk assessment purposes (41, 54, 55). Despite the many advances that have been made in the last decade, predictive toxicology that relies on SAR has not reached the stage that can replace animal testing under current regulatory statutes. Increasing understanding of the mechanistic basis of chemical carcinogenesis and of the relationship between molecular structure and carcinogenic activity appears to be the key to making SAR predictions compelling enough to warrant their use in regulatory decision-making. The predictive value of SAR models for human carcinogens, like that of empirical models, will probably always be open to question:

> Models can be confirmed by the demonstration of agreement between observation and prediction, but confirmation is inherently partial. Complete confirmation is logically precluded by the fallacy of affirming the consequent and by incomplete access to natural phenomena. Models can only be evaluated in relative terms, and their predictive value is always open to question. The primary value of models is heuristic (56).

REFERENCES

1. JD McKinney, A Richard, C Waller, MC Newman, F Gerberick. The practice of structure activity relationships (SAR) in toxicology. Toxicol. Sci. 56:8–17, 2000.
2. EC Miller. Some current perspectives on chemical carcinogenesis in humans and experimental animals: Presidential address. Cancer Res 38:1479–1496, 1978.
3. MD Shelby. The genetic toxicity of human carcinogens and its implications. Mutat Res 204:3–15, 1988.
4. Y-T Woo, JC Arcos, DY Lai. Metabolic and chemical activation of carcinogens: An overview. In: P Politzer, FJ Martin, eds. Chemical Carcinogens: Activation Mechanisms, Structural and Electronic Factors, and Reactivity. Amsterdam: Elsevier, 1998, pp 1–31.
5. J Ashby, RW Tennant. Chemical structure, Salmonella mutagenicity and extent of carcinogenicity as indicators of genotoxic carcinogenesis among 222 chemicals tested in rodents by the U.S. NCI/NTP. Mutat Res 204:17–115, 1988.
6. J. Ashby, RW Tennant. Definitive relationships among chemical structure, carcinogenicity and mutagenicity for 301 chemicals tested by U.S. NTP. Mutat Res 257: 229–306, 1991.
7. J Ashby, D Paton. The influence of chemical structure on the extent and sites of carcinogenesis for 522 rodent carcinogens and 55 different human carcinogen exposures. Mutat Res 286:3–74, 1993.
8. JC Arcos, MF Argus. Chemical Induction of Cancer. Structural Bases and Biologic Mechanism. Vol. IIA, Polynuclear Aromatic Hydrocarbons. New York: Academic Press, 1974.
9. JC Arcos, MF Argus. Chemical Induction of Cancer. Structural Bases and Biologic Mechanism. Vol. IIB, Aromatic Amines and Related Compounds. New York: Academic Press, 1974.
10. JC Arcos, Y-T Woo, MF Argus (with the collaboration of DY Lai). Chemical Induction of Cancer. Structural Bases and Biologic Mechanism. Vol. IIIA, Aliphatic carcinogens. New York: Academic Press, 1982.
11. Y-T Woo, DY Lai, JC Arcos, MF Argus. Chemical Induction of Cancer. Structural Bases and Biologic Mechanism. Vol. IIIB, Aliphatic and Aromatic Halogenated Carcinogens. Orlando, Fla: Academic Press, 1984.
12. Y-T Woo, DY Lai, JC Arcos, MF Argus. Chemical Induction of Cancer. Structural Bases and Biologic Mechanism. Vol. IIIC, Natural, Metal, Fiber and Macromolecular carcinogens. San Diego, Calif: Academic Press, 1988.
13. C Ioannides, DFV Lewis, J Trinick, S Neville, NN Sertkaya, M Kajbaf, JW Gorrod. A rationale for the non-mutagenicity of 2- and 3-aminobiphenyls. Carcinogenesis 10:1403–1407, 1989.
14. FA Beland, WB Melchior Jr, LGL Mourato, MA Santos, MM Marques. Arylamine-DNA adduct conformation in relation to mutagenesis. Mutat Res 376:13–19, 1997.
15. DY Lai, Y-T Woo, MF Argus, JC Arcos. Cancer risk reduction through mechanism-based molecular design of chemicals. In: SC DeVito & RL Garrett, eds. Designing Safer Chemicals. Green Chemistry for Pollution Prevention ACS Symposium Series No. 640. Washington, DC: American Chemical Society, 1996, pp 62–73.

16. AM Richard, Y-T Woo. A CASE-SAR analysis of polynuclear aromatic hydrocarbon carcinogenicity. Mutat Res 242:285–303, 1990.
17. R Jung, D Steinle, R Anliker. A comparison of genotoxicity and carcinogenicity data on aromatic aminosulfonic acids. Fd Chem Toxicol 30:635–660, 1992.
18. BL Van Duuren. Prediction of carcinogenicity based on structure, chemical reactivity and possible metabolic pathways. J Environ Pathol Toxicol 3:11–34, 1980.
19. JC Arcos. Criteria for selecting chemical compounds for carcinogencity testing. J Environ Pathol Toxicol 1:433–458, 1978.
20. Y-T Woo, JC Arcos, DY Lai: Structural and functional criteria for suspecting chemical compounds of carcinogenic activity: state-of-the-art of predictive formalism. In: HA Milman & EK Weisburger, eds. Handbook of Carcinogen Testing. Park Ridge, NJ: Noyes Publ, 1985, pp 1–25.
21. Y-T Woo, JC Arcos. Role of structure-activity relationship analysis in evaluation of pesticides for potential carcinogenicity. In: NN Ragsdale, RE Menzel, eds. Carcinogenicity and Pesticides. ACS Symposium Series No. 414, Washington, DC. American Chemical Society, 1989. pp 175-200.
22. C Hansch. A quantitative approach to biochemical structure-activity relationships. Acct Chem Res 2:232–239, 1969.
23. C Hansch. Structure-activity relationships of chemical mutagens and carcinogens. Sci Tot Environ 109/110:17–29, 1991.
24. G Klopman. Computer automated structure evaluation of organic molecules. J. Am Chem Soc 106:7315–7324, 1984.
25. G Klopman, HS Rosenkranz. Structure-activity relationships: Maximizing the usefulness of mutagenicity and carcinogenicity databases. Environ Health Perspect 96:67–75, 1991.
26. G Klopman. MULTICASE 1. A hierarchical computer automated structure evaluation program. Quant Struct Act Rel 11:176–184, 1992.
27. K Enslein, HH Borgstedt, ME Tomb, BW Blake, HB Hart. A structure-activity prediction model of carcinogenicity based on NCI/NTP assays and food additives. Toxicol Indust Health 3:267–287, 1987.
28. K Enslein, VK Gombar, BW Blake. Use of SAR in computer-assisted prediction of carcinogenicity and mutagenicity of chemicals by the TOPKAT program. Mutat Res 305:47–61, 1994.
29. EJ Matthews, JF Contrera. A new highly specific method for predicting the carcinogenic potential of pharmaceuticals in rodents using enhanced MCASE QSAR-ES software. Regul. Toxicol. Pharmacol. 28:242–264, 1998.
30. R Benigni, C Andreoli, A Giuliani. Structure-activity studies of chemical carcinogens: Use in an electrophilic reactivity parameter in a new QSAR model. Carcinogenesis 10:55–61, 1989.
31. R Benigni, C Andreoli, A Giuliani. Quantitative structure-activity relationships: Principles and applications to mutagenicity and carcinogenicity. Mutat Res 221:197–216, 1989.
32. DV Parke, C Ioannides, DFV Lewis. Metabolic activation of carcinogens and toxic chemicals. Hum Toxicol 7:397–404, 1988.
33. C Ioannides, DFV Lewis, DV Parke. Computer modelling in predicting carcinogenicity. Eur J Cancer Prev 2:275–282, 1993.

34. Y-T Woo, DY Lai, MF Argus, JC Arcos. Development of structure-activity relationship rules for predicting carcinogenic potential of chemicals. Toxicol Lett 79:219–228, 1995.
35. Y-T Woo, DY Lai, MF Argus, JC Arcos. Carcinogenicity of organophosphorus pesticides/compounds: An analysis of their structure-activity relationships. Environ Carcino Ecotoxicol Revs C14:1-42, 1996.
36. Y-T Woo, DY Lai, JC Arcos, MF Argus. Mechanism-based structure-activity relationship (SAR) analysis of carcinogenic potential of 30 NTP test chemicals. Environ Carcino Ecotoxicol Revs C15:139–160, 1997.
37. DM Sanderson, CG Earnshaw. Computer prediction of possible toxic action from chemical structure. The DEREK system. Hum Exp Toxicol 10:261–279, 1991.
38. CA Marchant, and the DEREK Collaborative Group. Prediction of rodent carcinogenicity using the DEREK system for 30 chemicals currently being tested by the National Toxicology Program. Environ Health Perspect 104 (suppl 5):1065–1073, 1996.
39. DW Bristol. Summary and recommendations: Activity classification and structure-activity relationship modeling for human health risk assessment of toxic substances. Toxicol Lett 79:265–280, 1995.
40. DW Bristol, JT Wachsman, A Greenwell. The NIEHS predictive-toxicology evaluation project. Environ Health Perspect 104 (suppl 5):1001–1010, 1996.
41. JT Wachsman, DW Bristol, J Spalding, M Shelby, RW Tennant. Predicting chemical carcinogenesis in rodents. Environ Health Perspect 101:444–445, 1993.
42. AM Richard. Application of SAR methods to non-congeneric data bases associated with carcinogenicity and mutagenicity: Issues and approaches. Mutat Res 305:73–97, 1994.
43. Environmental Protection Agency. Proposed test rules for hazardous air pollutants. Federal Register 61:33177–33200, 1996.
44. Y-T Woo, D Lai, J McLain, M Manibusan, V Dellarco, M Cox. Structure-activity relationship analysis of drinking water disinfection byproducts for potential carcinogenicity. In press.
45. DL Morgan, JK Dunnick, T Goehl, MP Jokinen, HB Matthews, E Zeiger, JH Mennear. Summary of the National Toxicology Program benzidine dye initiative. Environ Health Perspect 102 (suppl 2):63–78, 1994.
46. U.S. Environmental Protection Agency. Benzidine-based chemical substances: Significant new uses of certain chemical substances. Federal Register 61:52287–52297, 1996.
47. DY Lai, Y-T Woo, MF Argus, JC Arcos. Carcinogenic potential of organic peroxides: Prediction based on structure-activity relationships (SAR) and mechanism-based short-term tests. Environ Carcino Ecotoxicol Revs C14:63–80, 1996.
48. U.S. Environmental Protection Agency. Hazardous Waste Management System, Identification and listing of hazardous waste; land disposal restrictions, organobromine production waste. 40 CFR, Part 261, 1998.
49. CM Auer, DH Gould. Carcinogenicity assessment and the role of structure-activity relationship (SAR) analysis under TSCA section 5. Environ Carcino Ecotoxicol Revs C5:29–71, 1987.
50. CM Auer, JV Nabholz, KP Baetcke. Mode of action and the assessment of chemical hazards in the presence of limited data: Use of structure-activity relationship (SAR) under TSCA section 5. Environ Health Perspect 87:183–197, 1990.

51. "Pollution Prevention (P2) Framework." Office of Pollution Prevention and Toxics, U.S. Environmental Protection Agency. EPA-748-B-00-001, June, 2000.
52. U.S. Environmental Protection Agency. Guidelines for carcinogen risk assessment. Federal Register 51:33992–34003, 1986.
53. U.S. Environmental Protection Agency. Proposed guidelines for carcinogen risk assessment. Federal Register 61:17960–18011, 1996.
54. DY Lai, KP Baetcke, VT Vu, JA Cotruvo, SL Eustis. Evaluation of reduced protocols for carcinogenicity testing of chemicals: Report of a joint EPA/NIEHS workshop. Regul Toxicol Pharmacol 19:183–201, 1994.
55. National Toxicology Program Workshop. Mechanism-Based Toxicology in Cancer Risk Assessment: Implications for Research, Regulation, and Legislation. Chapel Hill, N.C: National Toxicology Program, National Institute of Environmental Health Sciences, January 11–13, 1995.
56. N Oreskes, K Shrader-Frechette, K Belitz. Verification, validation, and confirmation of numerical models in the earth sciences. Science 263:641–646, 1994.

5
Regulatory Genetic Toxicology Tests

Wai Nang Choy
Schering-Plough Research Institute, Lafayette, New Jersey

I. INTRODUCTION

The regulatory genetic toxicology tests are a series of well-defined mutagenicity tests designed to detect chemical and physical agents capable of inducing mutations. Mutation is defined as a DNA sequence change that leads to inheritable alteration of gene function. Agents that change DNA sequence are "toxic" to the gene, and thus designated as "genotoxic." Because mutations are often associated with cancers and birth defects, the two most fearsome human diseases, the genotoxicity of a commercial or environmental chemical is critical information for regulatory agencies for the assessment of human risks.

The justification for using genetic toxicology tests to predict carcinogenicity is based on the early somatic mutation theory of carcinogenesis, which simply postulates that cancer can be caused by mutations in the somatic cells (1). An increasing number of carcinogens are proposed to be nongenotoxic based on negative findings in genetic toxicology tests (2–4), but evidence of the somatic mutation theory of carcinogenesis remains strong (5, 6). In fact, almost all known human carcinogens listed by the International Agency for Research on Cancer (IARC) are mutagens (except for fiber, hormone, and lifestyle) (7). Recent human cancer genetic studies also have shown repeatedly that multiple genetic changes are required for the development of human cancers (8, 9), and that the two most intensively studied mechanisms of human carcinogenesis, activation of oncogenes and inactivation of tumor suppressor genes (6), are often mediated by mutations (5, 6). In addition, cancer cytogenetic studies provide retrospective evidence that all cancer cells are heteroploid (10), many with chromosome changes specific for the cancer type (11). It is, therefore, irrefutable that genetic changes are essential for neoplastic transformation and that the concern of genotoxic and nongenotoxic mechanisms of carcinogenesis is, in reality, on whether the chemical or physical agent directly initiates the first mutational event.

The genotoxicity of an agent theoretically can be evaluated by any genetic systems that detect mutations. More than a 100 mutagenicity tests were proposed in the 1970s when the field of genetic toxicology began. Most were based on existing research systems in bacterial or mammalian somatic cell genetics (12–14). However, many tests gradually disappeared over time because sufficient support for further evaluation and development was lacking. Fewer than 10 assays are routinely performed currently, among which is an assortment of in vitro and in vivo assays acceptable by worldwide regulatory agencies (Tables 1 and 2).

The acceptance of these assays was based on their sensitivity and reliability of detecting mutagens and carcinogens. Numerous validation studies have been preformed in the past twenty years. The results of these studies, as related to the prediction of rodent carcinogens, varied, depending on the chemical class and the cancer bioassay database used for analysis (15–17). Contrary to initial belief, various combinations of multiple genetic toxicology tests did not improve predictability (18–25). Recent correlation studies indicated that the most reliable assay for the prediction of rodent carcinogens is the Ames test, which predicted 60% to 70% rodent carcinogens in a diverse database (21, 24). The latest analysis is discussed in Chapter 2.

TABLE 1. Common *in vitro* Genetic Toxicology Tests Required by Regulatory Testing Guidelines

Bacteria	
Gene mutation	*Salmonella typhimurium* (Ames test) (4–5 strains, his^-)
	Escherichia coli (1–2 strains, trp^-)
Mammalian cells	
Gene mutation	Mouse lymphoma cells (L5178Y/TK)
	Chinese hamster ovary cells (CHO/HGPRT)
	Chinese hamster ovary AS52 cells (CHOAS52/XPRT)
Chromosome aberration	Human peripheral blood lymphocytes (HPBL)
	Chinese hamster ovary cells (CHO)
	Chinese hamster lung fibroblasts (CHL)

TABLE 2. Common *in vivo* Genetic Toxicology Tests Required by Regulatory Testing Guidelines

Rodents	
Somatic cells	
Chromosome aberration	Rat bone marrow cells
Micronucleus	Mouse bone marrow erythrocytes
	Mouse peripheral blood erythrocytes
	Rat bone marrow erythrocytes
	Rat peripheral blood erythrocytes
DNA repair	Rat hepatocyte unscheduled DNA synthesis (UDS)
Germ cells (OECD/EPA)	
Chromosome aberration	Rat testicular tissue
Sister chromatid exchange	Rat testicular tissue
DNA breaks	Alkaline elution (DNA breaks)
DNA repair	Rat testicular UDS
Gene mutation	Mouse dominant lethal test
	Mouse specific locus tests
	Mouse heritable translocation test

Several regulatory testing guidelines have been established for routine tests (26–29). These guidelines are very specific in the test battery, test species, genetic endpoints, treatment duration, test procedures, and data evaluations. They guidelines provide consistency and predictability for regulatory compliance. Importantly, based on the results of these tests, they also provide a "definition" albeit dogmatic, of genotoxicity for regulatory purposes to classify chemicals into genotoxic and nongenotoxic categories for hazard identification and risk characterization for risk assessments (Chapters 13 and 14).

II. REGULATORY TESTING GUIDELINES

Historically, numerous guidelines have been established in the United States, Canada, Europe, United Kingdom, Australia, Japan, and the Nordic countries. The major ones are guidelines of the U.S. Food and Drug Administration for food additives (30, 31), U.S. Environmental Protection Agency (US EPA; Toxic Substance Control Act [TSCA] and Federal Insecticide, Fungicide and Rodenticide Act [FIFRA]) (32–36), European Economic Community Council (37,38), United Kingdom (39–41), Organization for Economic Cooperation and Development [OECD] (28), and Japan Ministry of Health and Welfare (42, 43). These guidelines define the battery of tests; the OECD guidelines, especially, also provide detailed procedures for the conduct of these assays.

The basic requirements of the major guidelines are similar, but there are variations. These are mostly attributed to the scientific expertise and mission of the agency. In recent years, major efforts have been made to harmonize these guidelines, which resulted in the adoption of two major guidelines, presumably applicable worldwide. They are the International Conference on Harmonization (ICH) (26, 27) guidelines and the OECD guidelines (28). The ICH guidelines are for pharmaceuticals (44). The OECD guidelines were codeveloped with USEPA, mostly for environmental and agricultural chemicals.

The basic test requirements and strategies of these two guidelines are similar. The OECD guidelines, however, contain more tests for germ cell mutagens. These two guidelines are discussed in great detail in Chapters 9 and 10, respectively. Briefly, both guidelines require the *Salmonella* bacterial mutagenicity assay, an in vitro chromosome aberration assay and an in vivo cytogenetic assay. The in vitro chromosome aberration assay often uses Chinese hamster cells (CHO, CHL) or human lymphocytes, and the in vivo cytogenetic assay is commonly the rodent bone marrow micronucleus assay or bone marrow chromosome aberration assay. In addition, the OECD guideline requires a mammalian cell mutation assay, preferably in mouse lymphoma cells.

If an equivocal or positive response is observed in the above assays, additional in vitro, but often in vivo, tests would be required to assess genotoxicity further. This requirement also extends to chemicals of novel structure, chemicals that belong to a chemical class associated with carcinogenicity, or chemicals containing "structural alert" function groups (18). The requirements for additional tests are discussed Chapters 9 and 10, and described in Chapter 6.

With the development of good laboratory practice (GLP) guidelines, all tests intended for regulatory submissions are performed in compliance with GLP, as specified by the respective agencies (U.S. FDA, 21 CFR Part 58; U.S. EPA-FIFRA, 40 CFR, Part 160; U.S. EPA-TSCA, 40 CFR, Part 792; EEC Council Directive, 90/18/EED, and JMHW (Japan) Notification No. 313).

III. THE PRIMARY GENETIC TOXICOLOGY ASSAYS

The primary assays are the bacterial mutagenicity assay, the in vitro/in vivo cytogenetic assays, and the mammalian cell mutagenicity assay. The most common complementary assay is the unscheduled DNA synthesis assay. The procedures of these "routine" assays have been standardized (29) and published in the OECD guidelines (28). The principles of these assays are described below, and the complementary assays are described in Chapter 6.

A. Bacterial Mutagenicity Assay

The bacterial mutagenicity assay is routinely performed in *S. typhimurium* (15, 45, 46) and sometimes in *Escherichia coli* (47). The *Salmonella* assay is generally referred to as the "Ames test," named after its originator Dr. Bruce Ames at the University of California at Berkeley. The *Salmonella* assay is by far the most recognized and widely performed genetic toxicology assay. During the almost 20 years since its introduction in the mid-1970s, the test procedures have been firmly established, and test results been extensively validated (15 48). Although the reliability of this assay for the detection of mutagens and carcinogens varied with the chemical and toxicity database, this test has sometimes been conveniently used as the only test to determine the "genotoxicity" of chemicals, although such practice is not quite justifiable.

The *Salmonella* assay tests the ability of a test article to induce mutation(s) in the bacterial genes involved in histidine biosynthesis. Several *Salmonella* strains, each carrying a mutation in one of the histidine biosynthesis genes (collectively designated as *his*$^-$), are used in the assay. With existing mutations in the *his*$^-$ loci, the tester strains are auxotrophic for histidine and are unable to grow in the absence of exogenous histidine. The assay is to detect the potential of the test article to induce reverse mutations in the histidine loci (*his*$^-$) back to wild type (*his*$^+$), which enable the bacteria to regain de novo histidine biosynthesis, and be able to grow in the absence of histidine. Mutants generated by the reverse mutations (reversions) are identified in agar medium deficient in histidine. Because this assay detects reverse mutations, it is also referred to as "reverse mutation assay," and the mutants are called "revertants."

In addition to the *his*$^-$ genotype, several genetic changes were introduced into the *Salmonella* strains to enhance their susceptibility to mutagenesis. They are a mutation in excision DNA repair (*uvr*B), a mutation in the cell wall to increase cell permeability (*rfa*), and the introductions of plasmids (pKM101, pAQ1). Plasmid pKM101 enhances error-prone DNA repair and the multicopies of plasmid pAQ1 carrying the *his*$^-$ gene to increase the target size for mutagenesis, and thus the sensitivity, of the assay. Four *Salmonella* strains are commonly used in the assay: TA1535, TA100, TA1537, and TA98. Strains TA1535, TA100, and TA102 detect base-pair substitution mutations, and TA98, TA1537, TA97, and TA97a detect frameshift mutations. Strains TA1537, TA97, and TA97a are used interchangeably (49). Strain TA102 detects mutations in the AT base-pair (48, 50), as compared with other strains that detect mutations in the GC regions, and it is often added as the fifth strain. The genotypes of these *Salmonella* tester strains are shown in Table 3.

Because the existing mutations of the *his*$^-$ tester strains are well characterized and specific mutations (i.e., base-pair substitute or frameshift) are required for their reversion to wild type, the type of mutation induced by a mutagenic test article can be elucidated by examining the pattern of reversion response in several tester strains. This valuable information can be obtained from the *Salmonella* tests, but not from other routine genetic toxicology tests without DNA sequence analysis.

TABLE 3. Genotypes of Bacteria Strains for the Bacterial Mutagenicity Assay

Bacterial strains	Gene affected	Additional mutations			Type of mutation detected
		DNA repair	LPS	R factor	
Salmonella typhimurium					
TA 1535	*his* G46	*uvrB*	*rfa*	—	Base-pair substitution
TA 100	*his* G46	*uvrB*	*rfa*	pKM101	Base-pair substitution/frameshift
TA97	*his* D6610 *his* 01242	*uvrB*	*rfa*	pKM101	Frameshift
TA97a	*his* D6610 *his* 01242	*uvrB*	*rfa*	pKM101	Frameshift
TA98	*his* D3052	*uvrB*	*rfa*	pKM101	Frameshift
TA1537	*his* C3076	*uvrB*	*rfa*	pKM101	Frameshift
TA102	*his* G428	+	*rfa*	pKM101 pAQ1	Base-pair substitution
Escherichia coli					
WP2uvrA	*trp*	*uvrA*	+	—	Base-pair substitution
WP2uvrA (pKM101)	*trp*	*uvrA*	+	pKM101	Base-pair substitution

The *Escherichia* assay sometimes is added as a complement to the *Salmonella* assay. Similar to the *Salmonella* strain TA102, the *Escherichia* strains detect AT base-pair mutations (51, 52). Therefore, only one of two strains is required for the detection of A-T mutations. The *Escherichia* assay is also a reversion assay, but the target genes are involved in tryptophan biosynthesis. The tester strains thus are auxotrophic mutants (*trp*⁻) that require tryptophan to grow. Reverse mutations in the *trp*⁻ loci back to wild type (*trp*⁺) enable the bacteria to grow in the absence of tryptophan. Revertants are selected in agar medium deficient in tryptophan. Two *Escherichia* tester strains are commonly used in this assay. Both the WP2*uvr*A and WP2*uvr*A(pKM101) strains are defective in DNA repair (*uvr*A), and one strain also carries the plasmid pKM101. The genotypes of the *Escherichia* tester strains are shown in Table 3.

As for all in vitro genetic toxicology assays, an exogenous metabolic activation system is included in the assays to investigate the possibility that the test article would produce mutagenic metabolite(s) after metabolism. At present, the standard metabolic activation system for routine in vitro assays is the supernatant of Aroclor 1254-induced rat liver homogenates commonly referred to as S9 (microsomal fraction after centrifugation of liver homogenates at 9000xg) (48). S9 and its enzyme cofactors are typically added to the test system at the time of test article exposure. Common modifications of this metabolic activation system are the use of hamster S9, human S9, and rat S9 induced by enzyme inducers other than Aroclor 1254, such as phenobarbital or naphthoflavone (53).

Two common procedures are used to conduct the *Salmonella* assay: the plate incorporation assay (48) and the preincubation assay (54). The plate incorporation assay is the most widely used. It is conducted by exposing the test article, with or without metabolic activation, to the bacteria in soft agar medium before plating the bacteria onto the agar plate. Although the study protocols may vary, a typical assay consists of a dose range-finding assay and a mutagenicity assay using four to five tester strains (49, 55). Each study includes one assay without S9 metabolic activation and one assay with S9 activation at four to five dose levels. A repeat trial is recommended if the first trial does not show mutagenic response. For the second trial, a different exposure method, such as the preincubation assay, or a different metabolic activation system may be used (26, 27).

The preincubation method is similar to the plate incorporation assay, except that the test article is exposed to the bacteria in liquid medium for a predetermined period before plating. The preincubation method is reported to be more sensitive than the plate incorporation method, but its sensitivity depends on the variations of exposure conditions.

After dosing, bacteria are plated onto histidine-deficient agar plates and are incubated at 37°C for approximately 2 to 3 days. The revertant colonies are scored manually or by an automated counter. The number of revertants in test article-treated cultures is compared with those of the solvent control.

An increase of revertants over the solvent control constitutes a positive response. Several statistical methods have been proposed for data analysis (56), but the most commonly used method is direct comparison for fold increases. Except for strains such as TA1535, TA1537, and TA1538 that have low spontaneous revertant counts that require a threefold increase for a positive response, for other strains, an increase of revertant counts approximately twofold above the concurrent solvent control is generally considered a positive response (55).

Procedures of the *Escherichia* mutagenicity assay are similar to that of the *Salmonella* assay, except the selection medium is tryptophan-deficient agar. If the *Escherichia* mutagenicity assay is used to complement the *Salmonella* assay, only one strain is required (26, 27).

B. Mammalian Cell Mutagenicity Assays

Mutagenicity assays that use mammalian cells are theoretically more relevant than bacteria assays for the evaluation of mutagenicity of test articles in mammals. The most widely accepted mammalian mutation assays are the mouse lymphoma L5178 $TK^{+/-}$ assay (mouse lymphoma cell L5178Y using the heterozygous thymidine kinase gene $TK^{+/-}$ as the genetic marker) (57–61), the CHO/HGPRT assay (Chinese hamster ovary cell using the hypoxanthine-guanine phosphoribosyltransferase gene as the genetic marker) (62, 63), and less frequently, the CHOAS52/XPRT assay (Chinese hamster ovary cell AS52 strain using the xanthine-guanine phosphoribosyltransferase gene as the genetic marker) (64–66). All the above assays are forward mutation assays to detect mutations from wildtype to the mutant phenotype.

All three genetic markers are enzymes involved in nucleic acid biosynthesis. These assays were derived from somatic cell genetic studies in the 1960s when the most common genetic markers for mutation research were drug resistance to nucleotide analogues. Mutations in the HGPRT, XPRT, and TK genes abolish the enzyme functions and the ability to incorporate the respective nucleotides to the DNA. Because of this defect, mutants are also unable to incorporate toxic nucleotide analogues and are resistant to these analogs. Mutants defective in HGPRT and XPRT are resistant to 6-thioguanine (6-TG), and mutants defective in TK are resistant to trifluorothymidine (TFT) or bromodeoxyuridine (BUdR).

It should be noted that the target cells commonly used for mammalian cell mutagenicity assays are heteroploid established cell lines (e.g., CHO, CHOAS52) or tumor cell line (mouse lymphoma L5178Y), simply because they are easy to culture and to clone.

The procedures of the mammalian cell mutagenicity assay, although varying with different cell types, follow a basic plan (67, 68). A typical assay consists of a dose range-finding assay and a mutagenicity assay, often in two independent trials. Cells are typically treated with the test article for 3 to 6 hours in the presence or absence of Aroclor 1254-induced rat liver S9 for metabolic activation. Treated cells are allowed to multiply for a few days, usually about 7 days, for phenotypic expression before mutant selection. A phenotypic expression period is necessary to eliminate the pre-existing gene product and to allow the expression of a mutant recessive phenotype. Mutants are selected in the presence of their respective selection agent. Mutant colonies are quantified and compared with the controls. Statistical analyses are usually used for data evaluation. Several common pair-wise comparison and dose-response trend tests are acceptable.

For the mouse lymphoma assay, the size of the mutant colonies is also measured, and the mutants are classified into small-colony mutants and large-colony mutants. Large-colony mutants are presumably generated by single gene mutations, and small-colony mutants are believed to also associate with chromosome aberrations (60). This observation, although still somewhat controversial, led to the proposal that the mouse lymphoma assay detects both gene mutation and cytogenetic changes.

The mouse lymphoma assay is thought to be more sensitive than the other mammalian cell mutagenicity assays, simply because more test articles are positive in this assay. Correlation studies of the results of the mouse lymphoma assay, as compared with results of animal bioassays, also showed more false-positive results (61). For the lack of a good scientific explanation that carcinogenicity should correlate to mutagenicity in mouse lymphoma cells, many positive correlations may be only coincidental. In fact, several in vitro cell culture conditions, such as changes in the pH and osmolality, are known to affect the results of this assay (69–71).

The mouse lymphoma assay has become the favorite mammalian cell mutagenicity assay of several regulatory agencies (i.e., OECD and ICH). The mouse lymphoma assay is believed to be able to detect both gene mutation and chromosome changes, as demonstrated in the small-colony mutants. Also, the mouse lymphoma assay is considered more sensitive than the CHO/HGPRT assay because the location of the TK gene is on an autosome. The TK gene in mouse lymphoma cells is located on chromosome 11 (72), and the XPRT gene in CHOAS52 cells is located on chromosome 6 or 7 (73). In contrast, the HGPRT gene in the CHO cells is located on the X chromosome. As only one X chromosome is active in a diploid cell, it was postulated that test articles that induce large DNA deletions cannot be detected in the CHO/HGPRT assay because a deletion that extends beyond the HGPRT gene into the neighboring "essential" gene, in the absence of a complement active copy, is not compatible with cell survival. Mutations could, therefore, be underestmated because dead cells do not form colonies and would not register as mutants. Although this theoretical assumption has not been demonstrated experimentally, the mouse lymphoma assay has been selected by regulatory agencies as the preferred assay for mammalian cell mutations.

For the mouse lymphoma assay, selection of mutants has been performed in soft agar. A recent modification of this plating method is to clone and quantify mutants in multiwell microtiter plates (74, 75). This method appears to be advantageous to the plating method because of its increased cloning efficiency and the sensitivity of the assay. The microtiter method for the mouse lymphoma assay is currently under validation and is described in Chapter 7.

C. Cytogenetic Assays

The cytogenetic assays are used to detect test articles that damage chromosomes (clastogens) and concomitantly, test articles that interfere with normal chromosome segregation (aneuploid inducers). Chromosome aberration is a valid genetic marker for carcinogen identification because all cancer cells are heteroploid with constant change of their chromosome compositions. Indeed, heteroploid conversion from the diploid karyotype is an inevitable step for the establishment of immortal cell lines that may lead to cancer cells (10). Chromosome aberrations are tested in both in vitro and in vivo systems.

1. In Vitro Chromosome Aberration Assays

The in vitro chromosome aberration assays detect the ability of the test agent to induce chromosomal damage. The most common aberrations examined are chromosome breaks and chromosome gaps, but more complex chromosome changes such as translocations, endoreduplications, and polyploidy are also evaluated. The cells commonly used in this assay are Chinese hamster lung cells (CHL) (76, 77), CHO (78, 79) and human peripheral blood lymphocytes (HPBL) (80, 81).

The procedures of the chromosome aberration assay vary slightly with different cell types (82–84). A typical assay consists of a dose range-finding assay and a chromosome aberration assay, usually in two trials. Cells are treated with the test article, with or without Aroclor 1254-induced rat S9 metabolic activation for 3 to 6 hours, and harvested at approximately 1.5 times the cell cycle after treatment. Variations to this procedure include prolonged treatment times without activation, multiple harvests, or prolonged harvests for up to two to three cell cycles. These variations are developed to increase the sensitivity of the assay, especially for certain classes of chemicals, such as nucleotides (43) and nitrosamides (77). Harvested cells are fixed with methanol and acetic acid on microscope slides, stained, and examined under the microscope for chromosome aberrations. The chromosome aberration assays are relatively sensitive to nonphysiological cell culture conditions, such as pH and osmolality of the culture medium, test agent precipitates, and severe cytotoxicity traumatic to the cells (70, 85–88). Several commonly used statistical methods are acceptable for data analysis.

2. In Vivo Cytogenetic Assays: The Micronucleus Tests

The micronucleus test is the most widely used in vivo assay for the detection of clastogens and agents that induce aneuploidy (abnormal chromosomal segregation). This test was initially developed in mouse bone marrow erythrocytes (89), but it is also commonly conducted in rats (90) and, less frequently, in hamsters (91) and monkeys (92). The current micronucleus tests are mostly conducted in mouse or rat bone marrow erythrocytes (93, 94). Micronucleus assays of peripheral blood erythrocytes in mouse and rats are also acceptable for regulatory compliance.

The micronuclei are small nuclei that arise from chromosome fragments as a result of chromosome breaks (double-stranded DNA breaks) or whole chromosome detached from microtubules. In the micronucleus test, the target cells are the bone marrow erythroblasts. Induced micronuclei in the erythroblasts are retained in the erythrocytes after the extrusion of the main nuclei from the cells during maturation. The micronuclei can be scored in polychromatic erythrocytes (PCE; young erythrocytes) in the bone marrow or in the peripheral blood. They can also be scored in normochromatic erythrocytes (NCE, matured erythrocytes) in the peripheral blood after prolonged exposures. The micronucleus frequency presumably will reach a steady-state level if the exposure duration is equal to or greater than the normal lifespan of the NCE in the peripheral blood. An increase of micronuclei in the erythrocytes indicates that chromosome damage in the erythroblasts has occurred, resulting from the clastogenicity of the test article.

The origination of the micronuclei, whether by chromosome breaks or by nondisjunction, can be distinguished immunochemically for the presence of the kinetochore protein in the micronuclei (95), or by in situ DNA hybridization of the centromeric sequence in the micronuclei (96). The presence of kinetochore protein or the centromeric DNA sequence indicates the micronuclei were derived from whole chromosomes as a result of aneuploidy induction.

The procedures of the micronucleus test can vary slightly, depending on the type of data desired for the study (97). For a typical rodent bone marrow micronucleus assay, there are at least three acceptable dosing and bone marrow harvest schedules: one dose and three daily harvests (24-, 48-, and 72-hour harvests); two daily dosing and two daily harvests (24 and 48 hour harvests after the last dose); and three daily dosing and one harvest (24 hour harvest after the last dose). These multiple-dosing and harvest schedules were developed to accommodate genotoxic responses to test articles with different metabolism characteristics and to capture the window of maximum micronuclei occurrence in bone marrow PCE (98, 99). A typical micronucleus test consists of a dose range-finding assay and a micronucleus assay. Male (and female) animals are dosed with the test article, usually by intraperitoneal injection or oral gavage, although other routes of dosing are acceptable. Toxicity is monitored by clinical signs, animal death and bone marrow suppression. Bone marrow suppression is monitored by a decrease in ratio of PCE to NCE, or PCE to total erythrocytes (RBC, PCE+NCE), which are commonly referred to as PCE/NCE or PCE/RBC ratio. Dosing can be a single dose or daily doses for 2 to 3 days. Bone marrow cells are usually harvested from the femurs of the rodent at 24, 48, and/or 72 hours after the last dosing, depending on the dosing protocol. Bone marrow smears are prepared on microscope slides, stained with Giemsa (89) or nucleic acid-specific fluorescent dyes (i.e., acridine orange) (100) and scored for micronucleated PCE. An increase of the micronucleated PCE indicates clastogenicity of the test article. Commonly used statistical methods are acceptable for data analysis.

Micronuclei can also be scored in the PCE in mouse peripheral blood (101, 102). With the acridine orange staining technique (103), an interlaboratory study was conducted recently in Japan for the validation of the peripheral blood micronucleus test in the mouse and rats (104). One significant advantage of the peripheral blood assay is less invasive blood sampling, which allows for repetitive sampling from the same animal over time for kinetic studies of micronuclei induction. The peripheral blood micronucleus tests was originally thought to be only applicable to the mouse because the mouse spleen is known to be incapable of removing the micronucleated erythrocytes in the blood. However, recent studies in the rats also showed that newly formed micronucleated PCE (also referred to as reticulocytes) are also detectable in the peripheral blood before they are removed by the spleen (105). Furthermore, the ability to score micronuclei in NCE, which accumulate in the peripheral blood, permits the evaluation of clastogenicity of a test article in standard multiple-dose toxicology studies. Indeed, retrospective evaluation of micronuclei in NCE of peripheral blood have been performed in several mouse cancer bioassays in the National Toxicology Program (NTP) (106, 107). Such studies provide examples that the micronucleus test can be incorporated into routine chronic animal bioassays for the assessment of genotoxicity under the same exposure conditions as the animal bioassays.

IV. COMPLEMENTARY REGULATORY GENETIC TOXICOLOGY ASSAY

A. Unscheduled DNA Synthesis Assay

The unscheduled DNA synthesis (UDS) assay measures repairable DNA damage induced by the test article. The UDS assay is customarily conducted in hepatocytes, both in vitro and in vivo. Metabolic activation of the test article is believed to occur in the hepatocytes proximal to DNA, which enhances the sensitivity of mutagen detection. The in vivo-in vitro UDS assay has been proposed by the ICH guidelines to be an additional assay for the clarification of equivocal in vitro test results.

1. In Vitro Assay

Routine in vitro UDS assays are performed in primary cultures of rat hepatocytes (108–110), but assays in mouse, hamster, monkey, and human hepatocytes were also reported (111).

The procedures of the in vitro UDS assay in rat hepatocytes do not vary much (112, 113). In a typical assay, primary rat hepatocyte cultures are exposed to the test article simultaneously with radioactive thymidine (usually ^{3}H-thymidine), a radioactive precursor of DNA synthesis. Damaged DNA that is undergoing DNA repair, referred to as unscheduled DNA synthesis (as compared with DNA synthesis in cell division), will incorporate ^{3}H-thymidine into the repaired regions of the DNA, which in turn is detected by autoradiography. Repaired DNA appears as dark grains in the nuclei in autoradiographs. An increase of grain counts in the nuclear region indicated DNA repair and, therefore, DNA damage by the test article.

2. In Vivo-In Vitro Assay

The procedures of the in vivo-in vitro UDS assay is similar to that of the in vitro UDS assay, except the test article is administered to the animals. The UDS assay itself is conducted in cultured hepatocytes derived from dosed animals (114–116). This assay is commonly conducted in rats, but studies in mice have also been reported (117, 118).

There are a few variations in the procedures of the in vivo-in vitro UDS assay (113). A typical assay is to dose male (and female) rats with the test article, usually by a single oral gavage dose, and to isolate hepatocytes at two harvests, usually 2 hours and 16 hours after dosing. Hepatocytes are obtained from the liver of the dosed animals, and primary hepatocytes are immediately exposed to ^{3}H-thymidine. The amount of ^{3}H-thymidine incorporation—the indicator of UDS—is detected by autoradiography as grain counts. An increase of grain counts in the nuclear region indicates DNA repair.

Both the in vitro and the in vivo-in vitro UDS assays are most useful in detecting liver carcinogens and less reliable in detecting carcinogens of other organs (23, 119). The in vivo-in vitro UDS assay is considered to be more reliable than the in vitro UDS assay, which is known to produce more false-positive results than the in vivo-in vitro assay, as compared with results of rodent cancer bioassays (115, 116, 120, 121).

V. SUMMARY

All current routine genetic toxicology tests are mutation tests. The pivotal tests for regulatory compliance are described in this chapter. They are the bacterial mutagenicity assay (Ames test), in vitro chromosome aberration assays, in vitro mammalian cell mutation assays, and the in vivo cytogenetic assay (micronucleus test). New testing methodologies are constantly proposed, and advanced tissue culture, biochemical or recombinant DNA technologies, are often introduced. Extensive validations of these new tests are required before they can be used for regulatory purposes.

As related to cancer risk assessment, results of the genetic toxicology tests are used for the identification of mutagens. Conversely, if an agent is known to be a carcinogen, knowledge of its genotoxicity provides information on the mechanism of carcinogenesis (122, 123), which is important information for the selection of risk characterization methodologies, as described in Chapters 14 and 15.

For regulatory genetic toxicology, a major achievement is the recent completion of two internationally harmonized guidelines, the ICH and the OECD/EPA guidelines. These guidelines provide the scientific basis of the genetic toxicology tests, the rationale of the selection of the test batteries, the test procedures, and the guidance for data interpretation. These guidelines are essential for ascertaining credibility and consistency in making risk assessment decisions.

VI. REFERENCES

1. Boveri. The Origin of Malignant Tumors. Baltimore: Williams & Wilkins, 1929.
2. BE Butterworth, TJ Slaga. Nongenotoxic Mechanism of Carcinogenesis. Cold Spring Harbor, NY: Cold Spring Harbor Laboratory Press, 1987.
3. SM Cohen, L Ellwein. Cell proliferation in carcinogenesis. Science 249: 1007–1011, 1990.
4. JC Barrett. A multistep model for neoplastic development: Role of genetic and epigenetic changes. In: JC Barrett, ed Mechanism of Environmental Carcinogenesis. Vol. 2. Boca Raton, FL: CRC Press, 1987, pp 117–126.
5. AG Knudson. Genetics of human cancer. Ann Rev Genet 20:231–251, 1986.
6. AG Knudson. Antioncogenes and human cancer. Proc Natl Acad Sci U S A 90:11914–11921, 1993.
7. IARC. International Agency for Research on Cancer. IARC Monographs on the Evaluation of Carcinogenic Risks to Humans. Overall Evaluation of Carcinogenicity to Humans. Lyon, France, 1997.
8. B Volgelstein, ER Feareson, SR Hamilton, SE Kern, AC Preisinger, M Leppert, Y Nakamura, R White, AMM Smits, JL Bos. Genetic alternations during colorectal-tumor development. N Engl J Med 319:525–529, 1988.
9. SE Kern, B Vogelstein. Genetic alternations in colorectal tumors. In: J Brugge, T Curran, E Harlow, F McCormick, eds. Origins of Human Cancer. Cold Spring Harbor, NY, Cold Spring Harbor Laboratory Press, pp 577–585, 1991.
10. JW Littlefield. Variation, Senescence and Neoplasia in Culture Somatic Cells. Cambridge, MA: Harvard University Press, 1976.
11. F Mitelman. Catalog of Chromosome Aberrations in Cancer. New York: Alen Liss Inc. 1988.

12. AP Li, RH Heflich. Genetic Toxicology. Boca Raton, FL: CRC Press, 1991.
13. D Brusick. Principles of Genetic Toxicology, 2nd ed. New York: Plenum Press, 1987.
14. WN Choy. Genetic Toxicology Testing. In: AM Fan, LW Chang, eds. Toxicology and Risk Assessment: Principles, Methods, and Applications. New York: Marcel Dekker, 1996, pp 153–169.
15. BN Ames, J McCann, E Yamasaki. Methods for detecting carcinogens and mutagens with the *Salmonella*/mammalian-microsome mutagenicity test. Mutat Res 31:347–364, 1975.
16. J McCann, E Choi, E Yamasaki, BN Ames. Detection of carcinogens as mutagens in the *Salmonella*/microsome test: Assay of 300 chemicals. Proc Natl Acad Sci U S A 72:5135–5139, 1975.
17. IFH Purchase, E Longstaff, J Ashby, JA Styles, D Anderson, PA Lefevre, FR Westwood. An evaluation of 6 short-term tests for detecting organic chemical carcinogens. Br J Cancer 37:873–903, 1978.
18. J Ashby, RW Tennant. Chemical structure, *Salmonella* mutagenicity and extent of carcinogenicity as indicators of genotoxic carcinogenesis among 222 chemicals tested in rodents by the U.S. NCI/NTP. Mutat Res 204:17–115, 1988.
19. LS Gold, TH Slone, BR Stern, L Berstein. Comparison of target organs of carcinogenicity for mutagenic and non-mutagenic chemicals. Mutat Res 286:75–100, 1993.
20. BA Fetterman, BS Kim, BH Margolin, JS Schildcrout, MG Smith, SM Wagner, E Zeiger. Predicting rodent carcinogenicity from mutagenic potency measured in the Ames *Salmonella* assay. Environ Mol Mutagen 29:312–322, 1997.
21. E Zeiger. Identification of rodent carcinogens and noncarcinogens using genetic toxicity tests: Premises, promises, and performance. Regul Toxicol Pharmacol 28:85–95, 1998.
22. E Zeiger, B Anderson, S Haworth, T Lawlor, K Mortelmans. *Salmonella* mutagenicity tests: V. Results from the testing of 311 chemicals. Environ Mol Mutagen 21:2–141, 1992.
23. RW Tennant, JW Spalding, S Stasiewicz, WD Caspary, JM Mason, MA Resnick. Comparative evaluation of genetic toxicity patterns of carcinogens and noncarcinogens: Strategies for predictive use of short-term assays. Environ Health Prospect 75:87–95, 1987.
24. E Zeiger, JK Hasman, MD Shelby, BH Margolin, RW Tennant. Evaluation of four *in vitro* genetic toxicity tests for predicting rodent carcinogenicity: Confirmation of earlier results with 41 additional chemicals. Environ Mol Mutagen 16:1–14, 1990.
25. MD Shelby, E Zeiger. Activity of human carcinogens in *Salmonella* and rodent bone marrow cytogenetic tests. Mutat Res 234:257–261, 1990.
26. ICH. International Conference on Harmonization. Guidance on Specific Aspects of Regulatory Genotoxicity Tests for Pharmaceuticals. Fed Reg 61:18198–18202, 1996.
27. ICH. International Conference on Harmonization. Guidance on Genotoxicity: A Standard Battery for Genotoxicity Testing of Pharmaceuticals. Fed Reg 62:62472–62475, 1997.

28. OECD. Organisation for Economic Co-operation and Development. OECD Guidelines for the Testing of Chemicals. TG nos. 471-486, Adopted 1984, 1986, 1997, Paris, France.
29. SM Galloway, ed. Report of the international workshop on standardization of genotoxicity test procedures. Mutat Res 312:195–322, 1994.
30. USFDA. United States Food and Drug Administration. Toxicological Principles for the Safety Assessment of Direct Food Additives and Color Additives Used in Foods, "Redbook I." Bureau of Foods, 1982.
31. USFDA. United States Food and Drug Administration. Toxicological Principles for the Safety Assessment of Direct Food Additives and Color Additives Used in Foods, "Redbook II." Center for Food Safety and Applied Nutrition, Draft.
32. USEPA. United States Environmental Protection Agency. Health Effects Testing Guidelines, Part 798, Subpart F-Genetic Toxicity. Fed Reg 50:39435–39458, 1985.
33. USEPA. United States Environmental Protection Agency. Guidelines for Mutagenicity Risk Assessment. Fed Reg 51:34006–34012, 1986.
34. USEPA. United States Environmental Protection Agency. Revision of TSCA Test Guidelines. Fed Reg 52:19078–19081, 1987.
35. KL Dearfield, AE Aulletta, MC Cimino, MM Moore. Considerations in the U.S. Environmental Protection Agency's testing approach for mutagenicity. Mutat Res 258:259–283, 1991.
36. AE Auletta, KL Dearfield, MC Cimino. Mutagenicity test schemes and guidelines: US EPA Office of Pollution Prevention and Toxics and Office of Pesticide Program. Environ Mol Mutagen 21:38–45, 1993.
37. CPMP. Commission of the European Communities: The Rules Governing Medicinal Products in the European Community. Vol. III. Guidelines on the quality, safety and efficacy of medicinal products for human use, 1989, p 103.
38. CPMP. CPMP Working Party on Safety of Medicinal Products. Note for guidance. Recommendations for the development of nonclinical testing strategies III.58.89-EN, Draft No. 7, July 5, 1990.
39. DH. Department of Health United Kingdom. Guidelines for the Testing of Chemicals for Mutagenicity. Department of Health Report on Health and Society Subjects, No. 35, London: HMSO, 1989.
40. UKEMS. United Kingdom Environmental Mutagen Society. Basic Mutagenicity Tests: UKEMS Recommended Procedures. JD Kirkland, ed. Cambridge, UK: Cambridge University Press, 1990.
41. DJ Kirkland. Genetic toxicology testing requirements: Official and unofficial views from Europe. Environ Mol Mutagen 21:8–14, 1993.
42. MHW. Ministry of Health and Welfare. Japan Guidelines for Nonclinical Studies of Drug Manual. Japan: Yakuji Nippo Ltd, 1995.
43. T Sofuni. Japanese guidelines for mutagenicity testing. Environ Mol Mutagen. 21:2–7, 1993.
44. L Muller, Y Kikuchi, G Probst, L Schechtman, H Shimada, T Sofuni, D Tweats. ICH-harmonized guidelines on genotoxicity testing of pharmaceuticals: Evolution, reasoning and impact. Mutat Res 436:195–225, 1999.
45. BN Ames. Identifying environmental chemicals causing mutations and cancer. Science 204:587–593, 1979.

46. BN Ames, WE Durston, E Yamasaki, FD Lee. Carcinogens are mutagens: A simple test system combining liver homogenate for activation and bacterial for detection. Proc Natl Acad Sci U S A 70:2281–2285, 1973.
47. MHL Green. Mutation testing using Trp$^+$ reversion in *Escherichia coli*. In: BJ Kilbey, M Legator, W Nichols, C Ramel eds, Handbook of Mutagenicity Test Procedures 2nd ed. Amsterdam: Elsevier, 1984, pp 161–187.
48. DM Maron, BN Ames. Revised methods for the *Salmonella* mutagenicity tests. Mutat Res 113:173–215, 1983.
49. D Gatehouse, S Haworth, T Cebula, E Gocke, L Kier, T Matsushima, C Melcion, T Nobmi, T Ohta, S Venitt, E Zeiger. Recommendations for the performance of bacterial mutation assays. Mutat Res 312:217–223, 1994.
50. DEI Levin, M Hollstein, MF Charistman, EA Schwiers, BN Ames. A new *Salmonella* test strain (TA102) with A-T base pairs at the site of mutation detects oxidative mutagens. Proc Natl Acad Sci U S A 79:7445–7449, 1982.
51. P Wilcox, A Naidoo, DJ Wedd, DG Gatehouse. Comparison of *Salmonella typhimurium* TA102 with *Escherichia coli* WP2 tester strains. Mutagenesis 5:285–291, 1990.
52. K Watanabe, K Sakamoto, T Sasaki. Comparisons on chemically-induced mutation among four bacterial strains, *Salmonella typhimurium* TA102 and TA2638, and *Escherichia coli* WP2/pKM101 and WP2uvrA/pKM101: Collaborative study II. Mutat Res 412:17–31, 1998.
53. TE Johnson, DR Umbenhauer, SM Galloway. Human liver S-9 metabolic activation: Proficiency in cytogenetic assays and comparison with phenobarbital/beta-naphthoflavone or Aroclor 1254 induced rat S-9. Environ Mol Mutagen 28:51–59, 1996.
54. T Matsushima, T Sugimura, M Nagao, T Yahagi, A Shirai, M Sawamura. Factors modulating mutagenicity in microbial tests. In: KH Norpoth, RC Garner eds. Short-Term Test Systems for Detecting Carcinogens. New York: Springer, pp 273–285, 1980.
55. LE Kier, DJ Brusick, AE Auletta, ES von Halle, MM Brown, VF Simmmon, V Dunkel, J McCann, K Montelmans, M Prival, TK Rao, V Ray. The *Salmonella typhimurium*/mammalian microsomal assay: A report of the U.S. Environmental Protection Agency Gene-Tox Program. Mutat Res 168:69–240, 1986.
56. BS Kim, BH Margolin. Statistical methods for the Ames *Salmonella* assay: A review. Mutat Res 436:113–122, 1999.
57. D Clive, R McCuen, JFS Spector, C Piper, KH Mavournin. Specific gene mutations in L5178Y cells in culture: A report of the U.S. Environmental Protection Agency Gene-Tox Program. Mutat Res 115:225–256, 1983.
58. D Clive, W Caspery, PE Kirby, R Krehl, M Moore, J Mayo, TJ Oberly. Guide for performing the mouse lymphoma assay for mammalian cell mutagenicity. Mutat Res 189:143–156, 1987.
59. WJ Caspary, YJ Lee, S Poulton, BC Myhr, AD Mitchell, CJ Rudd. Evaluation of the L5178Y mouse lymphoma cell mutagenesis assay: Quality-control guidelines and response categories. Environ Mol Mutagen 12:19–36, 1988.
60. WF Blazak, FJ Los, CJ Rudd, WJ Caspary. Chromosome analysis of small and large L5178Y mouse lymphoma cell colonies: Comparison of trifluorothymidine-

resistant and unselected cell colonies from mutagen-treated and control cultures. Mutat Res 224:197–208, 1989.
61. AD Mitchell, AE Auletta, D Clive, PE Kirby, MM Moorem, BC Myhr. The L5178Y/tk$^{+/-}$ mouse lymphoma specific gene and chromosomal mutation assay. A phase III report of the U.S. Environmental Protection Agency Gene-Tox Program. Mutat Res 394:177–303, 1997.
62. AW Hsie, DA Casciano, DB Couch, DF Krahn, JP O'Neill, BL Whitfield. The use of Chinese hamster ovary cell to quantify specific locus mutation and to determine mutagenicity of chemicals: A report of the U.S. Environmental Protection Agency Gene-Tox Program. Mutat Res 86:193–214, 1981.
63. AP Li, JH Carver, WN Choy, AW Hsie, RS Gupta, KS Loveday, JP O'Neill, JC Riddle, LF Stankowski, LL Yang. A guide for the performance of the Chinese hamster ovary cell/hypoxanthine guanine phosphoribosyltransferase gene mutation assay. Mutat Res 189:135–141, 1987.
64. LF Stankwoski, AW Hsie. Quantitative and molecular analyses of radiation-induced mutation in AS52 cells. Radiat Res 105:37–48, 1986.
65. LF Stankowski Jr, KR Tindall. Characterization of the AS52 cell line for use in mammalian cell mutagenesis studies. In: MM Moore, DM DeMarini, FJ de Serres, KR Tindell eds. Mammalian Cell Mutagenesis, Banbury Report 28. Cold Spring Harbor, NY: Cold Spring Harbor Laboratory Press, 1987, pp 71–79.
66. KR Tindall, LF Stankowski Jr. Deletion mutations are associated with the differential induced mutant frequency response of the AS52 and CHO-K1-BH4 cell lines. In: MM Moore, DM DeMarini, FJ de Serres, KR Tindall, eds. Mammalian Cell Mutagenesis, Banbury Report 28. Cold Spring Harbor Laboratory Press, Cold Spring Harbor, NY: 1987, pp 283–292.
67. ER Nestmann, RL Brillinger, JPW Gilman, CJ Rudd, SHH Swierenga. Recommended protocols based on a survey of current practice in genotoxicity testing laboratories: II. Mutation in Chinese hamster ovary, V79 Chinese hamster lung and L5178Y mouse lymphoma cells. Mutat Res 246:255–284, 1991.
68. CS Aaron, G Bolcsfoldi, HR Glatt, M Moore, Y Nishi, L Stankowski, J Thesis, E Thompson. Mammalian cell gene mutation assays working group report. Mutat Res 312:235–239, 1994.
69. D Brusick. Genotoxicity effects in cultured mammalian cells produced by low pH treatment conditions and increased ion concentration. Environ Mutagen 8:879–886, 1986.
70. D Brusick, ed. Genotoxicity produced in cultured mammalian cell assays by treatment conditions. Mutat Res 189:71–179, 1987.
71. MA Cifone, B Myhr, A Eiche, G Bolcsfoldi. Effect of pH shifts on the mutant frequency at the thymidine kinase locus in mouse lymphoma L5178Y TK+/– cells. Mutat Res 189:39–46, 1987.
72. CA Kozak, FH Ruddle. Assignment of the gene for thymidine kinase and galactokinase to *Mus musculus* chromosome 11 and the preferential segregation of this chromosome in Chinese hamster/mouse somatic cell hybrids. Somatic Cell Genet 3:121–133, 1977.
73. KC Michaelis, LM Helvering, DE Kindig, ML Garriott, KK Richardson. Localization of xanthine guanine phosphoribosyl transferase gene (gpt) of *E. coli* in AS52

metaphase cell by fluorescence in situ hybridization. Environ Mol Mutagen 24:176–180, 1994.
74. J Cole, DB McGregor, M Fox, J Thacker, RC Garner. Gene mutation assays in cultured mammalian cells. In: DJ Kirkland, ed. Basic Mutagenicity Tests. Avon, UK: Cambridge University Press, 1990, pp 87–114.
75. TJ Oberly, DL Yount, ML Garriot. A comparison of the soft agar and microtiter methodologies for the L5178Y tk+/− mouse lymphoma assay. Mutat Res 388:59–66, 1997.
76. M Ishidate Jr, T Sofuni. The *in vitro* chromosomal aberration test using Chinese hamster lung (CHL) fibroblast cells in culture. In: J Ashby, FJ de Serres, M Draper, M Ishidate Dr, BH Margolin, BE Matter, MD Shelby, eds. Evaluation of Short-Term Tests for Carcinogens. Report of the International Programme on Chemical Safety's Collaborative Study on *in vitro* Assays. Amsterdam: Elsevier, 1985, pp 427–432.
77. M Ishidate Jr, MC Harnois, T Sofuni. A comparative analysis of data on the clastogenicity of 951 chemical substances tested in mammalian cell cultures. Mutat Res 195:151–213, 1988.
78. SM Galloway, AD Bloom, M Resnick, BH Margolin, F Nakamura, P Archer, E Zeiger. Development of a standard protocol for *in vitro* cytogenetic testing with Chinese hamster ovary cells: Comparison of results for 22 compounds in two laboratories. Environ Mutagen 7:1–51, 1985
79. SM Galloway, MJ Armstrong, C Reuben, S Colman, B Brown, C Cannon, AD Bloom, F Nakamura, M Ahmed, S Duk, J Rimpo, BH Margolin, MA Resnick, B Anderson, E Zeiger. Chromosome aberration and sister chromatid exchanges in Chinese hamster ovary cells: Evaluation of 108 chemicals. Environ Mol Mutagen 10:1–109, 1987.
80. RJ Preston, W Au, MA Bender, JG Brewen, AV Carrano, JA Heddle, A McFee, S Wolff, JS Wassom. Mammalian *in vivo* and *in vitro* cytogenetic assays: A report of the U.S. Environmental Protection Agency Gene-Tox Program. Mutat Res 87:143–188, 1981.
81. RJ Preston, JR San Sebastian, AF McFee. The *in vitro* human lymphocyte assay for assessing the clastogenicity of chemical agents. Mutat Res 189:175–183, 1987.
82. SHH Swierenga, JA Heddle, EA Sigal, JPW Gilman, RL Brillinger, GR Douglas, ER Nestmann. Recommended protocol based on a survey of current practice in genotoxicity testing laboratories IV. Chromosome aberration and sister-chromatid exchange in Chinese hamster ovary, V79 Chinese hamster lung and human lymphocyte cultures. Mutat Res 246:301–322, 1991.
83. SM Galloway, MJ Aardema, M Ishidate Jr, JL Ivett, DJ Kirkland, T Morita, P Mosesso, T Sofuni. Report from working group on *in vitro* tests for chromosomal aberrations. Mutat Res 312:241–261, 1994.
84. M Ishidate Dr, KF Miura, T Sofuni. Chromosome aberration assays in genetic testing *in vitro*. Mutat Res 404:167–172, 1998.
85. SM Galloway, DA Deasy, CL Bean, AR Kraynak, MJ Armstrong, MO Bradley. Effects of high osmotic strength on chromosome aberration, sister chromatid exchanges and DNA strand breaks, and the relation to toxicity. Mutat Res 189:15–25, 1987.

86. D Scott, SM Galloway, RR Marshall, M Ishidate Jr, D Brusick, J Ashby, BC Myhr. Genotoxicity under extreme culture conditions. A report form ICPEMC TASk Group. Mutat Res 257:147–204, 1991.
87. MJ Armstrong, CL Bean, SM Galloway. A quantitative assessment of the cytotoxicity associated with chromosomal aberration detection in Chinese hamster ovary cell. Mutat Res 265:45–60, 1992.
88. T Morita, T Nagaki, I Fukuda, K Okumura. Clastogenicity of low pH to various cultured mammalian cells. Mutat Res 268:297–305, 1992.
89. W Schmid. The micronucleus test for cytogenetic analysis. In: A Hollander, ed. Chemical Mutagens: Principles and Methods for Their Detection. Vol. 4. New York: Plenum Press, 1976, pp 31–53.
90. E George, AK Wootton, DG Gatehouse. Micronucleus induction by azobenzene and 1,2-dibromo-3-chloropropane in rat: Evaluation of a triple-dose protocol. Mutat Res 324:129–134, 1990.
91. A Basler. Aneuploid-inducing chemicals in yeast evaluated by the micronucleus test. Mutat Res 174:11–13, 1996.
92. WN Choy, PR Henika, CC Willhite, AF Tarantal. Incorporation of a micronucleus study into a developmental toxicology and pharmacokinetic study of L-selenomethionine in nonhuman primates. Environ Mol Mutagen 21:73–80, 1993.
93. JA Heddle, M Hite, B Kirkhart, K Mavournin, JT MacGregor, GW Newell, MF Salamone. The induction of micronuclei as a measure of genotoxicity: A report of the U.S. Environmental Protection Agency Gene-Tox Program. Mutat Res 123: 61–118, 1983.
94. KH Mavournin, DH Blakey, MC Cimino, MF Salamone, JA Heddle. The *in vivo* micronucleus assay in mammalian bone marrow and peripheral blood: A report of the U.S. Environmental Protection Agency Gene-Tox Program. Mutat Res 239: 29–80, 1990.
95. R Gudi, SS Sandhu, RS Athwal. Kinetochore identification in micronuclei in mouse bone-marrow erythrocytes: An assay for the detection of aneuploidy-inducing agents. Mutat Res 234:263–268, 1990.
96. M Hayashi, J Marki-Paakkanen, H Tanabe, M Honma, T Suzuki, A Matsuoka, H Mizusawa, T Sofuni. Isolation of micronuclei from mouse blood and fluorescence in situ hybridization with a mouse certromeric DNA probe. Mutat Res 307: 245–251, 1994.
97. H Tinwell, ed. Serial versus single dosing protocols for the rodent bone marrow micronucleus assay. Mutat Res 234:111–261, 1990.
98. JT MacGregor, JA Heddle, M Hite, BH Margolin, C Ramel, MF Salamone, RR Tice, D Wild. Guidelines for the conduct of micronucleus assay in mammalian bone marrow erythrocytes. Mutat Res 189:103–122, 1987.
99. M Hayashi, RR Tice, JT MacGregor, D Anderson, DH Blakey, M Kirsh-Volder, FB Oleson Jr, F Pacchierotti, F Romagna, H Shimada, S Sutou, B Vannier. *In vivo* rodent erythrocyte micronucleus assay. Mutat Res 312:293–304, 1994.
100. M Hayashi, T Sofuni, M Ishidate Jr. An application of acridine orange fluorescent staining to the micronucleus tests. Mutat Res 120:241–247, 1983.

101. JT MacGregor, CM Wehr, DH Gould. Clastogen-induced micronuclei in peripheral blood erythrocytes: The basis of an improved micronucleus test. Environ Mutagen 2:509–514, 1980.
102. JT MacGregor, R Schlegel, WN Choy, CM Wehr. Micronuclei in circulating erythrocytes: A rapid screen for chromosomal damage during routine toxicity testing in mice. In: AW Hayes, RC Schnell, TS Miya, eds. Development in the Science and Practice of Toxicology. Amsterdam: Elsevier, 1983, pp 555–558.
103. M Hayashi, T Morita, Y Kodama, T Sofuni, M Ishidate Jr. The micronucleus assay with mouse peripheral blood reticulocytes using acridine orange-coated slides. Mutat Res 245:245–249, 1990.
104. CSGMT. The Collaborative Study Group for the Micronucleus Test of the Mammalian Mutagenesis Study subgroup of the Environmental Mutagen Society of Japan. Micronucleus test with rodent peripheral blood reticulocytes by acridine orange supravital staining. Mutat Res 278:81–213, 1992.
105. HE Holden, JB Majeska, D Studwell. A direct comparison of mouse and rat bone marrow and blood as target tissues in the micronucleus assay. Mutat Res 391: 87–89, 1997.
106. WN Choy, JT MacGregor, MD Shelby, RR Maronpot. Induction of micronuclei by benzene in B6C3F1 mice: Retrospective analysis of peripheral blood smears from the NTP carcinogenesis bioassay. Mutat Res 143:55–59, 1985.
107. JT MacGregor, CM Wehr, PR Henika, MD Shelby. The *in vivo* erythrocyte micronucleus test: Measurement as steady state increases assay efficacy and permits integration with toxicity studies. Fundam Appl Toxicol 14:513–522, 1990.
108. AD Mitchell, DA Casciano, ML Meltz, DE Robinson, RHC San, GM Williams, ES von Halle. Unscheduled DNA synthesis test: A report of the U.S. Environmental Protection Agency Gene-Tox Program. Mutat Res 123:363–410, 1983.
109. GM Williams, C Tong, SV Brat. Tests with the rat hepatocyte primary culture/DNA repair tests. J Ashby, FJ de Serres, M Draper, M Ishidate Jr, BH Margolin, BE Matter, MD Shelby, eds. In: Evaluation of Short-Term Tests for Carcinogens. Report of the International Programme on Chemical Safety's Collaborative Study on *in vivo* Assays. Amsterdam: Elsevier, 1985, pp 341–345.
110. BE Butterworth, J Ashby, E Bermudez, D Casciano, J Mirsalis, G Probst, GM Williams. A protocol and guide for the *in vitro* rat hepatocyte DNA repair assay. Mutat Res 189:113–121, 1987.
111. KL Steinmetz, CE Green, JP Bakke, DK Spak, JC Mirsalis. Induction of unscheduled DNA synthesis in primary cultures of rat, mouse, hamster, monkey and human hepatocytes. Mutat Res 206:91–102, 1988.
112. SHH Swierenga, JA Bradlaw, RL Brillinger, JPW Gilman, ER Nestmann, RC San. Recommended protocols based on a survey of current practice in genotoxicity testing laboratories: I. Unscheduled DNA synthesis assay in rat hepatocyte cultures, Mutat Res 246:235–253, 1991.
113. S Madle, SW Dean, U Andrae, C Brambilla, B Burlinson, DJ Doolittle, C Furihata, T Hertner, CA McQueen, H Mori. Recommendations for the performance of UDS tests *in vitro* and *in vivo*. Mutat Res 312:263–285, 1994.

114. JC Mirsalis, BE Butterworth. Detection of unscheduled DNA synthesis in hepatocytes isolated from rats treated with genotoxic agents: An *in vivo-in vitro* assay for potential carcinogens and mutagens. Carcinogenesis 1:621–625, 1980.
115. JC Mirsalis, CK Tyson, BE Butterworth. Induction of DNA repair in hepatocytes from rats treated *in vivo* with genotoxic agents. Environ Mutagen 4:553–562, 1982.
116. JC Mirsalis. Summary report on the performance of the *in vivo* DNA repair assay. In: J Ashby, FL de Serres, MD Shelby, BH Margolin, M Ishidate Jr, GC Becking, eds. Evaluation of Short-Term Tests for Carcinogens. Report of the International Programme on Chemical Safety's Collaborative Study on *in vivo* Assays. Cambridge, UK: Cambridge University Press, 1988, pp 1.345–1.351.
117. JC Mirsalis, CK Tyson, EN Loh, JP Bakke, CM Hamilton, KL Steinmetz. An evaluation of the ability of benzo[a]pyrene, pyrene, 2- and 4-acetylaminofluorene to induce unscheduled DNA synthesis and cell proliferation in the liver of male rats and mice treated *in vivo*. In: J Ashby, FL de Serres, MD Shelby, BH Margolin, M Ishidate Jr, GC Becking, eds. Evaluation of Short-Term Tests for Carcinogens. Report of the International Programme on Chemical Safety's Collaborative Study on *in vivo* Assays. Cambridge, UK: Cambridge University Press, 1988, pp 1.361–1.366.
118. J Ashby, PA Lefevre, T Shank, J Lewtas, JE Gallagher. Relative sensitivity of ^{32}P-postlabelling of DNA and autoradiographic UDS assay in the lives of mice exposed to 2-acetylaminofluorene (2-AAF). Mutat Res 252:259–268, 1991.
119. JC Mirsalis, CK Tyson, KL Steinmetz, EN Loh, CM Hamilton, JP Bakke, JW Spalding. Measurement of unscheduled DNA synthesis and S-phase synthesis in rodent hepatocytes following *in vivo* treatment: Testing of 24 compounds. Environ Mutagen 14:155–164, 1989.
120. JC Mirsalis. *In vivo* measurement of unscheduled DNA synthesis and hepatic cell proliferation as an indicator of hepatocarcinogenesis in rodent. Cell Biol Genet Toxicol 3:165–173, 1987.
121. JC Mirsalis. KL Steinmetz, JP Bakke, CK Tyson, EKN Loh, CM Hamilton, MJ Ramsey, J Spalding. Genotoxicity and tumor promoting capabilities of blue hair dyes in rodent and primate liver. Environ Mutagen 8:55–56, 1986.
122. VL Dellarco, JA Wiltse. US Environmental Protection Agency's revised Guidelines for Carcinogen Risk Assessment: Incorporating mode of action data. Mutat Res 405:273–277, 1998.
123. WN Choy. Principles of Genetic Toxicology. In: AM Fan, LW Chang, eds. Toxicology and Risk Assessment: Principles, Methods, and Applications. New York: Marcel Dekker, 1996, pp 25–36.

6
Complementary Genetic Toxicology Assays

B. Bhaskar Gollapudi
The Dow Chemical Company, Midland, Michigan

I. INTRODUCTION

Genetic toxicology plays a critical role in assessing potential health risks from exposure to xenobiotic agents. This area of toxicology can also help us define critical cellular events that lead to the transformation of an apparently normal somatic cell into a genetically initiated cell with the potential to induce tumors. Short-term in vitro genetic toxicology screening assays have greatly facilitated research activity directed toward the identification of possible genotoxic agents. However, experience with these assays over the last two decades has shown that some chemicals identified in vitro as having genotoxic potential failed to elicit a similar activity in vivo. For obvious reasons, the in vitro systems are limited in many ways in mimicking the fate and biological consequences of the chemical in the whole animal. For this reason, the standard battery of genetic toxicology tests in tier I invariably includes at least one in vivo test. The bone marrow micronucleus test is frequently the assay of choice among genetic toxicologists in tier I because of its simplicity. However, there are a number of situations in which the micronucleus test may not be appropriate to address the in vivo genotoxic potential of a material. For example, this assay cannot provide any information on the genotoxicity that occurs only at the portal of entry tissue, such as the gastrointestinal tract after oral administration, the skin in dermal application studies, and the upper respiratory tract in inhalation studies. A number of other in vivo tests complement the results obtained from the basic set of screening tests. The need for these assays and the principles underlying various aspects of the commonly used complementary tests are discussed below.

II. NEED FOR COMPLEMENTARY GENETIC TOXICOLOGY ASSAYS

The need for complementary genetic toxicology assays is usually driven by the response observed in the initial screening battery, to meet regulatory requirements, or both. Additionally, complementary genotoxicity assays are often used to understand the mode of action in chemical carcinogenesis. Each of these situations is examined below.

A. Supplement to Screening Battery

During hazard identification process, the response observed in the initial battery of screening tests requires further evaluation before the risk posed by the test agent can be established. For example, a unique positive result in the bacterial reverse mutation assay raises a red flag on the potential genotoxicity of the material in vivo. The in vivo bone marrow micronucleus test may not necessarily be appropriate to address this concern if the material in question induces primarily point mutations, such as single base substitutions. Similarly, an initial battery of screening tests is unlikely to predict the genotoxicity of those materials that are activated in vivo through the complex interplay of hepatic and extrahepatic metabolism. Only an appropriately designed in vivo study can address the mutagenic potential of such materials.

B. Regulatory Studies

Most of the agencies charged with the regulation of chemicals and pharmaceutical substances have adopted a tiered testing approach to assess genotoxic potential. The Office of Pesticide Programs and Toxic Substances at the United States Environmental Protection Agency (EPA) requires follow-up studies if any of the tests in tier I elicit a positive response (1). For example, positive results in an in vitro assay for gene mutations in tier I could trigger an in vivo assay for interaction with gonadal cell DNA in tier II. Similarly, a test for dominant lethal mutations in the germ cells might be requested of a chemical that displays cytogenetic activity in bone marrow cells. From the above triggers, it is apparent that the type of test requested in tier II is dictated by the endpoint responding in tier I. The European Economic Council (EEC) directive for the regulation of plant protection chemicals, on the other hand, requires in vivo follow-up studies in addition to the bone marrow test if any of the in vitro tests in tier I for gene mutations give positive results (2). The International Commission on Harmonization (ICH) of technical requirements for registration of pharmaceuticals for human use took a similar approach (3).

C. Studies to Support Mode of Action in Carcinogenicity

The contribution of genotoxicity to a tumorigenic response in an animal bioassay is less contentious with chemicals that exhibit clear and biologically meaningful positive results in tier I test systems. In such cases, it is generally believed that the observed tumorigenic response is a consequence of genotoxicity. However, it is an overwhelming task to propose a mechanistic basis for tumor induction with certain chemical carcinogens. Included in this class are agents that induce tumors at sites other than the bone marrow and for which the evidence for genotoxicity is equivocal, that is, some in vitro positive results combined with negative bone marrow data. In such cases, it is imperative that evidence for genotoxicity should be gathered from the target tissue(s) and not from a surrogate tissue, such as the bone marrow.

III. GENERAL CONSIDERATIONS IN ASSAY PERFORMANCE

A. Selection of Species

The decision regarding the species of choice should be based on the experimental objective. Examples of species-specific genotoxicity have appeared in the literature (4, 5). If the study is undertaken to obtain mode of action data to explain tumor findings, then the relevant species and strain should be selected for the in vivo genotoxicity assay. For routine screening studies, the species of choice can be based on any prior toxicology data such as subchronic, chronic, and metabolism studies, taking into account the expected similarity to human situations. Several commonly used strains of laboratory rats (e.g., Sprague-Dawley, Fischer-344) and mice (e.g., CD-1, B6C3F1, Swiss) are suitable for these studies. In addition, hamsters are also used in these investigations, albeit less frequently. Historically, germ cell cytogenetic analyses have been conducted primarily in the mouse testicular cells, and very few studies using the rat have appeared in the literature.

B. Sex

The objective of the study usually determines the selection of the sex. In cases in which there is no remarkable difference in toxicity between the sexes, males are usually recommended for routine screening studies. Obviously drugs and pharmaceuticals targeted for a given sex (e.g., hormones) should be evaluated in the relevant sex. For technical reasons, germ cell cytogenetic effects are usually studied in males only. In female mammals, several stages of germ cell progression take place during fetal development, and any induced effects on these cell types (e.g., oogonia) cannot be readily investigated.

C. Animal Husbandry

It is needless to state that good animal husbandry practices are essential in obtaining scientifically meaningful results. Physical and biological stress factors, such as overcrowding, food/water deprivation, and infections can have adverse impact on the study and may lead to variable/spurious results.

Animals should be obtained from a reliable source. Animals used in the study must be individually identified so that all the events pertaining to a given animal can be tracked throughout the study. Several methods of uniquely identifying animals are practiced, including simple ones such as tail markings and metal ear tagging to the state of the art electronic transponders implanted subcutaneously. Before treatments, animals should be randomly assigned to various treatment groups to minimize wide variations in health status across the groups. This is usually done based on body weights and a randomization program. Animals should be observed at least twice daily for clinical signs, and these observations should be extended if the treatment induces overt signs of toxicity.

D. Number of Animals/Treatment Group

The sample size is largely dictated by statistical considerations that take into account, among other things, the background incidence of the type of genetic damage under investigation, its variability among individual animals, and the desired power of the assay. Some of these parameters have been well worked out for several of the commonly used tests.

E. Selection of Dose Levels

In toxicology studies, the criteria to be used for selecting dose levels are highly controversial. The regulatory guidelines for the conduct of in vivo genetic toxicology studies require dose levels up to the maximum tolerated dose (MTD) or up to 2 g/kg for relatively nontoxic test materials. The MTD is usually defined by the survival of the treated animals or by some other measures such as clinical signs of toxicity or cytotoxicity in the target tissue, among others. An extreme example of this recommendation is the use of a dose that might lead to the death of the treated animals after 4 days so long as the duration of the study in question is only 2 days. The recommendation for the use of sufficiently high doses in these studies was generally based on the premise that such doses compensate for the low statistical power of the commonly used study designs and short exposure duration. The MTD doses used in these studies often tend to be several orders of magnitude higher than the expected human exposure levels. It has been argued that such doses are irrelevant for safety assessment and risk evaluation because they may overwhelm or saturate the body's defense and protective mechanisms, such as the detoxification pathways, DNA repair mechanisms, and others. In certain cases, administration of unusually high doses may lead to the saturation of primary detoxification pathways, which in turn may lead to the formation of unusual metabolites through alternate metabolic pathways—a situation unlikely to be encountered during normal human exposure

Ideally, the highest dose level selection should be based on sound scientific judgement that takes into account any available pharmacokinetic data as well as the expected human exposure levels. When the in vivo genotoxicity assay is being conducted to follow-up on tumor findings, then the dose levels selected may closely mimic those used in the 2-year cancer bioassay with some upward adjustment for the reduced dosing duration. Because the highest dose level used in most of the carcinogenicity studies are already unreasonably high, it is important to limit doses to levels that are expected to yield linear pharmcokinetics.

F. Route of Administration

The following factors are taken into account when the route of exposure is selected: expected route of human exposure, bioavailability, test material characteristics, and technical limitations. Oral gavage (p.o.) and intraperitoneal (i.p.) administrations are two commonly used methods for acute studies. The use of i.p. method is highly contentious (6, 7). Proponents of this method argue that it optimizes the likelihood of bioavailability of the test agent to the target site and hence should be the choice in those instances in which absorption data from a relevant human exposure route is not available. However, this method has been criticized as irrelevant for obtaining human risk assessment data because normal, relevant biological processes are bypassed when the material is injected into the abdomen. Inhalation exposure is usually used for gaseous and highly volatile compounds, as well as in those instances in which the anticipated human exposure is by inhalation. For studies of longer duration, commonly used methods include dietary, drinking water, and oral gavage administrations. Intravenous (i.v.) administration is usually the method of choice for drugs that are administered to humans by this method.

Bioavailability of the administered dose is a critical consideration in designing these studies and selecting the route of administration. For example, it is a mistake to make a generalized statement that a test material is not genotoxic in vivo based on the results of an oral gavage bone marrow cytogenetic analysis when the administered material is not absorbed from the gastrointestinal tract. Results from such studies, although they preclude genotoxicity to a vast majority of target tissues because of the lack of bioavailability, do not address site of contact effects, in this case on the gastrointestinal tract. Similarly, one cannot conclude, based on a bone marrow test, that a chemical is not genotoxic when there is a significant level of first pass metabolism in the liver and the target tissue is not exposed to the parent compound at meaningful levels. Such studies fail to address the effect of the parent molecules on portal of entry tissue(s).

Test materials are usually administered in a vehicle and volume that do not upset the normal physiology of the experimental animal. The commonly used organic solvent dimethylsulfoxide should be avoided for in vivo studies because of its toxicity as well as its adverse impact on the normal metabolism.

G. Duration of Exposure

The common practice of acutely exposing animals to the agent in genotoxicity studies stems more from convenience and less from scientific merits. Repeated, subchronic dosing regimens are more relevant to human exposure situations and offer the advantage of analyzing genotoxicity at its steady-state levels. In many instances, such a treatment regimen also negates the need for multiple post-treatment tissue sampling times. Krishna et al (8) recently reviewed published studies in which repeated dosing protocols were successfully used in rodent genotoxicity evaluation.

H. Selection of Target Tissue for Analysis

The target tissues available for analyzing genotoxicity depend on the objective of the study and the type of assay used. Bone marrow and liver are the two most commonly used tissues for these analyses. Bone marrow, by virtue of its high mitotic rates, has been a favorite for cytogenetic analysis. Another feature that makes this tissue attractive for in vivo studies is the fact that it is a highly perfused tissue and blood-borne test materials and their metabolites can be assumed to reach this target tissue. The liver is an attractive tissue because, in a majority of cases, orally administered chemicals are delivered to this tissue through portal clearance. It is also a principle site of metabolic activation and detoxification, which enables the examination of the effects induced by reactive metabolites. Some of the methods currently available, such as the transgenic mutation assays, comet assay, and DNA adduct analysis can analyze genotoxicity virtually in any tissue of interest, whereas others, such as cytogenetic analysis, are usually restricted to those tissues that are actively dividing.

Prior toxicity data will be extremely valuable in selecting the target tissue for analysis. For example, if kidney is identified as a target organ in a subchronic dietary toxicity study or nasal mucosa in an inhalation study, it makes scientific sense to examine genotoxicity in these tissues and, if desired, in one or more non-target tissues. Tissue-specific metabolic pathways could lead to the generation of mutagenic metabolites that might be missed when a nontarget tissue is analyzed, as exemplified by the unique mutagenicity of the bladder carcinogen p-cresidine in the urinary bladder but not in other tissues (9). Similarly, the site of contact mutagen l-chloromethylpyrene elicited a positive result in the skin but not in other tissue (10). These examples underscore the limitations of the surrogate tissues in detecting organ-specific genotoxic events.

IV. COMMONLY USED IN VIVO CYTOGENETIC TESTS

A. Bone Marrow Chromosomal Analysis

The assay detects the effect of treatments on the structural integrity of chromosomes of mitotically active bone marrow cells. These effects are usually analyzed at the metaphase stage of the cell cycle through microscopy, and they consist of gross chromosomal structural alterations, such as chromosome/chromatid breaks and exchanges. The type of aberrations analyzed in this assay is usually lethal to the cell. However, the ability of a compound to induce such changes is predictive of its ability to induce other types of changes (small deletions, balanced translocations) that are compatible with cell survival and transmittable to future generations.

The assay involves treatment of animals with the test material and collection of bone marrow samples at one or more intervals after treatment. A spindle poison such as colchicine (4 mg/kg, i. p.) is administered a few hours before sacrifice (2–4 hours) to arrest cells at metaphase stage of the cell cycle. The bone marrow samples are collected into an isotonic medium (e.g., Hanks balanced salt solution) and the cells are subsequently treated with a hypotonic solution (e.g., 0.075 M potassium chloride, 10–20 minutes) to aid in the swelling of the slides and spreading of the chromosomes during slide preparation. At the end of the hypotonic treatment, the cells are exposed to several changes of a fixative (e.g., 3:1 methanol:acetic acid). Finally, the cells are suspended in a small volume of the fixative. The metaphase spreads are prepared by dropping the cell suspension onto cold slides, air-drying, then warming them to approximately 40–50° C to facilitate the drying process. The slides are usually stained in 5% Giemsa for chromosome examinations.

The sampling intervals are dictated by the average cell cycle time of the target cells and are selected to assure the recovery of the damage induced at various stages of the cell cycle. Multiple sampling times also facilitate the recovery of cells from cell cycle delay and accommodate such factors as the delayed metabolism of the test chemical. It is critical for these assays that the damage is assessed at the first posttreatment metaphases, because structural chromosomal damage often leads to reproductive death of the affected cells. As most of the chemically induced aberrations require an intervening S-phase, the sampling times selected should be sufficiently long for the cell cycle to progress through an intervening S-phase after treatment. For routine studies, two sampling times are recommended (11, 12) the first being 1.5 times the normal cell cycle time (e.g., $12 \times 1.5 = 18$ hours for rat) and a later sampling time of 24 hours.

The assay, as conducted routinely in most laboratories, is suitable for analyzing only structural aberrations such as breaks, exchanges, chromosome pulverization, among others. Numerical chromosomal changes, except endoreduplication, are difficult to assess unequivocally, as the air drying technique used to prepare the slides is prone to introduce artifacts. For example, it is difficult to establish whether a metaphase lacking a chromosome or two is real aneuploid or one in which chromosomes are lost due to cell breakage during slide preparation. Similarly, great care should be taken when classifying polyploidy since cospreading of chromosomes from two separate cells could theoretically result in an apparent polyploid cell.

The advantage of this assay lies in the fact that this target tissue is highly vascularized and, hence, toxicants and their metabolites are expected to reach the target tissue. Additionally, the high level of mitotic activity of the bone marrow facilitates the recovery of adequate numbers of cells for chromosomal analysis. Preparation of specimens for chromosomal analysis is relatively simple; however, analysis of slides for chromosome aberrations is technically demanding and requires extensive experience and patience.

In general, results from the bone marrow chromosome analyses correlate very well with those of the less labor intensive micronucleus test. Therefore, there is no need to conduct both these assays on any given compound. However, if an unexpected positive finding is seen in the micronucleus test when the in vitro cytogenetics result is negative, then chromosomal analyses will help to establish whether the observed effect is the result of clastogenicity. Similarly, a bone marrow metaphase analysis can be used as a complementary assay in those instances in which an in vitro clastogen yields negative bone marrow micronucleus test results.

Various regulatory bodies (11, 12) have published guidelines for the performance of this assay. Additional information on key methodological aspects has also been discussed by Preston et al (13) and Richold et al. (14).

B. Spermatogonial Metaphase Analysis

The spermatogonial metaphase analysis is one of the few germ cell assays that can be performed with relative ease. This assay measures the ability of an administered test material to reach the germ cells and cause clastogenicity. The blood-testes barrier usually prevents the exposure of germ cells to a number of toxicants. However, chemicals such as ethylnitrosourea, ethylmethane sulfonate, and metabolites of cyclophosphamide readily cross this barrier and cause germ cell cytogenetic damage. Because all the germ cell mutagens identified so far also cause somatic cell genotoxicity (15), this test is usually conducted only as a follow-up to a positive in vitro or in vivo somatic cell cytogenetic result.

The primary focus of this assay is to evaluate the clastogenic activity of the administered agent on various classes of spermatogonial cells that are undergoing waves of mitotic divisions. The cell cycle time of these spermatogonia is approximately 36 hours in mice and rats. Accordingly, the treated animals are sacrificed at various intervals after treatment such as 24, 36, and 48 hours to enable the sampling of cells treated at various stages of cell cycle, as well as to account for any cell cycle delay. Before they are killed, animals are treated with a spindle poison to arrest cells at the metaphase stage of the cell cycle. Preparation of cells for microscopy is similar to that described for bone marrow, with the exception of an extra step that involves the digestion of the seminiferous tubules with trypsin (0.1% for 15 minutes) to facilitate the release of spermatogonial cells.

From a theoretical point of view, this assay is extremely relevant for hazard identification. However, very little literature describes this test system and, hence, its methodological aspects and assay performance cannot be critically evaluated. Investigators wishing to use this assay may need to standardize the assay conditions, such as an appropriate positive control chemical and sampling times, in the species of interest. Guidelines for the performance of this assay have been published (11, 12).

V. UNSCHEDULED DNA SYNTHESIS ASSAY IN HEPATOCYTES

The occurrence of DNA synthesis in a cell that is not in the normal phase of replicating its genome, that is, cell division, is called unscheduled DNA synthesis (UDS). This is believed to be a manifestation of the DNA repair process in response to DNA damage induced by physical and chemical agents. The DNA repair process involves excision of damaged DNA or DNA adducts and resynthesis of the excised strand to restore normal DNA sequence. The repair process can be detected by measuring the incorporation of radiolabeled thymidine into nuclear DNA of cells that are not in the S-phase of the cell cycle. Thus, an increase in the number of cells undergoing UDS after treatment with a xenobiotic agent is an indication of the ability of the compound to induce DNA damage.

Williams (16) first proposed the use of rat hepatocytes for measuring UDS by autoradiographic method as an in vitro screening assay for identifying potential genotoxic and carcinogenic materials. These cells are virtually nondividing, and this feature greatly reduces the interference of replicative DNA synthesis in quantitating UDS. In addition, freshly isolated hepatocytes are capable of metabolically activating many promutagens and procarcinogens into their reactive forms, which eliminated the need for an exogenous metabolic activation system.

In this assay, freshly isolated rat hepatocytes grown on cover glasses are treated in vitro with the test material in the presence of radiolabeled thymidine [^{3}H thymidine] for 18 to 20 hours. At the end of the treatment, the cover glasses are processed for autoradiography, and the incorporation of the label into the DNA is determined by counting the silver grain in the nucleus. Nonspecific labeling is determined by counting silver grains in representative nuclear-sized areas of the cytoplasm, and a net nuclear grain count is determined by subtracting the cytoplasmic grain count from the nuclear count. The number of grains/nucleus and the number of cells in repair (arbitrarily defined as those cells having more than 5 grains) are the two parameters usually estimated from these studies. The in vitro rat hepatocyte UDS assay attracted considerable attention during the ensuing years as a screening tool in genotoxicity testing (17–20). Primary hepatocyte cultures from other species (mouse, hamster, and rabbit) have also been used successfully in studying species-specificity in genotoxic response (21).

An in vivo/in vitro version of the hepatocyte UDS assay was proposed by Mirsalis and Butterworth (22), in which the animals are treated with the test materials and the extent of UDS in hepatocytes is determined in vitro. This version of the assay takes into consideration the metabolism of the chemical in a whole animal, including the involvement of gut and cecal microflora in bioactivation of chemicals such as nitroaromatics and azo compounds (23). Typically, the test material is administered once and the animals are sacrificed at two intervals, that is, 2 to 4 hours and 12 to 14 hours after treatment when hepatocytes are isolated. Multiple sampling times are needed because the peak UDS response is chemical specific, for example, 1 hour for dimethylnitrosamine and 12 to 16 hours for 2-acetylaminofluorene (24). Hepatocytes are cultured in vitro in the presence of radiolabeled thymidine for 4 hours to facilitate the incorporation of the nucleotide into the DNA of cells that are undergoing UDS. The cultures are subsequently incubated in unlabeled medium overnight, and the UDS is quantified by autoradiography.

Although the in vivo/in vitro UDS assay can be performed using cells obtained from a variety of tissues, such as pancreas, trachea, stomach, kidney, and spermatocytes (23), liver is usually the target tissue used in routine genetic toxicology screening tests. However, if tissue-specific genotoxicity information is desired in light of tumorigenic findings, examination of UDS in target tissues may provide useful information on the mechanism of carcinogenesis.

From a technical standpoint, the hepatocyte UDS assay is relatively simple to perform, and there are standardized guidelines for conducting this assay (11, 12). The assay detects repairable DNA damage induced by a wide variety of mutagens and carcinogens over the entire genome (19). The usefulness of this assay as a complement to the standard battery of genetic toxicology tests was demonstrated by Probst et al (17), who evaluated 218 compounds.

VI. DOMINANT LETHAL MUTATIONS IN MALE GERM CELLS

The dominant lethal mutation assay in male germ cells has been used as a screening tool in genetic toxicology in many laboratories around the world for more than 30 years. Dominant lethal mutations, as the name implies, lead to the death of the individual carrying the mutation in a hetero- or hemizygous condition. Dominant lethal mutations are primarily the result of chromosomal events, such as breakage or loss. Dominant lethal mutations that arise in germ cells result in the death of the embryos carrying the mutation. Embryonic death can occur either before implantation (preimplantation loss) or after implantation (postimplantation loss or resorptions). The preimplantation loss is generally not used as an endpoint in these assays, as failure of ova to implant could be caused by dominant lethal mutations as well as other reproductive factors, such as failure of fertilization resulting from inadequate sperm quality or quantity (25).

Tests for the induction of dominant lethal mutations are usually conducted in male animals, because the results are more amenable to interpretation in terms of genotoxicity than in females in whom hormonal and other physiological factors could confound the assay endpoint. Dominant lethal mutation studies can be designed using a single exposure to males, followed by several weeks of successive mating (8 weeks for mouse and 10 weeks for rats) to evaluate sperms exposed at various stages of the spermatogenic cycle. Alternatively, the animals can be exposed to the test substance for several weeks (8 for mice and 10 weeks for rats), followed by 1 or 2 successive weeks of mating (26). In this modified protocol, the sperm sampled represent the germ cells treated at various stages of the spermatogenic cycle. During the mating period, each male is cohoused with 2 to 3 virgin females for 1 week, after which the females are removed and sacrificed 10 (mouse) or 13 (rat) days later. The pregnancies of these females will vary in age from 10 to 17 days for the mouse and 13 to 20 days for the rat. The uterine horns and ovaries of the females exteriorized through an abdominal incision and the following data are recorded: (a) the number of corpora lutea, (b) the number of total implants in utero, (c) the number of live implants, and (d) the number of dead implants/resorptions. The uteri of apparently nonpregnant females are stained with a 10% aqueous solution of sodium bisulfide and examined for evidence of early resorptions. For each mating period, the preimplantation loss is calculated as the difference between the corpora lutea and total implantation and expressed relative to corpora lutea. Postimplantation loss is calculated as the ratio of total resorptions to the total implantations.

Experience with a number of chemical mutagens has shown that the postmeiotic germ cells are the most sensitive to chemically induced dominant lethals (the first 3 weeks of mating in a sequential mating protocol). However, exceptions to this generalization were reported for 6-mercaptopurine, which induced dominant lethals only in early spermatocytes, that is, fifth week of mating (27). Because dominant lethal mutations arise primarily as a result of chromosomal breakage and aneuploidy, it is expected that premeiotic germ cells that carry such events are unlikely to progress through the spermatogenic cycle to form viable/functional sperm, which may explain why only the postmeiotic germ cells are sensitive to dominant lethal induction. From a recent analysis of literature data, Shelby (15) concluded that all chemicals that induce dominant lethal mutations in male germ cells also induced cytogenetic damage in the bone marrow, which suggested that germ cell mutagenicity assays be conducted only for those compounds positive in bone marrow cytogenetic tests.

Dominant lethal testing is relatively simple from a technical aspect and does not require any expensive equipment. The assay has performed well in detecting mutagens that induce heritable genetic changes, for example, those giving positive results in heritable translocation assay and specific locus mutation assay (15). Various regulatory bodies have published guidelines for the conduct of this assay (11, 12), and an excellent discussion on various aspects of data presentation was presented by Ashby and Clapp (28).

VII. IN VIVO ASSAYS IN MULTIPLE TISSUES

Commonly used assays that belong to this group are (a) DNA adduct analysis, (b) single cell gel electrophoresis assay for DNA strand breaks, and (c) mutation analysis in transgenic mice and rats. These assays are extremely valuable in studying genotoxicity at the portal of entry/site of contact and in distal tissues that results from tissue-specific factors such as metabolic activation. The first two assays identify DNA lesions that could potentially lead to mutations and, as such, do not provide any information on mutagenicity per se. Analyses for DNA adducts and strand breaks can also be conducted in vitro through the use of various cell types. Cell lines derived from transgenic animals are also available for in vitro mutagenesis studies.

A. DNA Adduct Analysis

Several sites in DNA are potential targets for electrophilic attack by reactive chemicals that lead to the formation of covalent DNA adducts. The nucleophilicity of these sites and the reactivity of the electrophile determine the extent and spectrum of adducts formed after chemical treatment. The sites susceptible to adduct formation include nucleic acid bases, as well as the sugar-phosphate backbone. For example, there are 17 potential sites for reaction in DNA treated with N-nitroso alkylating agents, and of these sites, the phosphate and ribose moieties each have one nucleophilic site (29). The exocyclic oxygen and amino groups of nuclei acid bases, N1, N3, and N7 positions of purines, and N3 position of pyrimidines are frequently involved in adduct formation. The N3 and N7 positions of guanine are most susceptible to electrophilic attack, compared with the exocyclic oxygen (30).

Cells have the ability to repair DNA adducts by several mechanisms, such as nucleotide excision repair. This repair capacity is influenced by various factors, such as the type of adduct, the tissues in which the adduct is induced, the dose and treatment duration, the time between treatment and DNA sampling, the rate of cell proliferation in the tissue, and the transcriptional state of the gene. The DNA adducts, if not repaired, have the potential to interfere with normal DNA replication, which leads to base substitution/frameshift mutations and/or cell killing. Some DNA adducts are more mutagenic than others. For example, adducts induced by alkylating agents at the exocyclic oxygen of the nucleic acid bases are generally more mutagenic than the ring nitrogens (30).

The DNA adducts represent the delivery of dose to an important target in mutagenic and carcinogenic processes. This dosimeter is extremely valuable in comparing tissue, strain, and species differences in the manifestation of genotoxicity and carcinogenicity. However, observation of DNA adducts in a tissue per se does not constitute evidence for carcinogenicity in that tissue, because adducts are frequently detected in both the target and nontarget tissues.

Several methods are available for adduct identification and analysis. The simplest is the binding of a radiolabeled test material with the DNA which, although easy to perform, suffers from many disadvantages, such as difficulty in synthesizing the labeled compound and the loss of label through metabolism. The techniques frequently used with considerable success are the ^{32}P-postlabeling method (31, 32), the immunological method (33), and the gas chromatography/mass spectroscopic method (34, 35). These methods vary in their sensitivity from one adduct per 10^{10} nucleotides for the ^{32}P postlabeling method to one adduct in 10^8 nucleotides for the immunological methods. Immunoassay results can also be influenced by the cross-reactivity of the antibodies to other DNA-adducts of the same compound or related adducts of a different agent in complex mixture studies, making data interpretation difficult.

The ^{32}P postlabeling method has an advantage over the others in that prior structural knowledge is not required for adduct detection. Furthermore, this technique could be used for detecting adducts formed by chemicals of diverse structures, such as alkylating agents, polycyclic aromatic hydrocarbons, aromatic amines, and others. Potential pitfalls of the postlabeling technique include the lack of structural identification of the adduct, potential artifacts, and interference of the endogenous preexisting adducts in the identification of induced ones. Immunological methods for adduct identification are simple and inexpensive. However, this method requires a relatively large sample of DNA, and the amount of adducts are quantitated in relative terms. The mass spectroscopic methods provide high quality information on the adduct structural identification. However, this method is not as sensitive as the postlabeling method.

In the ^{32}P-postlabeling method (31, 32, 36), DNA is enzynmatically digested to 3´-monophosphates of normal deoxynucleosides (N_p) and adducts (X_p). Adducted nucleosides are enriched by extracting them with butanol or converting N_p to N by 3´-dephosphorylation with nuclease P1 treatment. The X_p are subsequently labeled enzymatically with a high specific activity ^{32}P-ATP at the 5´ end and identified through multidirectional thin layer chromatography, followed by autoradiography. A digest of the untreated DNA representing the total nucleotides is also end labeled and chromatographed. The radioactivity observed in the X_p spots is expressed relative to the radioactivity observed in the total nucleotides to calculate the relative adduct labeling. The X_p are identified as additional spots on the chromatogram, and their structural identification can be accomplished by cochromatography with reference adducts.

Not all X_p identified by the ^{32}P-postlabeling technique represents premutagenic lesions or those formed by the direct reaction of the test material or its metabolite with DNA. For example, a number of indigenous spots (I spots) have been identified in untreated tissues, and their levels vary with age, the time of the day, and exposure to nongenotoxic carcinogens (37). These X_p spots have been postulated to represent DNA modifications of unknown origin with a regulatory function in the cell. Cellular processes, such as lipid peroxidation, result in the generation of DNA reactive compounds such as malonaldehyde (38), and agents that stimulate these processes can indirectly induce DNA adducts. However, such agents tend to do so at relatively high doses, and DNA adducts induced by these indirect mechanisms are likely to display a threshold phenomenon.

Immunological methods for DNA-adduct identification include radioimmunoassay (RIA) and enzyme-linked immunosorbent assay (ELISA) (33). These assays use antibodies against specific DNA adducts conjugated to proteins that are raised either in rabbits or monoclonally. In RIA, the antibody competes for two forms of the adducts, the radiolabeled tracer supplied in a constant amount and an inhibitor that is either a known adduct (standard) or an unknown sample. Inhibitors compete for the binding sites in the antibody, which results in reduced binding of the tracer with the antibody. The antigen-antibody complex is precipitated through the use of a secondary antibody or other reagents and the amount of radioactivity determined. This assay has a sensitivity of one adduct per 10^6 nucleotides. The ELISA has the same basic principle except that the tracer is bound to the bottom of a microtiter plate and the antigen-anibody complex. An increased sensitivity (four adducts/10^8 nucleotides) is achieved in ELISA by complexing the secondary antibody to an enzyme that cleaves a substrate to produce either a calorimetric or fluorometric endpoint. In addition to the above methods, DNA adducts can also be localized in tissue sections by immunohistochemical methods.

In spectroscopic methods, DNA samples are hydrolyzed either chemically or enzymatically to individual bases. These samples are then directly analyzed by high performance liquid chromatography (HPLC) separation followed by detection by mass spectrometry. Alternately, the hydrolyzed DNA samples can be extracted, derivatized, and analyzed by gas chromatography (GC) mass spectrometry. To verify the structure of adducts conclusively, these methods require synthesized standards. These methods are sensitive and highly selective. However, they require costly instrumentation and technical expertise. A recently developed assay that provides the greatest sensitivity for detecting DNA binding (attomole to picomole range) is accelerator mass spectrometry (39). In this method, biological systems are treated with a low radioactive test material, then DNA samples are converted into filamentous graphite by an initial oxidation followed by reduction on cobalt, and these samples are analyzed on an accelerated mass spectrometer. This method requires significant effort in preparing the samples for analysis and can be performed only at a limited number of facilities because of the type of equipment needed.

Assays for DNA adducts are useful when information is sought on the basic question of whether the material or its metabolites has the ability to react with the DNA either in vitro or in vivo. However, caution must be exercised in interpreting these data because increased levels of endogenous DNA adducts or the formation of new adducts could also result at toxic dose levels through secondary events such as lipid peroxidation.

B. The Comet Assay

The comet assay measures DNA single and double strand breaks in individual cells (40–42). In this assay, single cells or nuclei are suspended in an agarose gel, layered on a microscope slide, lysed and electrophoresed in an alkaline or neutral buffer. These slides are examined under a microscope after staining with a DNA specific dye, such as ethidium bromide or Yo Yo-1. Special treatments with RNase and proteinase K are often used to facilitate the untangling of nuclear DNA from RNA and proteins, respectively and to aid in the detection of strand breaks (43). Under these conditions, the migration of the negatively charged broken ends of DNA from the cell to the anode is proportional to the extent of strand breakage. The DNA strand breaks are produced either directly by the treatment (e.g., X-rays) or during the excision repair of DNA adducts (44). The name of the assay was based on the appearance of cells, with the trailing DNA tails resembling comets. For the detection of single strand breaks, alkaline lysis and electrophoresis conditions (pH > 13) are recommended to facilitate the unwinding of double strand DNA and expression of alkali labile sites. Under conditions of neutral pH (approximately 9), on the other hand, the extent of DNA migration is proportional to the double strand breaks. Treatment of several parameters associated with the comet tail (tail length, DNA content, and tail moment, defined as the product of length and DNA content) are generally used as the endpoints in this assay.

The comet assay can be conducted in any tissue from which single cells or nuclei can be prepared. Either acute or chronic treatment protocols can be used with this assay, or alternatively, this endpoint can be "piggy-backed" on any subchronic or chronic studies. Several variables, such as the tissue sampling time, cell cycle position, and electrophoresis conditions (time and voltage) have significant impact on the results of this assay. Most notably, great care should be taken to assure that the DNA damage is not originating from cell toxicity, apoptosis, and necrosis. Chemicals that induce DNA interstrand cross-links or DNA-protein cross-links inhibit the expression of DNA strand breaks and may be classified as negative in the routine assays unless special techniques (e.g., exposure of the treated cells to X-rays) are used for their identification. In addition, chemicals that induce DNA lesions that are not alkali-labile are also unlikely to be detected by this assay.

It should be emphasized that the detection of DNA strand breaks in a given tissue does not necessarily implicate that tissue as being a potential target organ for carcinogenesis. For example, in an extensive evaluation of 30 aromatic amines (20 genotoxic carcinogens and 10 noncarcinogens) using the comet assay, Sasaki et al (42) observed carcinogen-induced DNA strand breaks in both the target and nontarget organs. However, none of the noncarcinogenic amines induced strand breaks in any of the tissues. Interestingly, two of the carcinogens evaluated (4,4´-methylenedianiline and 2,4-diaminotoluene) did not induce DNA strand breaks in the target organ (bone marrow) but did so in a number of nontarget tissues.

The comet assay is a versatile and relatively inexpensive method for genotoxicity assessment. It can be used with practically any cell type from any species. The biological basis of this assay, however, has not been firmly established. The length of comet tails, which is a measure of genotoxicity, has been shown to be influenced by various factors, such as the DNA-unwinding time before electrophoresis (45) and the electrophoretic conditions (43, 46). Hence great care should be taken in minimizing technical artifacts to avert the possibility of spurious results.

C. Transgenic Mutation Assays

With the advent of transgenic technology, it is now feasible to study the frequencies of transgene mutations in mice and rats in any tissue of interest. Of the several transgenic models, the Big Blue (46, 47) and MutaMouse models (48–51) have been evaluated extensively as to their utility in genetic toxicology research. The Big Blue system is especially attractive because it offers transgenic mice and

Complementary Assays

rats to complement the two commonly used cancer bioassay species. The genome of transgenic rodents harbors shuttle vectors containing target genes for studying mutagenesis. After treatment with xenobiotic agents, the target gene is recovered from the genomic DNA of the tissue of interest by cleaving and packaging the vector into infectious bacteriophage λ. Each packaged vector represents a single recovered target gene. The packaged phage is used to infect host *Escherichia coli* and plated on indicator agar that contains the chromogenic substrate X-gal.

The genome of Big Blue mice harbors approximately 40 copies/haploid genome of a λ phage shuttle vector (λLIZ) that is approximately 43 kb long. These mice are available as hemizygous (40 copies/cell) or homozygous (80 copies/cell) on the inbred C57B16 background and as hemizygous for the transgene on the hybrid B6C3F1 background. Big Blue Fischer 344 rats are homozygous for the transgene and contain approximately 30–40 copies of the transgene per cell. There are currently two characterized target sites in the shuttle vector of the Big Blue model for mutation analyses, that is, the *lacI* and *cII*. The *lacI* target gene (1080 base pairs long) codes for the production of the *lac* repressor protein, which binds to the *lac* operator and prevents the transcription of the α*lacZ* reporter gene in *E. coli*. Mutations in *lacI* gene lead to the production of a defective repressor protein that does not bind to the operator. Under such conditions, the transcription of the α*lacZ* reporter gene occurs with the resulting production of *LacZ* protein—the enzyme β-galactosidase, through α complementation of the α*lacZ* protein from the phage with the ω*LacZ* protein in the *E. coli* host. This enzyme cleaves the chromogenic substrate (X-gal) present in the agar to produce blue mutant plaques. Phage particles without mutations in the target gene do not produce the enzyme, and the plaques that result from these phage particles will be colorless. Thus, the ratio of blue plaques to the total plaques represents the mutant fraction.

The *cII* target gene (294 base pairs long) offers the advantage of a positive selection system for the recovery of induced mutations (52, 53). This is a resource and labor saving feature as compared with the *lacI* target, where both mutant and nonmutant plaques have to be evaluated to estimate the mutant fraction. In addition, the smaller size of this target gene compared to *lacI* is also attractive from the standpoint of being able to characterize the mutational spectrum by DNA sequencing. The *cII* gene codes for a protein that plays a critical in the commitment of lytic or lysogenic cycle made by λ phage upon infection of *E. coli*. Under conditions of high *cII* protein levels, the phage goes into lysogeny and no plaques will be formed on the bacterial lawn. However, when the *cII* protein levels are low or defective because of a mutation in the *cII* gene, the phage will go into lytic cycle and form plaques. When λ phage particles that contain the shuttle vector from the treated animals are used to infect host *E. coli* strain 1250 (*hfl*A and *hfl*B) growing at a temperature that favors lysogeny (24° C), phages containing wild-type *cII* gene enter lysogeny, whereas those containing *cII* mutations multiply through the lytic cycle and form plaques in the bacterial lawn.

The genome of MutaMouse harbors approximately 40 copies of λgt101lacZ shuttle vector (1.8 kb long) per haploid genome (80 copies per cell) as a head-to-tail concatamer on a hybrid CD2F1 background, which was subsequently bred to homozygosity (49). The target gene for mutagenesis in this assay is the *lacZ* (3126 base pairs), and mutations in this gene lead to the production of defective enzyme. When phage containing *lacZ* mutations are used to infect the host bacteria (*E. coli* C) in the presence of the chromogenic substrate X-gal, they form colorless plaques, whereas the wild-type with functional β-galactosidase produce blue plaques. The scoring of colorless plaques on plates that contain predominantly blue nonmutant plaques is somewhat technically challenging. To alleviate this problem and to cut down on the costs associated with performing the assay, a positive selection was subsequently introduced to the MutaMouse protocol (51). The proprietary strain of *E. coli* C *lac galE* was used as the host bacteria to infect with the recovered phage, and this strain of bacteria does not replicate in the presence of galactose. Phage containing defective *lacZ* cannot convert the phenyl-galactose (P-gal) substrate into galactose and can replicate and form clear plaques, whereas phage containing intact *lacZ* gene will convert the P-gal substrate into galactose and lead to bacterial death upon infection. Thus, only mutant phage can replicate on the selection plates containing P-gal. The total phage titer can be estimated by plating the infected bacterial cells on nonselective plates without P-gal.

One of the critical protocol aspects of the transgenic mutation assays is the interval between treatment and tissue sampling. This interval, known as "manifestation time" (54), should be sufficiently long to accommodate several biological processes. These processes include: (a) DNA damage, repair, and fixation of the damage into mutation; (b) transcription of the mutant gene, dilution of pre-existing nonmutant messenger RNA and protein; and (c) cellular expansion of the mutant clone in the tissue. The optimal manifestation time in these assays is somewhat tissue specific and should be chosen to yield the maximum recovery of induced mutations. It is currently believed that a manifestation time of 35 days is appropriate for the majority of tissues after acute treatments. Alternatively, a shorter manifestation time may be appropriate after multiple treatment (e.g., 28-day treatment).

Considerable data have been accumulated over the past 10 years using the Big Blue and MutaMouse for a variety of genotoxic and nongenotoxic carcinogens (10, 55, 56). In general, the performance of these assays in identifying genotoxic carcinogens has been very good. Transgenic mutation assays with non-genotoxic hepatocarcinogens, such as chloroform, phenobarbital, and methyl-clofenapate uniformly yielded negative results (58–60). Transgenic models have also been successfully used to examine species specificity in carcinogenesis with compounds such as aflotoxin B1 (48) and tamoxifen (60). However, one has to keep in mind that the observation of an increased mutation frequency in any given tissue is not necessarily predictive of tumorigenicity in that tissue. Significant levels of mutation induction have been reported in the tumor target, as well as nontarget tissues after treatment with genotoxic carcinogens (61, 62).

Mutation induction in male germ cells of Big Blue mice and MutaMouse was investigated in a collaborative study using ethyl nitrosourea (ENU), isopropyl methanesulphonate (iPMS), and methyl methanesulphonate (MMS) (63). Both ENU and iPMS induced mutations in DNA recovered from testicular tissue as well as epididymal sperm. However, MMS was not mutagenic under any conditions. This observation was somewhat surprising because MMS was previously shown to be positive in the dominant lethal, heritable translocation and specific locus assays. However, because the endpoints measured in the above assays were different from those detected in the transgenic model, the negative response may be related to the type of DNA lesions induced by this methylating agent and the limitation of the test system. It is interesting to note that MMS also gave a negative response for mutation induction in Big Blue mouse liver (64).

One of the concerns in the use of transgenic Big Blue and MutaMouse models in genetic toxicology relates to the appropriateness of using the transgene as a surrogate for the endogenous gene in view of its transcriptional inactivity and, in the case of *lacI*, high levels CpG sequences in the gene. Depending on the mutagen used, either similar or different levels of mutation induction were demonstrated in the endogenous hprt locus and the *lacI* by Skopek et al (65, 66). Cosentino and Heddle (67) did not find a remarkable difference between the endogenous *Dlb-1* locus and the *lacZ* in MutaMouse treated with a wide variety of chemical mutagens. However, mutagenicity in the transgene by agents that induce large deletions (e.g., X-rays) are expected to be underestimated because deletions that extend into λ-vector are unlikely to produce viable phage (68). Transgenes may not behave exactly like endogenous genes. However, these differences do not preclude their use in genetic toxicology research.

VIII. TESTS FOR HERITABLE MUTATIONS IN MICE

These tests address the question of whether a test material has the potential to induce genetic changes that are transmittable to future generations. Such assays are needed for risk assessment if positive results are observed in either the dominant lethal test or tests for germ cell DNA interactions. The mouse specific locus tests with visible (69) and biochemical markers (70) have been used to evaluate the potential of chemicals to induce heritable point mutations. In the specific locus test using visible markers, one of the parental strains is homozygous for a number of recessive marker alleles for coat and eye color. When this strain is crossed to a homozygous wild-type strain, any mutations that occur at the marker locus in the wild-type can be detected in the F_1 offspring. The method can also identify mutations that occur in a locus other than the marker locus but that have an effect on the marker locus. For technical reasons, homozygous wild-type males are usually treated with the test substance and mated to females with the recessive marker loci. The matings are continued for several weeks, with a change of females every week. This allows the sampling of various stages of germ cell cycle during the first 7 weeks and the spermatogonial stem cell effects during the subsequent weeks. The F1 progeny is visually examined for changes in the phenotype at approximately 3 weeks of age.

Mutations can also be detected biochemically. Adult, DBA/2J males are treated with the test material and mated to C57B16 females. The matings are continued for several weeks to sample all the germ cell stages. The F1 offspring is examined for variations in normal electrophoretically expressed gene products from blood and kidney samples. Presumed mutants can be confirmed by further breeding.

Assays for heritable chromosomal translocations can be performed either by the breeding method or by cytological methods (71). In the breeding method, males are treated with the test material and sequentially mated to females to sample various germ cell stages. F1 males carrying balanced translocations are viable. However, such a change will result in either sterility or semisterility because of the translocation. Sterility or semisterility among the F1 males is determined by subsequent breeding experiments. Any suspected translocation heterozygote can be confirmed by cytogenetic analysis. Alternatively, the F1 males could be cytologically examined for the presence of multivalent chromosomal associations (vs. bivalent association among normal chromosomal homologues) at meiotic metaphase I.

Tests for heritable genetic changes in mammals are simple but fairly laborious. Very few laboratories in the world have the type of historical database necessary to perform a valid mouse specific locus test.

IX. CONCLUSIONS

A number of in vitro and in vivo assay systems are currently available in addition to those described in this review. These assays have served the scientific community extremely well during the past three decades by identifying genotoxins among the existing chemicals, as well as by preventing the introduction of new mutagens. The molecular biology revolution has already provided us with newer and highly sensitive tools, such as the gene chips for the rapid screening of chemicals for biological activity including interactions with DNA. These in silico technologies, in conjunction with the traditional assays, are expected to provide us with better models for defining as well as understanding the dose-response relationships in genetic toxicology and carcinogenicity.

REFERENCES

1. KL Dearfield, AE Auletta, MC Cimino, MM Moore. Consideration in the U.S. Environmental Protection Agency's testing approach for mutagenicity. Mutat Res 258:259–283, 1991.
2. European Economic Communities (EEC). Commission Directive 94/79/EC concerning the placing of plant protection products on the market. Official J European Communities L354:16–31, 1994.
3. L Muller, Y Kikuchi, G Probst, L Schechtman, H Shimada, T Sofuni, D Tweats. ICH-Harmonized guidances on genotoxicity testing of pharmaceuticals: Evolution, reasoning, and impact. Mutat Res 436:195–225, 1999.
4. S Teramoto, R Saito, H Aoyama, Y Shirasu. Dominant lethal mutation induced in male rats by 1,2-dibromo-3-chloropropane (DBCP). Mutat Res 77:71–78, 1980.
5. WM Generoso, KT Cain, LA Hughes. Tests for dominant lethal effects of 1,2-dibromo-3-chloropropane (DBCP) in male mice. Mutat Res 156:103–108, 1985.
6. J Ashby. Is there a continuing role for the i.p. injection route of exposure in short term rodent genotoxicity assays? Mutat Res 156:239–243, 1985.
7. MD Shelby. A case for the continued use of the intraperitoneal route of exposure. Mutat Res 170:169–171, 1986.
8. G Krishna, G Urda, J Theiss. Principles and practices of integrating genotoxicity evaluation into routine toxicology studies: A pharmaceutical industry perspective. Environ Mol Mutagen 32:115–120, 1998.
9. YF Sasaki, E Nishidae, YQ Su, N Matsusaka, S Tsuda, N Susa, Y Furukawa, S Ueno. Organ-specific genotoxicity of the potent rodent bladder carcinogen o-anisidine and p-cresidine. Mutat Res 412:155–160, 1998.
10. SW Dean, TM Brooks, B Burlinson, J Mirsalis, B Myhr, L Recio, V Thybaud. Transgenic mouse mutation assay systems can play an important role in regulatory mutagenicity testing in vivo for the detection of site-of-contact mutagens. Mutagenesis 14:141–151, 1999.
11. Organization for Economic Co-operation and Development (OECD). Guidelines for Testing of Chemicals. Paris Cedex: OECD Publications, 1997.
12. U.S. Environmental Protection Agency. Health Effects Testing Guidelines, 870 Series, Office of Prevention, Pesticides, and Toxic Substances, 1998.
13. RJ Preston, BJ Dean, S Galloway, H Holden, AF McFee, M Shelby. Mammalian in vivo cytogenetic assays. Mutat Res 189:157–165, 1987.
14. M Richold, J Ashby, J Bootman, A Chandley, DG Gatehouse, L Henderson. In vivo cytogenetics. In: DJ Kirkland, ed. Basic Mutagenicity Tests: UKEMS Recommended Procedures. New York: Cambridge Press, 1990, pp. 115–141.
15. M Shelby. Selecting chemicals and assays for assessing mammalian germ cell mutagenicity. Mutat Res 353:159-167, 1996.
16. GM Williams. Carcinogen-induced DNA repair in primary rat liver cell cultures: A possible screen for chemical carcinogens. Cancer Lett 1:231–236, 1976.
17. GS Probst, RE McMahon, LE Hill, CZ Thompson, JK Epp, SB Neal. Chemically-induced unscheduled DNA synthesis in primary rat hepatocyte cultures: A comparison with bacterial mutagenicity using 218 compounds. Environ Mutagen 3:11–32, 1981.

18. GM Williams, MF Laspia, VC Dunkel. Reliability of the hepatocyte primary culture/DNA repair test in testing of coded carcinogens and non-carcinogens. Mutat Res 97:359–370, 1982.
19. AD Mitchell, DA Casciano, ML Meltz, DE Robinson, RHC San, GM Williams, ES Von Holle. Unscheduled DNA synthesis tests: A report of the U.S. Environmental Protection Agency Gene-Tox Program. Mutat Res 123:363–410, 1983.
20. BE Butterworth, J Ashby, E Bermudez, D Casciano, J Mirsalis, G Probst, G Williams. A protocol and guide for the in vitro rat hepatocyte DNA-repair assay. Mutat Res 189:113–121, 1987.
21. CA McQueen, GM Williams. The hepatocyte primary culture/DNA repair test using hepatocytes from several species. Cell Biol Toxicol 3:209–218, 1987.
22. JC Mirsalis, BE Butterworth. Detection of unscheduled DNA synthesis in hepatocytes isolated from rats treated with genotoxic agents: An in vivo-in vitro assay for potential carcinogens and mutagens. Carcinogenesis 1:621–625, 1980.
23. JC Mirsalis. In vivo measurement of unscheduled DNA synthesis as an indicator of hepatocarcinogenesis in rodents. Cell Biol Toxicol 3:165–173, 1987.
24. BE Butterworth, J Ashby, E Bermudez, D Casciano, J Mirsalis, G Probst, G Williams. A protocol and guide for the in vivo rat hepatocyte DNA-repair assay. Mutat Res 189:123–133, 1987.
25. AJ Bateman. The dominant lethal assay in the male mouse. In: BJ Kilbey, M Legator, W Nichols, C Ramel, eds. Handbook of Mutagenicity Test Procedures. New York: Elsevier, 1977.
26. S Green, KS Lavappa, M Manandhar, C Sheu, E Whorton, JA Springer. A guide for mutagenicity testing using the dominant lethal assay. Mutat Res 189:167–174, 1987.
27. GM Genoroso, RJ Preston, JG Brewan. 6-Mercaptopurine, an inducer of cytogenetic and dominant lethal effects in pre-meiotic and early meiotic germ cells of male mice. Mutat Res 28:437–447, 1975.
28. J Ashby and MJL Clapp. The rodent dominant lethal assay: A proposed format for data presentation that alerts to pseudo-dominant lethal effects. Mutat Res 330:209–218, 1995.
29. B Singer. DNA damage: Chemistry, repair, and mutagenic potential. Regul Toxicol Pharmacol 23:2–13, 1996.
30. DK La. JA Swenberg. DNA adducts: Biological markers of exposure and potential applications to risk assessment. Mutat Res 365:129–146, 1996.
31. RC Gupta, MV Reddy, K Randerath. ^{32}P-Postlabeling analysis of non-radioactive aromatic carcinogen-DNA adducts. Carcinogenesis 3:1081–1092, 1982.
32. K Randerath, E Randerath, TF Danna, KL van Golen, KL Putman. A new sensitive ^{32}P-postlabeling assay based on the specific enzymatic conversion of bulky DNA lesions to radiolabeled dinucleiotide 5´-monophosphates. Carcinogenesis 10:1231–1239, 1989.
33. MC Poirier. Human exposure monitoring, dosimetry, and cancer risk assessment: The use of antisera specific for carcinogen-DNA adducts and carcinogen modified DNA. Drug Metab Rev 26:87–110, 1994.
34. PB Farmer, E Bailey, S Naylor, D Anderson, A Brooks, J Cushnir, JH Lamb, O Sepai, YS Tang. Identification of endogenous electrophiles by means of mass spec-

trometric determination of protein and DNA adducts. Environ Health Perspect 99:19–24, 1993.
35. KY Wu, A Ranssinghe, PB Upton, VE Walker, JA Swenberg. Molecular dosimetry of endogenous and ethylene oxide-induced N7-(2-hydroxyethyl)guanine formation in tissues of rodents. Carcinogenesis 20:1787–1792, 1999.
36. RC Gupta, G Spencer-Beach. Natural and endogenous DNA adducts as detected by ^{32}P-postlabeling. Regul Toxicol Pharmacol 23:14–21, 1996.
37. RG Nath, K Randerath, D Li, FL Chung. Endogenous production of DNA adducts. Regul Toxicol Pharmacol 23:22–28, 1996.
38. LJ Marnet, PC Burcham. Endogenous DNA adducts—Potential and paradox. Chem Res Toxicol 6:771–785, 1993.
39. ESC Kwok, BA Buchholz, JS Vogel, KW Turteltaub, DA Eastmond. Dose-dependent binding of *ortho*-phenylphenol to protein but not DNA in the urinary bladder of male F344 rats. Toxicol Appl Pharmacol 159:18–24, 1999.
40. DW Fairbairn, PL Olive, KL O'Neill. The comet assay: A comprehensive review. Mutat Res 339:37–59, 1995.
41. NP Singh, MT McCoy, RR Tice, EL Schneider. A simple technique for quantitation of low levels of DNA damage in individual cells. Exp Cell Res 175:184–191, 1988.
42. YF Sasaki, K Fujikawa, K Ishida, N Kawamura, Y Nishikawa, S Ohta, M Satoh, H Madarame, S Ueno, N Susa, N Matsusaka, S Tsuda. The alkaline single cell gel electrophoresis assay with mouse multiple organs: Results with 30 aromatic amines. Mutat Res 440:1–18, 1999.
43. NP Singh, RE Stephens. Microgel electrophoresis: Sensitivity, mechanism, and DNA electrostretching. Mutat Res 383:167–175, 1997.
44. G Speit, A Hartmann. The contribution of excision repair to the DNA effects seen in the alkaline single cell gel test (comet assay). Mutagenesis 10:555-559, 1995.
45. JE Yendle, H Tinwell, BM Ellitt, J Ashby. The genetic toxicity of time: Importance of DNA-unwinding time to the outcome of single-cell gel electrophoresis assays. Mutat Res 375:125–136, 1997.
46. GS Provost, PI Kretz, RT Hammer, CD Matthews, BJ Rogers, KS Lundberg, MJ Dycaico, JM Short. Transgenic systems for in vivo mutation analysis. Mutat Res 288:133–149, 1993.
47. MJ Dycaico, GR Stuart, GM Tobal, JG de Boer, BW Glickman, GS Provost. Species-specific differences in hepatic mutant frequency and mutational spectrum among lambda/*lacI* transgenic rats and mice following aflatoxin B1. Carcinogenesis 17:2347–2356, 1996.
48. JA Gossen, WJF de Leeuw, CHT Tan, EC Zwarthoff, F Berends, PHM Lohman, DI Knook, J Vijg. Efficient rescue of integrated shuttle vectors from transgenic mice: A model for studying mutations in vivo. Proc Natl Acad Sci U S A 86:7971–7975, 1989.
49. BC Myhr. Validation studies with MutaMouse: A transgenic mouse model for detecting mutations in vivo. Environ Mol Mutagen 18:308–315, 1991.
50. AJW Hoorn, IL Custer, BC Myhr, D Brusick, J Gossen, J Vijg. Detection of chemical mutagens using MutaMouse: A transgenic mouse model. Mutagenesis 8:7–10, 1993.

51. SW Dean, BC Myhr. Measurement of gene mutation in vivo using MutaMouse and positive selection for *LacZ* phage. Mutagenesis 9;183–185, 1994.
52. JL Jakubczak, G Merlino, JE French, WJ Muller, B Paul, S Adhya, S Garges. Analysis of genetic instability during mammary tumor progression using a novel selection-based assay for in vivo mutations in a bacteriophage λ transgene target. Proc Natl Acad Sci U S A 93:9073–9078, 1996.
53. BB Gollapudi, KM Jackson, WT Stott. Hepatic *lacI* and *cII* mutations in transgenic (λLIZ) rats treated with dimethylnitrosamine. Mutat Res 419:131–135, 1998.
54. JA Heddle. Mutant manifestation: The time factor in somatic mutagenesis. Mutagenesis 14:1–3, 1999.
55. V Morrison, J Ashby. A preliminary evaluation of the performance of the MutaMouse (*lacZ*) and Big Blue (*lacI*) transgenic mouse mutation assays. Mutagenesis 9:367–375, 1994.
56. NJ Gorelick. Overview of mutation assays in transgenic mice for routine testing. Environ Mol Mutagen 25:218–230, 1995.
57. BE Butterworth, MV Templin, AA Constan, CS Sprankle, BA Wong, LJ Pluta, JI Everitt, L Recio. Long-term mutagenicity studies with chloroform and dimethylnitrosamine in female *lacI* B6C3F1 mice. Environ Mol Mutagen 28:430–433, 1998.
58. D Gunz, SE Shephard, WK Lutz. Can nongenotoxic carcinogens be detected with *lacI* transgenic mouse mutation assay? Environ Mol Mutagen 21:209–211, 1993.
59. PA Lefevre, H Tinwell, SM Galloway, R Hill, JM Mackay, CR Elcombe, J Foster, V Randall, RD Callander, J Ashby. Evaluation of the genetic toxicity of the peroxisome proliferator and carcinogen methylclofenapate, including assay using MutaMouse and Big Blue transgenic mice. Hum Exp Toxicol 13:766–775, 1994.
60. R Davies, VIC Oreffo, S Bayliss, PA Dinh, KS Lilley, NH White, IL Smith, JA Styles. Mutation spectra of tamoxifen-induced mutations in the livers of *lacI* transgenic rats. Environ Mol Mutagen 28:430–433, 1996.
61. BB Gollapudi, WT Stott, BL Yano, JS Bus. Mode of action considerations in the use of transgenic animals for mutagenicity and carcinogenicity evaluations. Toxicol Lett 102-103:479-484, 1998.
62. A Hakura, Y Tsutsui, J Sonoda, Y Kai, T Imade, M Shimada, Y Sugihara, T Mikami. Comparison between in vivo mutagenicity and carcinogenicity in multiple organs by benzo(a)pyrene in the *lacZ* transgenic mouse (MutaMouse). Mutat Res 398:123–130, 1998.
63. J Ashby, NJ Gorelick, MD Shelby. Mutation assays in male germ cells from transgenic mice: Overview of study and conclusions. Mutat Res 388;111–122, 1997.
64. JC Mirsalis, JS Provost, CD Mathews, RT Hammer, JE Schindler, K Goloughlin, JT MacGregor, JM Short. Induction of hepatic mutations in *lacI* transgenic mice, Mutagenesis 8:265–271, 1993.
65. TR Skopek, KL Kort, DR Marino. Relative sensitivity of the endogenous *hprt* gene and the *lacI* transgene in ENU-treated Big Blue B6C3F1 mice. Environ Mol Mutagen 26:9–15, 1995.
66. TR Skopek, KL Kort, DR Marino, LV Mittal, DR Umbenhauer, GM Laws, SP Adams. Mutagenic response of the endogenous *hprt* gene and *lacI* transgene in benzo[a]pyrene-treated Big Blue B6C3F1 mice. Environ Mol Mutagen 28:376–384, 1996.

67. L Cosentino, JA Heddle. A comparison of the effects of diverse mutagens at the *lacZ* transgene and *Dlb-1* locus in vivo. Mutagenesis 14:113–119, 1999.
68. KS Tao, C Urlando, JA Heddle. Comparison of somatic mutations in transgenic versus host locus. Proc Natl Acad Sci U S A 90:10681–10685, 1993.
69. LB Russell, PB Selby, E von Halle, W Sheridan, L Valcovic. The mouse specific-locus test with agents other than radiations. Interpretation of data and recommendations for future work. Mutat Res 86:329–354, 1981.
70. SE Lewis. The biochemical specific-locus test and a new multiple-endpoint mutation detection system: Consideration for genetic risk assessment. Environ Mol Mutagen 18:303–306.
71. WM Generoso, JB Bishop, DG Gosslee, GW Newell, CJ Sheu, E von Halle. Heritable translocation test in mice. Mutat Res 76:191–215, 1980.

7

The Mouse Lymphoma Assay (MLA) Using the Microwell Method

Masamitsu Honma
National Institute of Health Sciences, Tokyo, Japan

Toshio Sofuni
Olympus Optical Co. Ltd., Tokyo, Japan

I. INTRODUCTION

The many genotoxicity tests that have been developed over the past years offer a variety of systems (prokaryotic, eukaryotic, in vitro, and in vivo) and endpoints (gene mutation, chromosome aberration, DNA damage). Those that use mammalian cells have the advantage of more closely approximating the human system and gene mutations that occur in mammalian cells and appear to be relevant for human genetic disorders, including cancers. The mouse lymphoma assay (MLA) quantifies genetic alterations that affect expression of the thymidine kinase (TK) gene (*tk*) in mouse L5178Y *tk+/−* lymphoma cells. Because the MLA can detect a wide range of genetic alterations (point mutations, larger scale chromosomal changes, recombination, mitotic nondisjunction, and others) (Fig. 1), it has been regarded as the most sensitive in vitro mammalian cell gene mutation assay (1–8). Most of the alterations detected by the test are found in human tumor cells and are presumably relevant for carcinogenesis (9). Compared with other genotoxicity tests, the MLA is capable of detecting mutagens and clastogens, which can be detected by the bacterial reverse mutation assay (Ames test) and the chro-

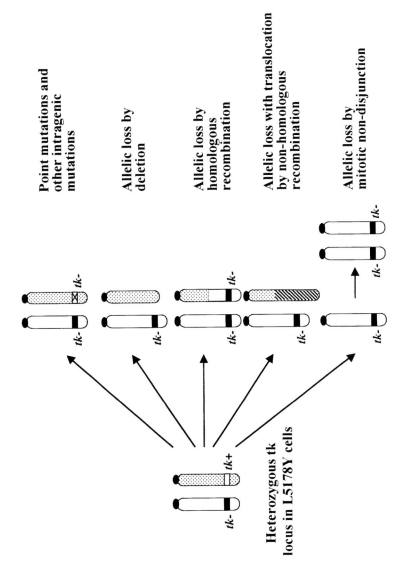

Fig. 1. Possible mutations at the tk locus. tk– and tk+ mean nonfunctional and functional tk allele, respectively.

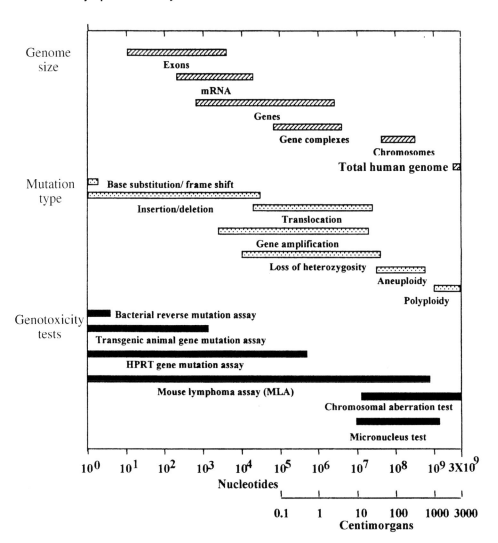

FIG. 2. Capacity of genotoxocity tests based on genome size and type of mutations.

mosome aberration test, respectively, as well as agents producing other genetic damage involving one chromosome (Fig. 2.).

The MLA was first developed by Drs. D. Clive and MM. Moore (10–12), and its protocol has since been standardized (13–24). There are two methodologies for performing the MLA: the original soft agar procedure and the newer microwell procedure (16, 25). Both procedures are acceptable, but the latter has several advantages: it is free of agar quality problems, has higher plating efficiencies, can detect small (tiny) colony mutants, and requires less subjectivity for classifying large and small colony mutants. This chapter is designed to assist in the setting up and performance of the MLA using the microwell method, as well as in analyzing and interpreting data (15, 26–28).

II. BASIC PROCEDURES

A. Cells and Culture Maintenance

Mouse lymphoma cells (L5178Y $tk^{+/-}$ clone 3.7.2C) are available from American Type Culture Collection (ATCC Number: CRL-9518). The cells are grown as suspension cultures in polystyrene tissue culture flasks with constant mixing. Logarithmic growth is normally maintained with population doubling times (PDT) of 9–11 hours. Cell density is determined by hemacytometer or a cell counting machine, and the cultures are routinely diluted each day to approximately 2×10^5 cells/ml to prevent overgrowth (more than 10^6 cells/ml). To prepare working cell stocks for gene mutation experiments, cultures are purged of pre-existing $tk^{-/-}$ mutants by exposure for 1 day to THMG medium (THMG medium is culture medium containing 3 μg/ml thymidine, 5 μg/ml hypoxanthine, 0.1 μg/ml methotrexate, and 7.5 μg/ml glycine), and then the cells are transferred to THG medium (THMG but without methotrexate) for 2 days. The purged cultures are frozen as working stocks in liquid nitrogen at a density of $1-5 \times 10^6$ cells/ml/tube in culture medium containing 5% dimethyl sulfoxide (DMSO). Each experiment is started from the working stock. Cells are usually used 3 or 4 days later after thawing, when they are growing logarithmically.

B. Media

RPMI 1640 medium and horse serum are used for the assay. The horse serum should be freshly thawed and then incubated at 56° C for 30 minutes to inactivate the enzymes that destroy trifluorothymidine (TFT). Basic medium (designated RPMI0) consists of RPMI 1640 medium supplemented with 200 μg/ml sodium pyruvate, 100 unit/ml penicillin, and 100 μg/ml streptomycin. Growth medium (designated RPMI10) is RPMI0 with 10% (v/v) heat-inactivated horse serum. The treatment medium is growth medium with 5% serum (RPMI5). The cloning medium (for 96-well microwell plates) is growth medium with 20% serum (RPMI20).

C. S9 Activation

Livers from phenobarbital- and 5,6-benzoflavon pretreated rats are used to prepare postmitochondrial supernatant fractions (S9) for exogenous metabolic activation (the S9 is also commercially available). An S9 mix (4 ml S9, 2 ml 180 mg/ml glucose-6-phosphate, 2 ml 25 mg/ml nicotinamide adenine dinucleotide phosphate (NADP), and 2 ml 150 mM KCl) is prepared just before use. The concentration of S9 in treatment cultures is 2%.

D. Test Chemicals

The test chemical is dissolved, as appropriate, dependent on solubility with RPMI0, saline, water, or DMSO. Treatment is initiated by adding dilutions of these solutions (10% maximum for aqueous solutions and 1% for organic solution) into treatment medium containing cells in suspension. Preliminary assessment should test concentrations up to 10 mM or 5 mg/ml.

E. Control Chemicals

Appropriate solvent and positive control cultures are included with each experiment. The positive control is usually 10 μg/ml methylmethansulfonate (MMS) in experiments without S9 mix, and 3 μg/ml cyclophosphamide (CP) in experiments with S9 mix. The MMS and CP should be freshly prepared with physiological saline (1 mg/ml and 0.3 mg/ml, respectively) for each experiment. Negative control is the solvent used for the testing chemicals.

F. Preliminary Experiments

Experiments to determine the solubility and appropriate dose range of the test chemical are conducted before performing the main experiment. The solvents, in order of preference, are RPMI0, physiological saline, distilled water, and DMSO. Suspensions are avoided, if possible. If the test chemical is insoluble, however, it should be suspended in RPMI0.

In the dose range-finding test, chemicals at concentrations of up to 5 mg/ml, usually regardless of the solubility, are tested for 3 hours with and without S9 mix. Relative suspension growth (RSG) is calculated by daily cell growth (DCG) during a 2-day expression period (see below). The highest dose selected for the main experiment should be one with a 10% to 20% RSG. There is no perceived need to test concentrations greater than 5 mg/ml.

G. Main Experiments

The MLA protocol is diagrammed in Figure 3. An experiment usually consists of one solvent control, one positive control, and three to six test chemical concentrations. Equal-difference or equal-ratio dilutions are serially prepared from the highest concentration to obtain graded concentration levels. Mutant and viable colonies are isolated in 96-microwell plates (flat bottom). The number of microwell plates required for each treatment is shown in Table 1.

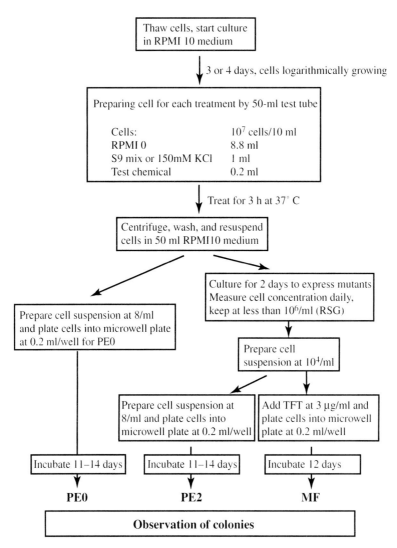

FIG. 3. A scheme of whole experimental procedure of the mouse lymphoma assay (MLA) using microwell plates. PE0, PE2, and MF mean the plating efficiency at day 0, the plating efficiency at day 2, and the mutation frequency, respectively.

Mouse Lymphoma Assay

TABLE 1. Number of Microwell Plates Required for Each Treatment Type

Treatment Group	Number of Treatment	Treatment Type		
		PE0	PE2	MF
Solvent Control	1	2	2	4
Dose 1	1	1	1	2
Dose 2	1	1	1	2
Dose 3	1	1	1	2
Dose 4	1	1	1	2
|	|	|	|	|
Positive Control	1	1	1	2

Abbreviations: *PE*, plating efficiency; *MF*, mutation frequency.

1. Treatment of Cell Cultures

Approximately 10^7 cells in 10 ml RPMI10 are placed in each of a series of sterile disposable 50-ml centrifuge tubes. The serum level in the treatment medium should be reduced to 5% (v/v). RPMI0, solvent, test chemical or positive control solution, and 1 ml S9 mix (with metabolic activation) or 150 mM KCl (without metabolic activation) are added (Table 2) to each tube to achieve a final volume of 20 ml.

The tubes are closed tightly and placed on a rocker platform in an incubator at 37° C for 3 hours. After treatment, the cells are centrifuged at low speed (approximately 1,000 r.p.m for 5 minutes), and the supernatant is discarded. Each culture is washed with RPMI0 once by resuspension and centrifugation. The cells are then resuspended in 50 ml RPMI10 at approximately 2×10^5 cells/ml. Cell densities are determined by hemacytometer or a cell-counting machine. Cells are

TABLE 2. Components of Cell Cultures in Treatments

	Volume	
Component	Without metabolic activation	With metabolic activation
Cell suspension (10^6/ml in RPMI 10)	10 ml (10^7 cells)	10 ml (10^7 cells)
RPMI0	8.8 ml	8.8ml
S9 mix	—	1 ml
150 mM KCl	1 ml	—
Test chemical	0.2 ml	0.2 ml

transferred to flasks for growth through the expression period or diluted to be plated for survival as described below.

2. Plating for Survival

A sample of each culture is diluted to 8 cells/ml in RPMI20. To accurately achieve this, a two-step dilution should be performed as shown in Table 3. This final 50 ml or 25 ml constitutes the survival cultures for the solvent controls and for the test chemicals and positive controls, respectively. Using a multichannel pipette, 0.2 ml of each culture is placed into each well of the 96-well microwell plates (two plates for each solvent control and one plate for each test chemical and the positive control). The plates are incubated at 37° C in a humidified incubator gassed with 5% CO_2 in air to allow cell growth. Colony growth is assessed after 11–13 days.

3. Expression Period

Treated cultures are maintained in tissue culture flasks without shaking for 45 hours, during which the TK^- phenotype will be expressed. A cell-counting machine or a hemacytometer is used after day 1 to measure cell density, and then each culture is diluted to approximately 2×10^5 cells/ml. On completion of the 45-hour expression time, cell densities are measured again, and the RSG is calculated from the day 1 and day 2 measurements (see below).

4. Plating for Viability

The cultures are adjusted to 10^4 cells/ml with RPMI20 to a volume of 100 ml for solvent controls and 50 ml for test chemicals and positive control for screening for TFT resistance. A sample of each culture is removed and diluted to 8 cells/ml. To achieve this accurately, a two-step dilution is made, as shown in Table 4. This final 50 ml or 25 ml constitutes the viability cultures for the solvent control and for the test chemical and the positive control, respectively. Each viability culture is then dispensed at 200 µl per well on 96-well microwell plates (two plates for each solvent control and one plate for each test chemical and the positive control). The plates are incubated at 37° C in a humidified incubator gassed with 5% CO_2 in air for 11–13 days.

TABLE 3. A Two-Step Dilution Procedure for Plating for Survival

	Initial Cell Conc. (A)	Step 1		Intermediate Cell Conc. (B)	Step 2		Final Cell Conc.
		A ml	Medium (ml)		B ml	RPMI 20 (ml)	
Survival	2×10^5/ml	0.1	9.9	2×10^3/ml	0.2 (0.1)	49.8 (24.9)	8/ml

TABLE 4. A Two-Step Dilution Procedure for Plating for Viability

	Initial Cell Conc. (A)	Step 1		Intermediate Cell Conc. (B)	Step 2		Final Cell Conc.
		A ml	Medium (ml)		B ml	RPMI 20 (ml)	
Viability	1×10^4/ml	0.5	9.5	5×10^2/ml	0.8 (0.4)	49.2 (24.6)	8/ml

5. Plating for TFT Resistance

After the dilutions are prepared, TFT is added to cultures (10^4 cells/ml) to a final concentration of 3 μg/ml (100 μl and 50 μl of 1000 times stock concentration [3 mg/ml] to 100 ml of the solvent control and 50 ml of the test chemical and positive control, respectively). Each TFT-treated culture is dispensed at 200 μl per well with a multichannel pipette onto 96 well-microwell plates (four plates for each solvent control and two plates for each test chemical and the positive control). The plates are incubated for 12 days at 37° C in a humidified incubator gassed with 5% CO_2 in air to allow colony development.

6. Colony Counting

The number of wells containing colonies is counted by naked eye or with the aid of a dissecting microscope at 10× magnification. A well containing no colonies is classified as negative. The number of negative wells (empty wells) per 96-well microwell plate are quantified for the survival (PE_0), viability (PE_2), and mutation (TFT) plates. For the TFT plates, large and small colonies are scored for information of the mechanism of action of the test chemical. Large and small colonies are characterized as follows:

Large colonies:	Size:	More than ¼ of well's diameter
	Morphology:	Diffused totally or in periphery
Small colonies:	Size	Less than ¼ of well's diameter
	Morphology:	Compact

H. Calculations

1. Survival and Viability

From the zero term of the Poisson distribution the probable number of clones/well (P) on microwell plates is given by

$$P = -\ln(EW/TW) \qquad [\text{EQ. 1}]$$

where EW is empty wells and TW is total wells (29).

The plating efficiency (PE) is

$$PE = P/\text{number of cells plated/well)} \qquad [EQ.\ 2]$$

When 1.6 cells/well are plated on average for all survival and viability plates

$$PE = P/1.6 \qquad [EQ.\ 3]$$

The relative survival (RS) in each test culture will therefore be determined by comparing plating efficiencies in test and control cultures.

$$\%RS = [PE(test)/PE(control)] \times 100 \qquad [EQ.\ 4]$$

2. Mutation Frequency

Mutation frequency (MF) expressed as mutants/10^6 viable cells is calculated as

$$MF = [PE(mutant)/PE(viable)] \times 10^6 \qquad [EQ.\ 5]$$

From the formula of PE and with knowledge that 2×10^3 cells are plated/well for mutation to TFT resistance.

$$PE(mutant) = P(mutant)/(2 \times 10^3)$$
$$PE(viable) = P(viable)/1.6 \qquad [EQ.\ 6]$$

With colony sizing, mutation frequency could be separately calculated as total mutation frequency, large colony mutation frequency, and small colony mutation frequency. Without colony sizing, the total mutation frequency should be conducted.

3. Relative Total Growth

Relative total growth (RTG) is the measure of the cytotoxicity of the test chemical. First, the RSG is calculated from the DCG.

$$RSG = [DCG1 \times DCG2\ (test)/[DCG1 \times DCG2\ (control)] \qquad [EQ.\ 7]$$

DCG is growth rate between day 0 and day 1 (DCG1) or between day 1 and day 2 (DCG2). Then the RTG is calculated as

$$RTG\ (\%) = RSG \times RV(\%) \qquad [EQ.\ 8]$$

RV (relative viability) is calculated by comparing plating efficiencies in the test and control culture at day 2.

4. Percentage of Small-Colony Mutants

Percentage of small colony mutants (%SC) is simply calculated as

$$\%SC = \text{(small colony mutation frequency)}/$$
$$\text{(total mutation frequency)} \times 100 \qquad [EQ.\ 9]$$

Alternatively,

$$\%SC = (\text{no. of wells containing small colonies})/(\text{no. of positive wells}) \times 100 \qquad [\text{EQ. 10}]$$

is possible.

I. Criteria for Acceptable Conditions

To demonstrate acceptable cell growth and maintenance throughout the experiment, the absolute plating efficiency for the solvent control has to be 60% to 140% of survival (PE_0) and 70% to 130% for viability (PE_2) according to the 1994 consensus agreements formed by the MLA workshop in Portland, OR (19). Mutation frequencies of solvent controls with and without S9 mix should ideally be within the range of 50–250×10^{-6}. When the mutation frequency of the positive control increases twofold or more over the concurrent negative control, the experimental sensitivity is acceptable.

To assess the mutagenic potential of a test chemical properly, a range of cytotoxic doses should be tested, with the awareness that excessive cytotoxicity can cause false-positive responses. According to the Portland agreement (19), the top concentration should show 10% to 20% RS or RTG, whichever is lower. Data obtained under excessively cytotoxic conditions (less than 10% RS or RTG) may be excluded from evaluation.

J. Statistics

All results should be subjected to appropriate statistical analysis. The UKEMS guideline recommends the method established by Robinson et al (30). This includes two procedures, one is a pair-wise comparison of each treatment with the concurrent solvent control, and the other is assessment of a linear trend between mutation frequencies and increasing doses. A statistical package (Mutant™; UKEMS, York, UK) for this analysis is commercially available. The data can also be statistically analyzed by another technique, with adjustment of the family-wise type I error (31). The procedure consists of elimination of data that show a downturn phenomenon using the Simpson-Margolin procedure, dose-response effect evaluation, and multiple comparisons with the concurrent control by a modified Dunnett's procedure.

K. Criteria for a Positive Response

A statistically significant dose-related increase in mutation frequency is required for a positive response, but data at concentrations eliciting excessive cytotoxicity (less than 10% RS or RTG) should be excluded.

L. Criteria for a Negative Response

A negative response is no significant increase in mutant frequency at any concentrations that show more than 10% RS or RTG. Data at concentrations exhibiting sufficient cytotoxicity (10% to 20% RS or RTG) are required for a negative judgement.

M. Data Sheets

An example of a set of data sheets for the MLA is shown in Table 5. Data in the sheets consist of preliminary experiments, main experiments, and summary.

TABLE 5.

I. Preliminary Experiments
I-A. Solubility test of the test chemical
    1) Test chemical no.:
    2) State: solid, liquid
    3) Solubility for solvents: RPM10, saline, water, DMSO
    4) Choice of solvent:
    5) Comments:

I-B. Dose range finding tests
    1) Lot of L5178Y cells:
    2) S9: +, −
    3) Date of experiment:

		Cell Concentration (× 10^5/ml)					Daily Cell Growth		
		Day 0		Day 1		Day 2			Day 1 ×
	Conc.	Adj. Conc.	Conc.	Adj. Conc.	Conc.	Day 1	Day 2	Day 2	RSG
Sol. ( )									
Dose 1									
Dose 2									
Dose 3									
Dose 4									
Dose 5									
Dose 6									
Dose 7									
Dose 8									
Pos. ( )									

    4) Dose of 10–20% of RSG and Comments:
II. Main Experiments
II-A. General information
    1. Test chemical
      a) Test chemical no.:

Mouse Lymphoma Assay

    b) Solvent:
    c) Dose: (Dose)
    d) S9: +, −
    e) Positive control: Dose: Lot No.:
2. Cells
    a) Lot of cells:
    b) Expected spontaneous MF:
3. Date of experiment:
4. Comments:

II-B. Cytotoxicy test (PE0, RS)

	Conc. (/ml)	No. of Positive Wells			Total Well No.	Negative Well No.	PE0 (%)	RS (%)
		Plate 1	Plate 2	Total				
Sol. ( )					192			
Dose 1					96			
Dose 2					96			
Dose 3					96			
Dose 4					96			
Dose 5					96			
Dose 6					96			
Dose 7					96			
Dose 8					96			
Pos. ( )					96			

Comments:

II-C. Cell growth during expression time

	Conc.. (/ml)	Cell Concentration ($\times 10^5$/ml)				Daily Cell Growth			RSG
		Day 0	Day 1		Day 2			Day 1 ×	
		Adj. Conc.	Conc.	Adj. Conc.	Conc.	Day 1	Day 2	Day 2	
Sol. ( )									
Dose 1									
Dose 2									
Dose 3									
Dose 4									
Dose 5									
Dose 6									
Dose 7									
Dose 8									
Pos. ( )									

Comments:

(Continued)

TABLE 5. (Continued)

II-D. Cell viability (PE2, RS2, RTG)

	Conc. (/ml)	No. of Positive Wells			Total Well No.	Negative Well No.	PE2 (%)	RS2 (%)	RTG (%)
		Plate 1	Plate 2	Total					
Sol. ( )					192				
Dose 1					96				
Dose 2					96				
Dose 3					96				
Dose 4					96				
Dose 5					96				
Dose 6					96				
Dose 7					96				
Dose 8					96				
Pos. ( )					96				

Comments:

II-E. Number of mutant colonies

	Conc. (/ml)	Number of Positive Wells											
		Plate 1			Plate 2			Plate 3			Plate 4		
		L	S	Total	L	S	Total	L	S	Total	L	S	Total
Sol. Con.													
Dose 1													
Dose 2													
Dose 3													
Dose 4													
Dose 5													
Dose 6													
Dose 7													
Dose 8													
Pos. Con.													

Comments:

Mouse Lymphoma Assay

II-F. Mutation frequency

	Conc. (/ml)	Total Positive Well No.			Total Well No.	Negative Well No.	PE2	Mutation Frequency ($\times 10^{-6}$)			% SC
		Large	Small	Total				L-MF	S-MF	T-MF	
Sol. Con.					384						
Dose 1					192						
Dose 2					192						
Dose 3					192						
Dose 4					192						
Dose 5					192						
Dose 6					192						
Dose 7					192						
Dose 8					192						
Pos. Con.					192						

Comments:

III. Summary data

	Conc. (/ml)	PE0 (%)	PE2 (%)	RS0 (%)	RSG (%)	RTG (%)	Mutation Frequency ($\times 10^{-4}$)	% SC	Statistical Analysis
Sol. Con.									
Dose 1									
Dose 2									
Dose 3									
Dose 4									
Dose 5									
Dose 6									
Dose 7									
Dose 8									
Pos. Con.									

Judgment:
Comments:

III. MODIFIED PROCEDURE

A. Duplicate Culture

Duplicate cultures are originally recommended for the microwell method in the MLA, because variability between cultures may be greater than that attributable to theoretical binominal variance, and a heterogeneity factor obtained from duplicate experiments may be required for reliable statistical analysis (22, 32). Duplicate experiments involve duplicating all treatments, including solvent and positive controls, and the data should be statistically analyzed by Robinson's method (30).

B. Long-Term Treatment

In the standard MLA, cells are treated with a test chemical for 3 or 4 hours with and without S9 mix. That treatment time, however, may be insufficient for detecting some clastogens. For example, the mutagenicities of nucleoside or base analogues and spindle poisons, which do not directly affect DNA but bring about mutations and chromosome alterations through nucleotide metabolism and chromosome segregation, appear to be cell-cycle dependent. For detecting the mutagenicities of such non-DNA targeting chemicals, extended treatment time (24 hours) is sometimes effective (24). When a test chemical yields negative responses in the short-term MLA with and without S9 mix, a long-term treatment trial may be conducted in the absence of S9 mix (24).

The long-term MLA is different from the standard MLA in treatment protocol, which is as follows: cultures of 50 ml at 2×10^5 cells/ml in RPMI medium supplemented with 10% horse serum are treated in flasks with a series of diluted test chemicals for 24 hours in a 37° C, 5% CO_2 humidified incubator. The cells are then centrifuged, washed once, and resuspended in fresh medium. They are transferred to new flasks and adjusted to 50 ml at 2×10^5 cells/ml for growth through the expression period (45 hours) or diluted to be plated for estimation survival. The rest of the procedures are the same as that used for the short-term treatments.

IV. BIOLOGICAL INTERPRETATION OF MUTATIONS IN THE MLA

The spontaneous mutation frequency in the microwell version is generally $50-200 \times 10^{-6}$. That is much higher than the spontaneous frequencies in other mammalian gene mutation assays, even considering that autosomal genes usually mutate at a higher rate than the X-linked hypoxanthine phosphorybosyl transferase (*hprt*) gene. Recently, the L5178Y cell line was shown to have a point mutation at codon 170 in the *p53* gene and to overproduce the mutant p53 protein

(33, 34). The p53 mutant cell lines are known to exhibit a repair-deficient phenotype that results in genomic instability and increased susceptibility to mutation, and the mutator phenotype of the MLA cells is probably the result of this p53 mutation (35, 36). In fact, this may be a desirable characteristic for a cell line used in in vitro mutagenicity and clastogenicity testing if the presence of the p53 mutation confers increased sensitivity to genotoxic agents. The high mutation frequency also reduces the number of cells required for the gene mutation assay. Because p53 mutant cells such as L5178Y fail to undergo apoptosis or cell-cycle arrest and may show increased survival, mutations may be apt to accumulate when those cells are treated with cytotoxic levels of test chemicals. Perhaps this is why loss of specificity and false-positive results sometimes occur under highly cytotoxic conditions (less than 10% RS or RTG) (33).

Two distinct phenotypic classes of TK-deficient mutants can be observed in the MLA: large-colony (lc) mutants and small-colony (sc) mutants. Generally, lc mutants are associated with small scale genetic changes, such as point mutations, small deletions, and small insertions, and sc mutants usually involve allelic loss and large-scale changes at the chromosomal level (2, 5, 37). Molecular and cytogenetic analysis of TK-deficient mutants, however, demonstrated that most lc mutants also result from allelic loss and sometimes involve chromosomal alteration (3, 4, 8). This discrepancy is also related to p53 mutations (35, 36). It is likely that a mutator phenotype caused by a p53 mutation would contribute to gross structural changes, such as chromosomal alterations, but not small DNA changes such as point mutations. L5178Y cells may easily generate chromosomal mutations spontaneously and by various kinds of genetic damage, including DNA strand breaks, alkylation, oxdation, among others. It is believed that the percentage of sc mutants (%SC) can predict the clastogenicity of test chemicals, but practically speaking, it is difficult to estimate clastogenicity by the %SC. The %SC data may be used as additional information or for quality control for the performance of the test, but not for the prediction of clastogenicity

V. MATERIALS AND EQUIPMENT

The following equipment, media, and chemicals have been used for the assay in our laboratories. This list does not necessarily constitute an endorsement of any specific product.

(a) Equipment and supplies
  Tissue culture flask (100 ml)   Corning No. 25120, Falcon No. 3024
  Tissue culture flask (50 ml)   Corning No. 25110, Falcon No. 3028
  Test tube (50 ml)   Corning No. 25339, Falcon No. 2027
  Test tube (15 ml)   Corning No. 25310, Falcon No. 2027
  96-well microwell plate   Corning No. 25861, Costar No. 4820

8-channel multimechanical pipette (200 μl)	Costar No. 4820
Reservoir (50 ml)	Costar No. 4870A
Coulter Counter Z2 model	Coulter Electronics
Burker-Turk hematocytometer	Erma, Tokyo
Rocker platform	TAITEC RP12, Tokyo
Low-magnifying microscope	Olympus TH3

(Other equipment and supplies conventionally used in a laboratory)

(b) Media components

RPMI 1640 medium (powder)	Gibco-BRL No. 31800
Penicillin-streptomycin solution (×100)	Gibco-BRL No. 15140
Sodium pyruvate	Gibco-BRL No. 11840
Horse serum	JRH BIOSCIENCES
S9	Kikkoman Co. Ltd, Chiba, Japan
Glucose-6-phosphate	Sigma No. G7250
Nicotinamide adenine dinucleotide phosphate	Oriental No. 309-50471

(c) Chemicals

Thymidine	Sigma No. T9250
Hypoxanthine	Sigma No. H9377
Methotrexate	Sigma No. M8407
Glycine	Sigma No. G6388
Trifluorotymidine (TFT)	Sigma No. T2255
Methylmethanesulfonate (MMS)	Aldrich No. 12,992-5
Cyclophosphamide (CP)	Aldrich No. 21,870-7

REFERENCES

1. MM Moore, D Clive, JC Hozier, BF Howard, AG Batson, NT Turner, J Sawyer. Analysis of trifluorothymidine-resistant (TFT) mutants of L5178Y/TK+/− mouse lymphoma cells. Mutat Res 151:161–174, 1985.
2. BF Blazak, FJ Los, CJ Rudd, WJ Caspary. Chromosome analysis of small and large L5178Y mouse lymphoma cell colonies: Comparison of trifluorothymidine-resistant and unselected cell colonies from mutagen-treated and control culture. Mutat Res 224:197–208, 1989.
3. ML Applegate, MM Moore, CB Broder, A Burrell, G Juhn, KL Kasweck, P-F Lin, A Wadhams, JC Hozier. Molecular dissection of mutations at the heterozygous thymidine kinase locus in mouse lymphoma cells. Proc Natl Acad Sci U S A 87:51–57, 1990.
4. D Clive, P Glover, M Applegate, J Hozier. Molecular aspects of chemical mutagenesis in L5178Y/tk+/− mouse lymphoma cells. Mutagenesis 5:191–197, 1990.
5. RD Combes, H Stopper, WJ Caspary. The use of L5178Y mouse lymphoma cells to assess the mutagenic, clastogenic and aneugenic properties of chemicals. Mutagenesis 10:403–408, 1995.

6. L-S Zhang, M Honma, A Matsuoka, T Suzuki, T Sofuni, M Hayashi. Chromosome painting analysis of spontaneous and methylmethanesulfonate-induced trifluorothymidine-resistant L5178Y cell colonies. Mutat Res 370:181–190, 1996.
7. M Liechty, Z Hassanpour, J Hozier, D Clive. Use of microsatellite DNA polymorphisms on mouse chromosome 11 for in vitro analysis of thymidine kinase gene mutations. Mutagenesis 9:423–427, 1994.
8. M Leichty, J Scalzi, K Sims, H Crosby Jr, D Spencer, L Davis, W Caspary, J Hozier. Analysis of large and small colony L5178Y tk+/− mouse lymphoma mutants by loss of heterozygosity (LOH) and by whole chromosome 11 painting: Detection of recombination. Mutagenesis 13:461–474, 1998.
9. J Hozier, M Applegate, MM Moore. In vitro mammalian mutagenesis as a model for genetic lesions in human cancer. Mutat Res 270:201–209, 1992.
10. D Clive, JFS Spector. Laboratory procedure for assessing specific locus mutations at the tk locus in cultured L5178Y mouse lymphoma cells. Mutat Res 31:17–29, 1975.
11. D Clive, KO Johnson, JFS Spector, AG Batson, MMM Brown. Validation and characterization of the L5178Y/TK+/− mouse lymphoma mutagen assay system. Mutat Res 59:61–108, 1979.
12. MM Moore-Brown, D Clive, BE Howard, AG Batson, KO Johnson. The utilization of trifluorothymidine (TFT) to select for thymidine kinase-deficient (TK−/−) mutants from L5178Y/TK+/− mouse lymphoma cells. Mutat Res 85:363–378, 1981.
13. MM Moore, D Clive. The quantitation of TK−/− and HGPRT-mutants of L5178Y/TK+/− mouse lymphoma cells at varying times post-treatment. Environ Mutagen 4:499–519, 1982.
14. MM Moore, BE Howard. Quantitation of small colony trifluorothymidine-resistant mutants of L5178Y/TK+/− mouse lymphoma cells in RPMI-1640 medium. Mutat Res 104:287–294, 1982.
15. J Cole, CF Arlett, MHL Green, J Lowe, W Muriel. A comparison of the agar cloning and microtitration techniques for assaying cell survival and mutation frequency in L5178Y mouse lymphoma cells. Mutat Res 111:371–386, 1983.
16. NT Turner, AG Batson, D Clive. Procedure for the L5178Y/TK+/− to TK−/− mouse lymphoma cell mutagenicity assay. In: BJ Kilbey, M Legator, W Nichols, C Ramel, eds. Handbook of Mutagenicity Test Procedures. Amsterdam: Elsevier Science Publishers, 1984, pp 239–268.
17. JB Majeska, DW Matheson. Development of an optimal S9 activation mixture for the L5178YTK+/− mouse lymphoma mutation assay. Environ Mol Mutagen 16:311–319, 1990.
18. DL Spencer, KC Hines, WJ Caspary. An in situ protocol for measuring the expression of chemically-induced mutations in mammalian cells. Mutat Res 312:85–97, 1994.
19. D Clive, G Bolcsfoldi, J Clements, J Cole, M Honma, J Majeska, M Moore, L Muller, B Myhr, T Oberly, MC Odelhkim, C Rudd, H Shimada, T Sofuni, V Thybaud, P Wilcox. Consensus agreement regarding protocol issues discussed during the mouse lymphoma workshop: Portland, Oregon, May 7, 1994. Environ Mol Mutagen 25:165–168, 1995.
20. ML Garriott, DA Casciano, LM Schechtman, GS Probst. International workshop on mouse lymphoma assay testing practices and data interpretations: Portland, Oregon, May 7, 1994. Environ Mol Mutagen 25:162–164,

21. T Sofuni, M Honma, M Hayashi, H Shimada, N Tanaka, S Wakuri, T Awogi, KI Yamamoto, Y Nishi, M Nakadate. Detection of in vitro clastogens and spindle poisons by the mouse lymphoma assay using the microwell method: Interim report of an international collaborative study. Mutagenesis 11:349–355, 1996.
22. T Sofuni, P Wilcox, H Shimada, J Clements, M Honma, D Clive, M Green, V Thybaud, RHC San, BM Elliot, L Muller, L. Meeting report on protocol issues regarding the use of the microwell method discussed in the mouse lymphoma workshop: Victoria B.C., Canada, March 27, 1996. Environ Mol Mutagen 29:434–438, 1997.
23. M Honma, M Hayashi, H Shimada, N Tanaka, S Wakuri, T Awogi, KI Yamamoto, K, N Ushio-Kodani, Y Nishi, M Nakadate, T Sofuni. Evaluation of the mouse lymphoma tk assay (microwell method) as an alternative to the in vitro chromosomal aberration test. Mutagenesis 14:5–22, 1999.
24. M Honma, L-S Zhang, H Sakamoto, M Ozaki, K Takeshita, M Momose, M Hayashi, T Sofuni. The need for long-term treatment in the mouse lymphoma assay. Mutagenesis 14:23–29, 1999.
25. J Cole, DB McGregor, M Fox, J Thacker, RC Garner. Gene mutation assays in cultured mammalian cells. In: DJ Kirkland, ed. Basic Mutagenicity Tests. Avon: Cambridge University Press, 1990, pp 87–114.
26. J Cole, FN Richmond, BA Bridges. The mutagenicity of a-amino-N6-hydroxyadenine to L5178Y tk+/– mouse lymphoma cells: Measurement of mutations to ouabain, 6-thioguanine and trifluorothymidine resistance, and the induction of micronuclei. Mutat Res 253:55–62, 1991.
27. P Clay, MF Cross. Microwell mutation assays: Evaluation of ethylmethansulfonate, benzo[a]pyrene and benzidine using the tk locus in L5178Y mouse lymphoma cells. Mutagenesis 5 (suppl):45–54, 1990
28. TJ Oberly, DL Yount, ML Garriot. A comparison of the soft agar and microtiter methodologies for the L5178Y tk± mouse lymphoma assay. Mutat Res 388:59–66, 1997.
29. EM Furth, WG Thilly, BA Penman, HL Liber, WM Rand. Quantitative assay for mutation in diploid human lymphoblasts using microtiter plates. Anal Biochem 110:1–8, 1981.
30. WD Robinson, MHL Green, J Cole MJR Healy, RC Garner, D Gatehouse. Statistical evaluation of bacterial/mammalian fluctuation test. In: DJ Kirkland, ed. Statistical Evaluation of Mutagenicity Test Data. Cambridge: Cambridge University Press, 1989, pp 102–140.
31. M Hayashi, T Ohmori, Y Honda, M Honma, Y Nishi, T Awogi, Y Kasahara, N Tanaka, Y Nakagawa, S Wakuri, A Hirose, H Shimada, T Morita, A Wakata, T Sofuni, I Yoshimura. Statistical evaluation of the mammalian mutation assay data by mouse lymphoma L5178Y/tk microtitration method. Mutagenesis, 1999, (in press).
32. CF Arlett, DM Smith, GM Clarke, MHL Green, J Cole, DB McGregor, JC Asquith. Mammalian cell gene mutation assay based upon colony formation. In: DJ Kirkland, ed. Statistical Evaluation of Mutagenicity Test Data. Cambridge: Cambridge University Press, 1989, pp 102–140.
33. RD Storer, AR Kraynak, TW McKelvey, MC Elia, TL Goodrow, JG DeLuca. The mouse lymphoma L5178YTK+/– cell line is heterozygous for codon 170 mutation in the p53 tumor supressor gene. Mutat Res 373:157–165, 1997.

34. LS Clark, DW Hart, PJ Vojta, K Harrington-Brock, JC Barrett, MM Moore, KR Tindall. Identification and chromosomal assignment of two heterozygous mutations in the Trp53 gene in L5178Y/Tk+/−3.7.2C mouse lymphoma cells. Mutagenesis 13:427–434, 1998.
35. M Honma, M Hayashi, T Sofuni. Cytotoxic and mutagenic responses to X-rays and chemical mutagens in normal and p53-mutated human lymphoblastoid cells. Mutat Res 374:89–98, 1997
36. M Honma, L-S Zhang, M Hayashi, K Takeshita, Y Nakagawa, N Tanaka, T Sofuni. Illegitimate recombination leading to allelic loss and unbalanced translocation in p53-mutated human lymphoblastoid cells. Mol Cell Biol 17:4774–4781, 1997.
37. D Clive, NT Turner, R Krehl, J Eyre. The mouse lymphoma assay may also be used as a chromosome aberration assay. Environ Mutagen 7 (suppl 3):33, 1985.

8

The In Vitro Micronucleus Assay

Marilyn J. Aardema
The Procter & Gamble Company, Cincinnati, Ohio

Micheline Kirsch-Volders
Vrije Universiteit Brussel, Brussels, Belgium

I. INTRODUCTION

The in vitro micronucleus assay is emerging in the field of genetic toxicology. Increased interest in this assay is the result of recent advances that have produced a method that provides an easy, quick, and inexpensive assessment of many types of cytogenetic damage, including clastogenic and aneugenic events. In addition, the in vitro micronucleus assay allows for the simultaneous assessment of several measures of toxicity, including cell proliferation and cell death (apoptosis), on the same slide. In this chapter, the in vitro micronucleus assay is described and particular attention is paid to how this emerging assay contrasts to the analysis of chromosomes in the standard in vitro metaphase assay, which has been used for decades and is currently required for regulatory purposes (1).

A. Principles of the Assay

1. Induction of Micronuclei

A micronucleus is a small, separate nucleus from the main nucleus in a cell. Micronuclei are formed during telophase of mitosis or meiosis, when the nuclear envelope is reconstituted around the chromosomes of the developing daughter cells. A chromosome fragment(s) lacking a centromere (acentric fragment), or lagging whole chromosome not incorporated into the main nucleus when the nuclear membrane reforms, can have a separate nuclear membrane form around it, which results in a micronucleus. Formation of micronuclei during mitosis is depicted in Figures 1 and 2. Because the DNA within a micronucleus is separate from the main nucleus, micronuclei represent "chromatin loss." Micronuclei result from either structural damage to the chromosome, which leads to formation of acentric chromosome fragments (Fig. 1), or damage to the mitotic targets (e.g. spindle apparatus), which leads to loss of a whole chromosome(s) (Fig. 2). In the standard in vitro metaphase assay, chromosome damage is assessed by the direct visual observation of the condensed chromosomes in metaphase (Figs. 1b & 2b).

It is important to note that micronuclei are formed during cell division, regardless of where the actual damage (either DNA damage or damage to mitotic targets) occurred during the cell cycle. Studies with synchronized cells or time course studies can help determine where in the cell cycle the damage that caused formation of micronuclei was induced. In contrast, chromosome aberrations can be formed during any of the cell cycle stages, and the specific types of chromo-

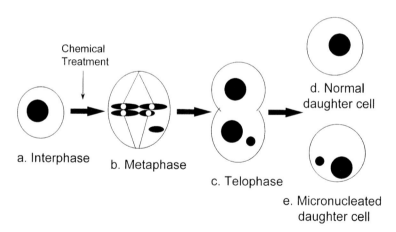

FIG. 1. Formation of a micronucleated cell containing an acentric chromatid fragment. Induction of structural chromosome damage in interphase (**a**) resulting in the formation of a chromatid fragment lacking a centromere (acentric) can be visualized when the chromosomes are condensed in metaphase (**b**) of mitosis (as is performed in the standard metaphase assay). Reformation of the nuclear membrane around the chromatid fragment produces a micronucleus (**c**) that can be quantitated in the daughter cells formed after division (**e**).

In Vitro Micronucleus Assay

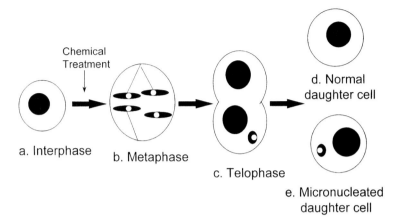

FIG. 2. Formation of a micronucleated cell containing a whole chromosome. A cell with damage to the spindle apparatus, can result in a whole chromosome that is not connected to the microtubules (**b**). As this cell undergoes division, this chromosome will lag during anaphase and will not be segregated properly to a daughter cell nucleus. Reformation of the nuclear membrane around this lagging chromosome produces a micronucleus (**c**) which is segregated into one of the daughter cells (**e**). Special staining techniques for centromeres can be used to identify the presence of the centromere in this micronucleus. This distinguishes a micronucleus formed from an acentric chromatin fragment (Fig. 1) from a micronucleus formed from a whole chromosome (or centric fragment).

some aberrations observed at metaphase may provide information as to whether DNA damage was induced in G0/G1 (chromosome-type aberrations) or S/G2 (chromatid-type aberrations). One recent exception to the long-held belief that micronuclei are formed only during cell division is the report by Shimizu et al (2) of the novel formation of micronuclei from amplified DNA during the S phase. Future studies may identify other novel mechanisms of formation of micronuclei that will need to be addressed; however, for the purposes of this chapter, only micronuclei formation during cell division will be considered.

Numerical chromosome changes are directly observed in metaphase cells, including polyploidy (extra copy[ies] of all chromosomes) and endoreduplication (reduplication of chromosomes from preceding S phase). Observation of these may provide information on damage to mitotic targets in the preceding mitosis (polyploidy) or damage in late S phase G2 leading to replication errors (endoreduplication). Because both micronuclei and chromosome aberrations can only be analyzed in dividing cells, their utility is, by definition, limited to dividing cells/tissues; they are of no use for examining damage induced in nondividing cells or tissues.

Because micronuclei are analyzed in daughter cells derived from treated cells, they represent genetic damage that is passed down, or inherited, through at least one cell division. This is in contrast to the standard analysis of chromosome aberrations in metaphase cells, where the optimal assessment of damage is made in cells in their first metaphase after chemical treatment, that is in cells that have not divided. To assess stable aberrations in metaphase that are transmitted to daughter cells (e.g., reciprocal translocations) requires the use of special staining methods like fluorescence in situ hybridization (FISH) chromosome painting. In contrast, reciprocal translocations cannot be assessed in micronuclei.

2. Factors That Affect the Frequency of Micronucleated Cells

Because a cell has to divide to form a micronucleus, the frequency of micronuclei is most accurately measured by the analysis of only those cells that have divided. Unlike the in vivo micronucleus assay in which recently divided cells are differentially stained, an easy identification of cells that have divided in vitro has not always been possible. In 1985, Fenech and Morley (3, 4) reported the development of a simple method, the cytokinesis block, to identify cells that have divided by adding cytochalasin B to the cells in culture. Cytochalasin B is a metabolite of the mold *Helminthosporium dematioideum* (5) that blocks cytokinesis but not nuclear division and results in the formation of binucleated cells from cells that have undergone nuclear division once (Fig. 3). Cytochalasin B inhibits cytokinesis by binding high-molecular-weight complexes in the plasma membrane that

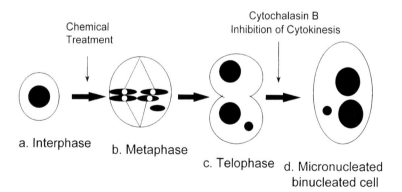

FIG. 3. Formation of a micronucleated binucleated cell. In the presence of cytochalasin B, the actin filaments involved in cleavage of the cytoplasm to form the two daughter cells are inhibited, leading to the formation of a binucleated cell (**d**). Measurement of micronuclei in binucleated cells allows the investigator to analyze cells that have divided only once. Cells that do not divide remain mononucleated, whereas cells that have divided two or more times are multinucleated (not shown).

In Vitro Micronucleus Assay

induce actin polymerization to form microfilaments that induce cleavage of the dividing cell into the two daughter cells (6). By quantifying micronuclei in binucleated cells, the investigator is assured that the cell has divided once. Thus the ratio of #micronucleated binucleated cells/#binucleated cells provides an accurate measure of the true frequency of induction of micronucleated cells, compared with the lower frequency that would be obtained in the absence of cytochalasin B, where the frequency would be calculated as: #micronucleated cells/total cells (including cells that did not divide and had no chance of forming micronuclei) (7). Currently, debate is ongoing regarding whether the use of cytochalasin B is necessary, especially for continuously cycling cell lines. The advantages and disadvantages of the use of cytochalasin B are discussed in Section II D.

Unlike the analysis of chromosomes in metaphase in which the specific types of chromosome aberrations are identifiable and the presence of extra chromosomes (e.g., polyploidy) is readily apparent, the chromosome content of micronuclei is not directly visualized. Complicating this is the fact that the nuclear membrane can reform around multiple chromosome(s) or fragments, therefore one micronucleus may contain multiple chromosome fragments. This is a disadvantage if one is trying to precisely quantify the amount of chromosome damage induced in a cell, as the relationship between the number of micronuclei does not necessarily equate to the degree of induced chromosome aberrations.

In addition, not all chromosome aberrations are expected to form a micronuclei. Those chromosome aberrations that form acentric fragments have a higher probability of forming micronuclei, because the acentric fragment has no means of attaching to the spindle and segregating appropriately. Those aberrations that do not result in acentric fragments, like stable symmetrical exchanges, would not be expected to result in micronuclei. Because of this, one of the potential criticisms of the micronucleus assay is that it detects unstable chromosome damage which may not be relevant for human disease.

Other factors that can affect the frequency of cells with micronuclei include:

1. Further division of cells containing micronuclei. Because the segregation of micronuclei into the daughter cells is random, if a parent cell containing one micronucleus divides, only one daughter cell will receive the micronucleus and the other daughter cell will not receive a micronucleus (8). This would result in an apparent lowering of the frequency of cells with micronuclei. In contrast, if multiple micronuclei are present in a parent cell, there is the possibility that both daughter cells will receive micronuclei, which would increase the frequency of micronucleated cells.
2. If the DNA contained within a micronucleus is essential to cellular function, further division of a micronucleated cell could be lethal. Dying cells clearly reduce the frequency of observed damage.

3. Reincorporation of micronuclei into the main nucleus. During subsequent division, there is a chance that the chromosome(s) or chromosome fragments from the micronuclei will be incorporated into the main nucleus during the next telophase. This would reduce the frequency of micronucleated cells in the population.
4. Elimination of micronuclei from a cell. It has been shown that some micronuclei, especially larger micronuclei that contain whole chromosomes, can be extruded from the cells, thereby lowering the frequency of micronucleated cells observed at a given time (9–11). Micronuclei containing amplified DNA (double minutes) have recently been reported to also be released from the cell (2).

Because of these factors, it is not surprising that the absolute frequency of micronucleated cells is often found to be less than the frequency of cells with aberrations in metaphase (8, 12, 13). This is an important consideration when one is trying to compare results quantitatively from an in vitro micronucleus assay to an in vitro metaphase assay.

3. Biological Consequence of Micronuclei

Because micronuclei are a result of chromatin loss, they may represent a risk for cancer, especially if the lost chromatin contains a gene(s) involved in the cancer process, such as a tumor suppressor gene. However, as there are few direct studies on the biological consequences of micronuclei, the strength of their utility as a biomarker of disease such as cancer is largely undefined at this time. Most is known about death of micronucleated cells. Early investigations conducted by Obe et al (14) compared the cell cycle stage of DNA in the micronuclei to the cell cycle stage of the DNA in the main nucleus using autoradiographic methods. The authors reported that some micronuclei can be in a different cell cycle stage relative to the main nucleus, as evidenced by the observation of cells in which the main nucleus incorporated tritiated thymidine during DNA synthesis, whereas the micronucleus was not labeled. Later investigations (15) confirmed this observation that some micronuclei can be out of synchrony with the main nucleus, although in the majority of cells in this study, the micronuclei were undergoing DNA synthesis at the same time as the main nucleus. It was proposed that DNA in micronuclei that are out of synchrony with the main nucleus undergoes premature chromosome condensation when the main nucleus enters mitosis, and the micronucleus is degraded, which leads to cell death (14). More recent evidence supporting the degradation of DNA in micronuclei comes from the observation of a late accumulation of p53 in micronuclei (i.e., later than the p53 accumulation in the main nucleus that occurs after DNA damage), presumably as a result of micronucleus DNA degradation (16, 17).

Other mechanisms leading to cell death of micronucleated cells are also likely. For instance, if the genetic information contained in micronuclei is not transcriptionally active and is essential to cellular function, then it can be predicted that cells with micronuclei would die. Cell death would certainly be expected for micronuclei that are completely extruded from the cell [9–11]. Eckert et al, (18) concluded that micronuclei containing whole chromosomes do not lead to an increase in mutation frequency in the L5178Y tk +/– assay, presumably because of lethality of these micronucleated cells. In addition, if a micronucleus that contains a whole chromosome is reincorporated into the main nucleus, there is the possibility (though perhaps remote) that there may be no adverse consequence as far as transmitted DNA damage.

B. Basis for Interest in the In Vitro Micronucleus Assay

The analysis of whether a cell contains a micronucleus involves a relatively easy "yes" or "no" visual evaluation. Thus, the quantification of the number of cells that contain micronuclei, along with recording the number of micronuclei each cell contains, is simple. Because of this simplicity, the analysis of micronuclei is fast, and therefore relatively inexpensive. It does not require the more highly trained technical personnel needed for analyzing and classifying the specific types of chromosome aberrations typically recorded in the in vitro metaphase assay. (It must be noted that the analysis of chromosome damage in the in vitro metaphase assay could be similarly simplified by recording "aberrant cell" or "normal cell.") The analysis of micronuclei can be automated, and several systems for the automated analysis of micronuclei have been reported (see below). Because of these benefits, interest has increased in the in vitro micronucleus assay as a method for screening chemicals for genotoxicity. Recent adaptations of the assay to a low-volume microwell method have increased its use even further, because it now can be used for prescreening numerous chemicals of limited quantity (19–21). This is possible because the population of cells available for analysis of micronuclei is so large (all interphase cells that have divided) compared with the relatively low percentage of cells in metaphase in a culture. Thus, in a microliter volume, a few thousand cells can be seeded, exposed to the test chemical, processed in situ, and analyzed for micronuclei. This method is described in more detail below.

In addition to its use in genotoxicity screening, there is increased interest in the in vitro micronucleus assay because of the mechanistic information it can provide. Because micronuclei can be formed from the loss of a whole chromosome, this assay provides a more definitive assessment of aneuploidy (instead of the indirect method of measuring polyploidy in the metaphase assay as a means of identifying aneugens). Confirmation that a micronucleus contains a whole chromosome, however, requires special staining methods, as described in more detail below. Other mechanistic information that can be obtained from the analysis of the main interphase nucleus is discussed below.

C. Mechanistic Information

Because the specific types of chromosome damage induced by a chemical or physical agent can provide clues as to the mechanism of action, several techniques have been developed to determine whether a micronucleus contains acentric chromosome fragments, which suggests chromosome breakage, or centromeric DNA, which may indicate damage to the spindle leading to loss of a whole chromosome (aneuploidy) rather than a DNA reactive event. The most common methods for identifying whether a micronucleus contains centromeric DNA involve use of fluorescent probes such as the CREST (calcinosis, Raynaud phenomenon, esophageal dismotility, sclerodactyly, and telangiectasia) anti-kinetchore antibodies from human CREST patients (22–24), centromere specific probes (25, 26), combined use of centromere and telomere probes (27, 28), or tandem labeling probes (29) to more accurately distinguish clastogenic from aneugenic events. The advantages and disadvantages of the different methods are discussed in detail by Kirsch-Volders et al (30). It has been proposed that if 70% to 100% of the micronuclei contain centromeric DNA, then a chemical can be concluded to induce aneuploidy (24).

In addition to identifying the contents of micronuclei, chromosome-specific fluorescent probes have been used in the cytokinesis-block assay to identify the contents of each of the two main nuclei in the binucleated cell. This powerful methodology allows for the investigation of two mechanisms of induction of aneuploidy, namely chromosome nondisjunction (missegregation of chromosomes to the nuclei of the daughter cells) compared with chromosome loss (formation of micronuclei) (30). Examples of how this methodology is used are provided in this chapter.

II. METHODOLOGY

A. General Comments

Many different methods for the conduct of the in vitro micronucleus assay have been published. Variations in the protocol include the type of cells used (cell lines versuss primary cells), how the cells are processed (in situ vs dropping cells onto slides), which cells are analyzed (asynchronous vs synchronized), and the use of cytochalasin B. Because there are so many variations, there is no agreed upon protocol, even for routine genotoxicity testing, at this time. To address this, a number of efforts are underway that are expected to result in the design of a common method for routine testing in the near future.

In Vitro Micronucleus Assay

The basic design of the in vitro micronucleus assay is similar to the in vitro metaphase assay (31). Briefly, cells grown in vitro are exposed to a chemical or physical agent in the presence or absence of exogenous metabolic activation. After exposure, the cells are processed at some interval(s) of time to allow cells to divide and convert the chromosome damage to micronuclei. The cells are stained and examined for the presence of micronuclei through use of a standard light or fluorescence microscope. Further details of the assay are provided below.

B. Cells

As with the in vitro metaphase assay, a variety of cells have been used in the in vitro micronucleus assay and all appear equally acceptable. These include established cell lines, the most common being Chinese hamster cells lines (32, 33). Human peripheral blood lymphocytes (whole blood or isolated lymphocytes) are commonly used (34). Fenech (35) provides a detailed review of the important variables in the in vitro micronucleus assay in cytokinesis-blocked human lymphocytes. The most appropriate cell type in human peripheral blood to analyze has been debated, and Wuttke et al, (36) reported an increased sensitivity of B-lymphocytes to radiation induced micronuclei compared to T-lymphocytes. An in vitro hepatocyte micronucleus assay has been described (37–41).

C. Treatment and Recovery Times

The appropriate treatment and recovery times for the in vitro micronucleus assay have not been defined at this time, and the literature contains numerous variations. For the purposes of defining whether a chemical is genotoxic (as opposed to a mechanistic study which may involve a very specific experimental design), a combination of treatment regimens is typically used. The most common protocol involves treating asynchronously growing cell lines, and for human lymphocytes, initiation of treatment 24 hours after phytohaemoglutamin growth stimulation, which results in the treatment of cells at the initiation of the first S phase. Current thinking is that the timing for the in vitro micronucleus assay should be based on the methods used in the in vitro metaphase assay in which cells are typically harvested at a time equivalent to 1.5 normal cell cycles after initiation of treatment. One common regimen involves a short exposure to the test chemical (3–6 hours), followed by washing and a recovery time in culture in the absence of the test article. For cells with a 12–14 hours normal cell cycle (1.5 cell cycles = 18–21 hours), a harvest time of at least 18–21 hours from the beginning of treatment has been frequently used. Results of ongoing studies will provide data to determine whether a longer recovery time to allow cells with damage to progress through mitosis and form micronuclei is advantageous. The appropriate recovery time may be cell type dependent.

A short 3- to 6-hour exposure is usually conducted in the presence and absence of metabolic activation. A short exposure helps maximize the concentration of the chemical that can be tested because cells can usually tolerate higher exposures to a toxic chemical for a shorter length of time. In addition to a short exposure, a continuous exposure up to the time of harvesting the cells (at 1.5 normal cell cycles) is used to maximize exposure to all cell cycle stages. Studies examining the timing of appearance of micronuclei are needed to understand more fully the appropriate treatment/harvest times for micronucleus induction. Time course studies have shown that an early induction of micronuclei is observed with spindle poisons as expected, based on their interaction with cells in G2/M, whereas later induction of micronuclei is observed with damage induced in S or G1 phases of the cell cycle (42).

D. Cytochalasin B

Currently the use of cytochalasin B in the in vitro micronucleus assay for cell lines is being debated. Fenech (7) reviewed both the advantages and disadvantages of the use of cytochalasin B. In this review, the benefits of the use of cytochalasin B appear to outweigh the disadvantages, the largest benefit being the ability to control for the influence of the kinetics of the cell population (including chemically induced delay) on the observed frequency of micronucleated cells in the cytokinesis-block method. The one major concern with the use of cytochalasin B is the interference with the induction of micronuclei by spindle poisons, as reported by Antoccia et al (43) and Minissi et al (44). For cell lines, the use of cytochalasin B has been questioned (32, 45) since the cells are continuously dividing, and there is the potential for cytochalasin B to interfere with induction of micronuclei, for instance by chemicals that also affect cytokinesis or microtubules, such as was observed with colchicine (43). Further studies comparing micronucleus induction with and without the cytokinesis-block method, using a variety of treatment/harvest times and including weak clastogens and aneugens acting via different mechanisms are needed.

For human lymphocytes, it is generally agreed that cytochalasin B should be used, because not all the cells are stimulated into division; however, there is still the concern for the potential interference of cytochalasin B on chemically induced micronuclei. Montero-Montoya et al (46) described a method using 5-bromodeoxyuridine to label proliferating lymphocytes and avoid any potential problems with cytochalasin B. This method does address the problem of nonproliferating cells, but it does not help identify those cells that have divided just once. At this time, insufficient data are available from studies that directly compared the micronucleus test with and without cytochalasin B to determine whether the benefits of being able to precisely identify cells that have divided once exceed the potential problems.

The appropriate concentration of cytochalasin B varies with the cell type. Typical concentrations are between 3 to 6 µg/ml and yield 50% to 80% binucleated cells. Contrary to what might be expected, if the concentration of cytochalasin B is too low, then the frequency of micronucleated binucleated cells can actually be elevated, because undamaged normally cycling cells can divide before the cytochalasin B has time to induce the cytokinesis-block, whereas slower cycling cells with chromosome damage are more likely to be affected by the cytokinesis-block(47).

There are numerous variations in the timing of addition of cytochalasin B to the cell culture relative to chemical exposure. Typically, for a dosing regimen that involves a short 3–6 hour exposure to the test article followed by washing, cytochalasin B is added after the washing. For the continuous exposure in cell lines, cytochalasin B is added at the same time as the test chemical. For human lymphocytes, one common protocol involves the addition of cytochalasin B at 44 hours (either after a short or long exposure to the test article), as this is when the cells are progressing into mitosis. Cytochalasin B is usually left in the cultures until the cells are harvested. However, recent unpublished results from some laboratories indicate that a 60- to 90-minute recovery in the absence of cytochalasin B before harvesting improves the quality of the appearance of the cells.

Some examples of common treatment regimens with human lymphocytes in the presence of cytochalasin B are shown in Figure 4. Examples of common treatment regimens with cell lines in the presence of cytochalasin B are shown in Figure 5. It is important to note that these are just examples, and a common protocol will be developed in the near future, once the various validation and harmonization efforts are complete.

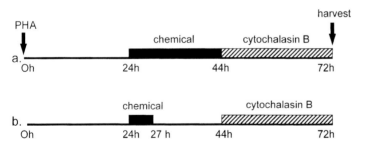

FIG. 4. Examples of treatment/harvest times for human lymphocytes in the cytokinesis-block method. Lymphocytes are stimulated with phytohaemagglutinin (PHA) to stimulate cell division. Twenty-four hours later, the test chemical is added to the cell culture. The cells are exposed to the chemical for a long time (**a**), such as 20 hours in this diagram, or for a short time (**b**), such as 3 hours in this diagram. Cytochalasin B is added for the last 28 hours until the cells are harvested at 72 hours.

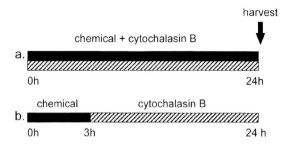

FIG. 5. Examples of treatment/harvest times for cell lines in the cytokinesis-block method. Cells are exposed to the test chemical for a long time (**a**) in the presence of cytochalasin B or for a short time (**b**) before the addition of cytochalasin B.

E. Preparation of Slides and Analysis

1. Fixation and Slide Preparation

In contrast to the metaphase assay in which the cells are treated with a hypotonic solution and fixative to free the chromosomes from the cytoplasm, in the in vitro micronucleus assay, the cytoplasm must be preserved to ensure the accurate detection of micronuclei. As such, suspension or trypsinized cells can be put onto microscope slides by cytocentrifugation, then fixed (e.g., 70% to 100% ethanol or methanol) and stained. Cells can be dropped onto slides (dry or wet slides) after a mild hypotonic treatment (e.g., 0.075 M KCl) and fixation (e.g., methanol:acetic acid, 3:1 to 25:1). A method for preparing slides for human lymphocytes was described (48), and a recent review of methods used by laboratories around the world was published (34). Another common method involves fixing cells in situ. Cells growing attached to the slides can be exposed to hypotonic (e.g., 1% Na citrate) and then fixed (e.g., methanol: acetic acid, 6:1). Surrallés et al (49) discussed the advantages of using a hypotonic treatment before fixation to make the cytoplasm less condensed and allow easier identification of micronuclei when a non-DNA-specific stain like Giemsa is used.

2. Staining

Standard Giemsa staining of slides is commonly used. However, it has been shown that DNA-specific stains like 6´-diamidino-2-phenylindole (DAPI) or acridine orange provide a more accurate detection of micronuclei, especially small micronuclei (49). As discussed above, other specific staining methods can be used, such as antikinetochore antibodies, to identify whether a micronucleus contains a chromosome fragment or a whole chromosome.

In Vitro Micronucleus Assay

3. Analysis

   a. Frequency of Micronuclei. The frequency of micronucleated cells is typically measured by analyzing at least 1000 cells. In prescreening assays, fewer cells may be analyzed. Micronuclei are generally classified by the criteria described by Heddle (50) and Fenech (51): size less than one third the diameter of the main nucleus, not linked with the nucleus, contained within the cytoplasm.

   In the cytokinesis-block method, the number of micronucleated binucleated cells in a total of at least 1000 binucleated cells is recorded. The criteria for selecting a binucleated cell to analyze are: the two nuclei should be of approximately equal size; the two nuclei may be attached by a nucleoplasmic bridge; the two nuclei may touch or overlap (51). In the cytokinesis-block method, cells that are mononucleated are concluded to be cells that have not divided, whereas cells with more than two nuclei (multinucleated) have undergone nuclear division more than once. Because some aneugens have been reported to increase the frequency of micronuclei in mononucleated cells in the cytokinesis-block method (52), some protocols also include a measure of the number of micronucleated mononucleated cells. Aneugens have also been reported to increase the frequency of multinucleated cells (53). Currently, all types of cells should be recorded until more is learned about the different types of damage and how they are detected in this assay.

   In addition to measuring the frequency of micronucleated cells, the number of micronuclei per cell is often recorded as a measure of the severity of damage on an individual cell basis.

   b. Cytotoxicity. One of the advantages of the cytokinesis-block method is that it provides an easy way to assess cytotoxicity on the same slide as that used for the measurement of micronuclei. The ratio of cells that have divided once (mononucleated), twice (binucleated), or more (multinucleated) gives a measure of the proliferation of the cells in the culture. A decrease in the proliferation of the cells in culture is one measure of cytotoxicity in the form of chemically induced cell cycle delay or cell death. Cells are analyzed for the number of mononucleated, binucleated, or multinucleated cells, and proliferation can be expressed in a variety of ways, including:

   - % binucleated cells: (#binucleated cells/total mono- + bi- + multinucleated cells) × 100
   - nuclear division index (NDI) = (#mono- +2(#bi-) + 3(# tri-) + 4(#quadranucleated cells)/total number of cells (54).

   In the absence of cytochalasin B, various measures of cytotoxicity have been reported. A cell count of a treated culture compared with a control culture can be used to calculate comparative cell growth. If cells are cultivated in a relatively low cell density, the number of clones consisting of one, two, four, and eight cell(s) can be assessed to determine proliferation (33). A proliferation index (PI) can be calculated according to the formula:

$$PI = \frac{(c1 \times 1) + (c2 \times 2) + (c4 \times 3) + (c8 \times 4)}{\text{No. of clones}} \qquad [EQ.\ 1]$$

where $c1$ = single cells, $c2$ = two-cell colonies, $c4$ = four-cell colonies, $c8$ = eight-cell colonies.

Along with the various measurements of proliferation described above, the presence of apoptotic and necrotic interphase nuclei can be assessed on the same slides used for measuring micronuclei (30).

III. ALTERNATIVE METHODOLOGIES

In addition to the general methods described above, a number of alternative methods have been described in the literature. These are:

A. Mitotic Shake-Off Method

One variation on the standard method is the mitotic shake-off method for detection of aneugens (33, 55). V79 cells growing asynchronously in culture are exposed to the test substance for 3 hours, mitotic cells are collected by shake-off from the flasks, then centrifuged and reseeded onto microscope slides in quadriperm dishes. Three hours later, after the cells have attached, the medium is removed, and cells are exposed to a mild hypotonic solution (1.5% trisodium citrate-2-hydrate) and fixed in situ. With this method, only cells treated in G2 or mitosis are analyzed, and specific effects, such as induction of micronuclei by spindle inhibitors, can be studied.

B. Microwell Method for Screening

For screening large numbers of chemicals or chemicals with limited quantity as is common during the early stages of drug development, a microwell method of the in vitro micronucleus assay was developed by Balwierz et al (19) and adapted for routine, contract lab use (20, 21). In this method, around 5000 cells are seeded into each chamber of an eight-well plastic chamber slide (Nunc) in a volume of 0.250 ml media. About 24 hours later, the cells are exposed to the test substance through use of treatment/recovery times discussed above. Cytochalasin B is typically used. The dosing of the test article volume can be reduced to 0.100 ml to save on the quantity of chemical used for the assay. After treatment, the cells are exposed in situ to mild hypotonic (1% sodium citrate) and fixed (methanol: glacial acetic acid, 6:1) in situ. With this method, up to eight different treatments can be assessed on the same slide. For an initial assessment, about 400 binucleated cells are analyzed for micronuclei. This method can be adapted to any size chambers.

C. Cytosine Arabinoside Assay

In this adaptation of the in vitro micronucleus assay, chemicals that induce excision repair can be distinguished from genotoxic chemicals that do not induce excision repair (7). Cytosine arabinoside is added to cells in culture to inhibit excision repair, thereby increasing the formation of DNA strand breaks that lead to acentric fragments and increased micronuclei. If a chemical is genotoxic via induction of strand breaks and acentric fragments directly, little or no increase is seen in the presence of cytosine arabinoside.

D. Bleomycin Amplification Assay

Snyder (56) described a method for assessing DNA intercalation using the in vitro micronucleus assay with V79 cells. In this method, cells are exposed to both a concentration of bleomycin shown to be negative for micronucleus induction and a test chemical (also at a concentration negative for micronucleus induction). The enchancement of micronucleus-induction and decreased proliferation are measured. If an enhancement of micronucleus induction is observed, it is interpreted as demonstrating that the test chemical interacts with DNA in such a way (e.g., intercalation) as to allow greater access of the minor groove of DNA to binding by bleomycin, leading to greater induction of bleomycin-induced DNA nicks and formation of micronuclei.

E. Automated Analyses of Micronuclei

To reduce the time involved in scoring micronucleated cells, methods for automating this analysis have been developed. These include flow cytometric (57, 58) and image analysis methods (59, 60). Flow cytometric methods provide the advantage of being able to analyze large numbers of cells very quickly and to determine the DNA content of micronuclei. However, as reviewed by Nusse and Marx (57), cell debris and apoptotic bodies complicate the analysis, and this method should not be used when apoptosis or a large number of dying cells are induced. Roman et al (58) described a new gating procedure that provided a better way to discriminate cell debris in a flow cytometric analysis, but the complication of apoptotic cells remains. Frieauff et al (60) concluded that flow cytometric analyses had advantages for testing single chemicals, but image analysis was a faster method for analyzing micronuclei from a large series of chemicals, as would be needed in routine screening.

IV. USES AND INTERPRETATION OF DATA

The uses of the in vitro micronucleus assay are too numerous to review, but recent examples are described briefly, below. One major use of the assay, not discussed here because it is beyond the scope of this chapter, is in assessing environmental and occupationally induced chromosome damage in vivo in humans.

A. Chemical-Specific Investigations

The in vitro micronucleus assay is used to study the genotoxicity of specific chemicals of interest. These type of studies often use the standard methods and cell types described above. However, some reports have shown how the in vitro micronucleus assay can be modified to address specific questions. Saranko and Recio (61) compared the ability of 1,2:3,4-diepoxybutane (DEB) (a metabolite of 1,3 butadiene, a genotoxic carcinogen) to induce micronuclei with mutations in Big Blue Rat 2 lambda/lacI transgenic fibroblasts. These studies were conducted to help define the role of genotoxicity of this metabolite in the overall carcinogenicity of 1,3 butadiene.

Another recent example of a chemical-specific study was reported for 1,4-dioxane (a presumed nongenotoxic rodent carcinogen) (62). The in vitro micronucleus assay was used in a battery of standard genotoxicity tests to confirm that 1,4-dioxane was not genotoxic in vitro, although a positive result in a liver micronucleus assay was obtained.

Ong et al (63) reported the induction of micronucleated and multinucleated cells by glass fibers of various types and sizes. The presence of kinetochore-positive staining in the micronuclei and the induction of multinucleated cells was concluded to indicate that these fibers induced aneuploidy and inhibited cytokinesis.

B. Structure Activity Relationships

Snyder (56) reported the results of the in vitro micronucleus assay to study a series of antihistamines. Results of this work, along with other mechanistic data, suggest that the genotoxicity of various antihistamines is caused by DNA intercalation, which was not expected based on the structure of these drugs. From a structure activity relationship (SAR) standpoint, Snyder was able to identify the need for a dimethylamino substituent for the genotoxic activity of these intercalating antihistamines (56).

Because the in vitro micronucleus assay detects both clastogenic and aneugenic activity, Tafazoli and Krisch-Volders (64) and Tafazoli et al (65) used the in vitro micronucleus assay in human lymphocytes to compare the genotoxicity of various chlorinated hydrocarbons. In these studies, the results of the micronucleus assay were compared with the alkaline single cell gel electrophoresis assay (Comet assay).

Quinolones are a class of antibacterial compounds that inhibit bacterial DNA gyrase and can cross-react with mammalian topoisomerase, which results in genotoxicity. The in vitro micronucleus assay has been used by different laboratories to investigate the SAR of different quinolone compounds to aid in the drug development process (66–68). Recently, Snyder and Cooper (69) reported the adaptation of the in vitro micronucleus assay for assessing the photogenotoxicity of structurally related fluoroquinolones.

The in vitro micronucleus assay has been used as part of a multiendpoint assessment. Again, because the in vitro micronucleus assay detects aneuploidy-inducing activity, Pfeiffer et al [70] used the in vitro micronucleus assay in V79 cells to study the aneuploidy-inducing potential of bisphenols. As part of the multiendpoints assessment, they measured disruption of cytoplasmic microtubules, disruption of the mitotic spindle, and induction of both metaphase arrest and micronuclei by various bisphenols.

C. Correlation to Other Endpoints

The transformation of primary Syrian hamster embryo (SHE) cells in vitro has been shown to be a useful tool for the identification of both genotoxic and nongenotoxic rodent carcinogens (71). One way to help determine whether a positive response in the SHE cell transformation assay is caused by a genotoxic or nongenotoxic mechanism is to measure micronucleus induction in SHE cells. As reviewed by Tsutsui and Barrett (72), an assessment of micronuclei in SHE cells treated with various estrogens has been useful in understanding the role of genotoxicity, particularly aneuploidy, in the transforming ability of these compounds. Gibson et al (73) reported a good agreement between the SHE cell micronucleus assay and SHE cell transformation assay for mutagenic chemicals, but a number of chemicals were negative in the SHE micronucleus assay that induced cell transformation, as would be expected because transformation can occur via a nongenotoxic mechanism.

Eckert et al (18) studied the relationship between induction of micronuclei containing chromosome 11 (containing the *tk* gene) and mutations in the L5178Y mouse lymphoma assay. Although micronuclei containing chromosome 11 were induced by various aneugens, stable, viable mutants were not recovered.

Van Goethem et al (74) compared induction of micronuclei in vitro and the alkaline single cell gel electrophoresis (Comet) assay with various cobalt compounds. Their results demonstrate how these two methods can be used to investigate more directly the relationship between induced DNA strand breaks and formation of micronuclei.

D. Thresholds

To obtain the statistical power needed to detect small differences above background and establish a no-effect-level, or threshold, for a genotoxic event requires the analysis of a large number of cells. Because the analysis of micronuclei is relatively easy and fast, it is practical for the study of thresholds for genotoxins. Elhajouji et al (75, 76) and Marshall et al (77) used the in vitro micronucleus assay in human lymphocytes to determine practical thresholds for a variety of spindle inhibitors. In these studies, different endpoints were measured to help establish a threshold, including the general induction of micronuclei (75–77), the induction of centromere-positive micronuclei (indicative of micronuclei containing whole chromosomes) that had been collected by flow cytometry (75), and the induction of nondisjunction of specific human chromosomes in the interphase nuclei of cytokinesis-blocked cells (76, 77). In addition to providing critical data to support the existence of thresholds for spindle inhibitors, these studies resulted in an important observation that the measure of micronuclei (that result from chromosome loss) is a less sensitive measure of aneuploidy, compared with the measurement of nondisjunction (the irregular distribution of chromosomes to the daughter nuclei).

E. Affect of Different Genotypes

The in vitro micronucleus assay provides an easy method for assessing the influence of different genotypes on induction of genotoxicity. These studies involve genetically engineered cell lines, as well as human and rodent genetic variants. To investigate the effect of metabolism on genotoxicity of nitroarenes and aromatic amines, sublines of Chinese hamster cells stably expressing human NAT1 or NAT2 N-acetyltransferases or *Salmonella typhimurium* O-acetyltransferase were compared for sensitivity to micronucleus induction (78). Cells expressing human NAT2 N-acetyltransferase had the highest sensitivity to chromosome damage induced by nitroarenes. Norppa (79) discussed how the in vitro micronucleus assay (among other cytogenetic methods) can be used to study the effect of human polymorphisms, such as in the detoxification enzyme glutathione S-transferase.

F. Biomarker of Human Cancer Predisposition, Treatment Outcome

Because increased sensitivity to radiation-induced chromosome damage is common in people predisposed to cancer, the in vitro micronucleus test has been investigated as a tool for identifying increased carcinogenic risk in humans. Scott et al (80, 81) reported increased radiation-induced micronuclei in peripheral blood lymphocytes from breast cancer patients. They proposed that this assay may be useful for identifying women with a predisposition for breast cancer. The use of the in vitro micronucleus assay for determining clinical outcome of various cancers was reported by Shibamoto et al (82). Because of the potential power of this method as a biomarker of human disease and treatment, a collaborative project was initiated to define the assay better (35).

V. CONCLUSIONS

The in vitro micronucleus assay is an exciting method that has emerged rapidly in the field of genetic toxicology over the last few years. It is a quick and easy method for assessing many types of chromosome damage. In the near future, the results of ongoing validation studies around the world will become available. With these data, the appropriate experimental design for genotoxicity testing will be determined, and the use of the assay in regulatory testing schemes will be defined. Continued innovations with the assay promise to broaden its usefulness in addressing novel research problems.

REFERENCES

1. M Kirsch-Volders. Towards a validation of the micronucleus test. Mutat Res 392:1–4, 1997.
2. N Shimizu, N Itoh, H Utiyama, GM Wahl. Selective entrapment of extrachromosomally amplified DNA by nuclear budding and micronucleation during S phase. J Cell Biol 140:1307–1320, 1998.
3. M Fenech, A Morley. Solutions to the kinetic problem in the micronucleus test. Cytobios 43:233–246, 1985a.
4. M Fenech, A Morley. Measurement of micronuclei in lymphocytes. Mutat Res 148:99–105, 1985b.
5. MA Ridler, GF Smith. The response of human cultured lymphocytes to cytochalasin B J Cell Sci 3:595–602, 1968.
6. CD Lin, S Lin. Actin polymerization induced by a high motility-related high-affinity cytochalasin binding complex from human erythrocyte membrane. Proc Natl Acad Sci U S A 76:2345–2349, 1979.
7. M Fenech. The advantages and disadvantages of the cytokinesis-block micronucleus method. Mutat Res 392:11–18, 1997.
8. JRK Savage. A comment on the quantitative relationship between micronuclei and chromosomal aberrations. Mutat Res 207:33–36, 1988.
9. JW Parton, ML Garriott, JE Beyers. Expulsion of demecolcin-induced micronuclei from mouse bone marrow polychromatic erythrocytes. Environ Mol Mutagen 17:79–83, 1991.
10. G Schwemmer, U Kliesch, I-D Adler. Extruded micronuclei induced by colchicine or acrylamide contain mostly lagging chromosomes identified in paintbrush smears by minor and major mouse DNA probes. Mutagenesis 12:201–207, 1997.
11. S Nito, F Ariyuki, A Okaniwa. Spontaneous expulsion of micronuclei by enucleation in the micronucleus assay. Mutat Res 207:185–192, 1988.
12. A Wakata, MS Sasaki. Measurement of micronuclei by cytokinesis-block method in cultured Chinese hamster cells: Comparison with types and rates of chromosome aberrations. Mutat Res 190:51–57, 1987.
13. M Hayashi, A Matsuoka, H Sakamoto, N Yamazaki, T Sofuni. Correlation between micronucleus induction and chromosomal aberrations. Environ Mol Mutagen 31:4, 1998.

14. G Obe, B Beek, VG Vaidya. The human leukocyte test system III. Premature chromosome condensation from chemically and x-ray induced micronuclei. Mutat Res 27:90–101, 1975.
15. J Kramer, G Schaich-Walch, M Nusse. DNA synthesis in radiation-induced micronuclei studied by bromodeoxyuridine (BrdUrd) labelling and anti-BrdUrd antibodies. Mutagenesis 5:491–495, 1990.
16. C Unger, S Kress, A Buchmann, M Schwarz. γ-Irradiation-induced micronuclei from mouse hepatoma cells accumulate high levels of the tumor suppressor protein p53. Cancer Res 54:3651–3655, 1994.
17. C Granetto, L Ottaggio, A Abbondandolo, S Bonatti. p53 Accumulates in micronuclei after treatment with a DNA breaking chemical, methylnitrosourea, and with the spindle poison, vinblastine. Mutat Res 352:61–64, 1996.
18. I Eckert, WJ Caspary, M Nusse, M Liechty, L Davis, H Stopper. Micronuclei containing whole chromosomes harbouring the selectable gene do not lead to mutagenesis. Mutagenesis 12:379–382, 1997.
19. PS Balwierz, RT Bunch. Validation of a high-throughput in vitro/in situ micronucleus assay for screening compounds for clastogenic/aneugenic activity. Environ Mol Mutagenesis 31 Suppl. 29:3, 1998.
20. P Curry, P Balwierz, JL Ivett. An in situ micronucleus assay: Implications for discovery toxicology. Environ Mol Mutagenesis 31 Suppl. 29:3, 1998
21. MJ Aardema, LL Crosby, DP Gibson, PT Curry, JL Ivett. The in situ micronucleus assay: Utility for genotoxicity screening. Environ Mol Mutagenesis 31 Suppl. 29:2, 1998.
22. BK Vig, SE Swearngin. Sequence of centromere separation: Kinetochore formation in induced laggards and micronuclei. Mutagenesis 1:461–465, 1986.
23. EJ Thomson, PE Perry. The identification of micronucleated chromosomes: A possible assay for aneuploidy. Mutagenesis 3:415–418, 1988
24. A Antoccia, F Degrassi, A Battistoni, P Ciliutti, C Tanzarella. In vitro micronucleus test with kinetochore staining: Evaluation of test performance. Mutagenesis 6:319–324, 1991.
25. JM de Stoppelaar, B de Roos, GR Mohn, B Hoebee. Analysis of DES-induced micronuclei in binucleated rat fibroblasts: comparison between FISH with a rat satellite I probe and immunocytochemical staining with CREST serum. Mutat Res 392:139–149, 1997.
26. M Schuler, DS Rupa, DA Eastmond. A critical evaluation of centromeric labeling to distinguish micronuclei induced by chromosomal loss and breakage in vitro. Mutat Res 392:81–95, 1997.
27. BM Miller, T Werner, HU Weier, M Nusse. Analysis of radiation-induced micronuclei by fluorescence in situ hybridization (FISH) simultaneously using telomeric and centromeric DNA probes. Radiat Res 131:177–185, 1992.
28. A Russo, G Priante, AM Tommasi. PRINS localization of centromeres and telomeres in micronuclei indicates that in mouse splenocytes chromatid non-disjunction is a major mechanism of aneuploidy. Mutat Res 72:173–180, 1996.
29. DA Eastmond, DS Rupa, HW Chen, LS Hasegawa. Multicolor fluorescence in situ hybridization with centromeric DNA probes as a new approach to distinguish chro-

mosome breakage from aneuploidy in interphase cells and micronuclei. In BK Vig, ed. Chromosome Segregation and Aneuploidy. NATO ASI series, 1993, pp 377–390.
30. M Kirsch-Volders, A Elhajouji, E Cundari, P Van Hummelen. The in vitro micronucleus test: A multi-endpoint assay to detect simultaneously mitotic delay, apoptosis, chromosome breakage, chromosome loss and non-disjunction. Mutat Res 392:19–30, 1997.
31. SM Galloway, MJ Aardema, M Ishidate Jr, JL Ivett, DJ Kirkland, T Morita, P Mosesso, T Sofuni. Report from working group on in vitro tests for chromosomal aberrations. Mutat Res 312:241–261, 1994.
32. B Miller, S Albertini, F Locher, V Thybaud, E Lorge. Comparative evaluation of the in invitro micronucleus test and the in vitro chromosome aberration test: Industrial experience. Mutat Res 392:45–59, 1997.
33. S Kalweit, D Utesch, W von der Hude, S Madle. Chemically induced micronucleus formation in V79 cells—Comparison of three different test approaches. Mutat Res 439:183–190, 1999.
34. J Surrallés, AT Natarajan. Human lymphocytes micronucleus assay in Europe. An international survey. Mutat Res 392:165–174, 1997.
35. M Fenech. Important variables that influence base-line micronucleus frequency in cytokinesis-blocked lymphocytes—A biomarker for DNA damage in human populations. Mutat Res 404:155–165, 1998.
36. K Wuttke, C Streffer, W-U Müller. Radiation induced micronuclei in subpopulations of human lymphocytes. Mutat Res 286:181–188, 1993.
37. T Alati, P Eckl, RL Jirtle. An in vitro micronucleus assay for determining the radiosensitivity of hepatocytes. Radiat Res 119:562–568, 1989.
38. K Müller, P Kasper, L Müller. An assessment of the in vitro hepatocyte micronucleus assay. Mutat Res 292:213–224, 1993.
39. K Müller-Tegethoff, P Kasper, L Müller. Evaluation studies on the in vitro rat hepatocyte micronucleus assay. Mutat Res 335:293–307, 1995.
40. K Müller-Tegethoff, B Kersten, P Kasper, L Müller. Application of the in vitro rat hepatocyte micronucleus assay in genetic toxicology. Mutat Res 392:125–138, 1997.
41. PM Eckl, I Raffelsberger. The primary rat hepatocyte micronucleus assay: General features. Mutat Res 392:117–124, 1997.
42. H Stopper, SO Müller. Micronuclei as a biological endpoint for genotoxicity: A mini review. Toxicol In Vitro 11:661–667, 1997.
43. A Antoccia, C Tanzarella, D Modesti, F Degrassi. Cytokinesis-block micronucleus assay with kinetochore detection in colchicine-treated human fibroblasts. Mutat Res 287:93–99, 1993.
44. S Minissi, B Gusstavino, F Degrassi, C Tanazeralla, M Rizzoni. Effect of cytochalasin B on the induction of chromosome missegregation by colchicine at low concentrations in human lymphocytes. Mutagenesis 14:43–49, 1999.
45. BM Miller, E Pujadas, E Gocke. Evaluation of the micronucleus test in vitro using Chinese hamster cells: Results for four chemicals weakly positive in the in vivo micronucleus test. Environ Mol Mutagen 26:240–247, 1995.
46. R Montero-Montoya, L Serrano, P Ostrosky-Wegman. In vitro induction of micronuclei in lymphocytes: The use of bromodeoxyuridine as a proliferation marker. Mutat Res 391:135–141, 1997.

47. J Surrallés, E Carbonell, R Marcos, F Degrassi, A Antoccia, C Transarella. A collaborative study on the improvement of the micronucleus test in cultured human lymphocytes. Mutagenesis 7:407–410, 1992.
48. P Van Hummelen, M Kirsch-Volders. An improved method for the "in vitro" micronucleus test using human lymphocytes. Mutagenesis 5:203–204, 1990.
49. J Surrallés, J Catalan, A Creus, H Norppa, N Xamena, R Marcos. Micronuclei induced by alachlor, mitomycin-C, and vinblastine in human lymphocytes: Presence of centromeres and kinetochores and influence of staining technique. Mutagenesis 10:417–423, 1995.
50. JA Heddle. A rapid in vitro test for chromosomal damage. Mutat Res 18:187–190, 1973.
51. MF Fenech. The cytokinesis-block micronucleus technique. In: GP Pfeifer, ed. Technologies for Detection of DNA Damage and Mutations. New York: Plenum Press, 1996.
52. A Elhajouji, M Cunha, M Kirsch-Volders. Spindle poisons can induce polyploidy by mitotic slippage and micronucleate mononucleates in the cytokinesis-block assay. Mutagenesis 13:193–198, 1998.
53. YG Liu, ZL Wu, JK Chen. Differential effects of aneugens and clastogens on incidences of multinucleated cells and of micronucleate cells in Chinese hamster lung(V79) cell line in vitro. Mutat Res 413:39–45, 1998
54. DA Eastmond, JD Tucker. Identification of aneuploidy-inducing agents using cytokinesis-blocked human lymphocytes and an antikinetochore antibody. Environ Mol Mutagen 13:34–43, 1989.
55. A Seelbach, B Fissler, A Strohbusch, S Madle. Development of a modified micronucleus assay in vitro for the detection of aneugenic effects. Toxicol In Vitro 7:185–193, 1993.
56. RD Snyder. A review and investigation into the mechanistic basis of the genotoxicity of antihistamines. Mutat Res 411:235–248, 1998.
57. M Nusse, K Marx. Flow cytometric analysis of micronuclei in cell cultures and human lymphocytes: Advantages and disadvantages. Mutat Res 392:109–115, 1997.
58. D Roman, F Locher, W Suter, A Cordier, M Bobadilla. Evaluation of a new procedure for the flow cytometric analysis of in vitro, chemically induced micronuclei in V79 cells. Environ Mol Mutagen 32:387–396, 1998.
59. H Thierens, A Vral, F De Scheerder, L De Ridder, A Tates. Semi-automated micronucleus scoring in cytokinesis-blocked lymphocytes after irradiation. Int J Radiat Biol 72:319–324, 1997.
60. W Frieauff, F Pötter-Locher, A Cordier, W Suter. Automatic analysis of the in vitro micronucleus test on V79 cells. Mutat Res 413:57–68, 1998.
61. CJ Saranko, L Recio. The butadiene metabolite, 1,2:3,4-diepoxybutane, induces micronuclei but is only weakly mutagenic at lacI in the Big Blue Rat2 lacI transgenic cell line. Environ Mol Mutagen 31:32–40, 1998.
62. T Morita, M Hayashi. 1,4-Dioxane is not mutagenic in five in vitro assays and mouse peripheral blood micronucleus assay, but is in mouse liver micronucleus assay. Environ Mol Mutagen 32:269–280, 1998.
63. T Ong, Y Liu, BZ Zhong, WG Jones, WZ Whong. Induction of micronucleated and multinucleated cells by man-made fibers in vitro in mammalian cells. J Toxicol Environ Health 50:409–414, 1997.

64. M Tafazoli, M Kirsch-Volders. In vitro mutagenicity and genotoxicity study of 1,2-dichloroethylene, 1,1,2-trichloroethane, 1,3-dichloropropane, 1,2,3-trichloropropane and 1,1,3-trichloropropene, using the micronucleus test and the alkaline single cell gel electrophoresis technique (comet assay) in human lymphocytes. Mutat Res 371:185–202, 1996.
65. M Tafazoli, A Baeten, P Geerlings, M Kirsch-Volders. In vitro mutagenicity and genotoxicity study of a number of short-chain chlorinated hydrocarbons using the micronucleus test and the alkaline single cell gel electrophoresis technique (Comet assay) in human lymphocytes: A structure-activity relationship (QSAR) analysis of the genotoxic and cytotoxic potential. Mutagenesis 13:115–126, 1998.
66. CV Ciaravino, MJ Suto, JC Theiss. High capacity in vitro micronucleus assay for assessment of chromosome damage: Results with quinolone/naphthyridone antibacterials. Mutat Res 298:227–236, 1993.
67. PT Curry, ML Kropko, JR Garvin, RD Fiedler, JC Theiss. In vitro induction of micronuclei and chromosome aberrations by quinolones: Possible mechanisms. Mutat Res 352:143–150, 1996.
68. DP Gibson, X Ma, AG Switzer, VA Murphy, MJ Aardema. Comparative genotoxicity of quinolone and quinolonyl-lactam antibacterials in the in vitro micronucleus assay in Chinese hamster ovary cells. Environ Mol Mutagen 31:345–351, 1998.
69. RD Snyder, CS Cooper. Photogenotoxicity of fluoroquinolones in Chinese hamster V79 cells: Dependency on active topoisomerase II. Photochem Photobiol 69:288–293, 1999.
70. E Pfeiffer, B Rosenberg, S Deuschel, M Metzler. Interference with microtubules and induction of micronuclei in vitro by various bisphenols. Mutat Res 390:21–31, 1997.
71. RJ Isfort, GA Kerckaert, RA LeBoeuf. Comparison of the standard and reduced pH Syrian hamster embryo (SHE) cell in vitro transformation assays in predicting the carcinogenic potential of chemicals. Mutat Res 356:11–63, 1996.
72. T Tsutsui, JC Barrett. Neoplastic transformation of cultured mammalian cells by estrogens and estrogenlike chemicals. Environ Health Perspect 105: suppl 3:619–624, 1997.
73. DP Gibson, R Brauninger, HS Shaffi, GA Kerckaert, RA LeBoeuf, RJ Isfort, MJ Aardema. Induction of micronuclei in Syrian hamster embryo cells: Comparison to results in the SHE cell transformation assay for National Toxicology Program test chemicals. Mutat Res 392:61–70, 1997.
74. F Van Goethem, D Lison, M Kirsch-Volders. Comparative evaluation of the in vitro micronucleus test and the alkaline single cell gel electrophoresis assay for the detection of DNA damaging agents: Genotoxic effects of cobalt powder, tungsten carbide and cobalt-tungsten carbide. Mutat Res 392:31–43, 1997.
75. A Elhajouji, P Van Hummelen, M Kirsch-Volders. Indications for a threshold of chemically-induced aneuploidy in vitro in human lymphocytes. Environ Mol Mutagen 26:292–304, 1995.
76. A Elhajouji, F Tibaldi, M Kirsch-Volders. Indication for thresholds of chromosome non-disjunction versus chromosome lagging induced by spindle inhibitors in vitro in human lymphocytes. Mutagenesis 12:133–140, 1997.

77. RR Marshall, M Murphy, DJ Kirkland, KS Bentley. Fluorescence in situ hybridisation with chromosome-specific centromeric probes: A sensitive method to detect aneuploidy. Mutat Res 372:233–245, 1996.
78. M Watanabe, A Matsuoka, N Yamazaki, M Hayashi, T Deguchi, T Nohmi, T Sofuni. New sublines of Chinese hamster CHL stably expressing human NAT1 or NAT2 N-acetyltransferases or Salmonella typhimurium O-acetyltransferase: Comparison of the sensitivities to nitroarenes and aromatic amines using the in vitro micronucleus test. Cancer Res 54:1672–1677, 1994.
79. Norppa (1997) discussed how the in vitro micronucleus assay (among other cytogenetic methods) can be used to study the effect of human polymorphisms such as in the detoxification enzyme glutathione S-transferase.
    Norppa H. Cytogenetic markers of susceptibility: Influence of polymorphic carcinogen-metabolizing enzymes. Environ Health Perspect 105:829–835, 1997.
80. D Scott, JB Barber, EL Levine, W Burrill, SA Roberts. Radiation-induced micronucleus induction in lymphocytes identifies a high frequency of radiosensitive cases among breast cancer patients: A test for predisposition? Br J Cancer 77:614–620, 1998.
81. D Scott, JB Barber, AR Spreadborough, W Burrill, SA Roberts. Increased chromosomal radiosensitivity in breast cancer patients: A comparison of two assays. Int J Radiat Biol 75:1–10, 1999.
82. Shibamoto, Oike, H Mizuno, T Fukuse, S Hitomi, M Takahashi. Proliferative activity and micronucleus frequency after radiation of lung cancer cells as assessed by the cytokinesis-block method and their relationship to clinical outcome. Clin Cancer Res. 4:677–682, 1995.

9

ICH Guidances on Genotoxicity and Carcinogenicity: Scientific Background and Regulatory Practice

Lutz Müller
Novartis Pharma AG, Basel, Switzerland

Peter Kasper
Federal Institute for Drugs and Medical Devices, Bonn, Germany

Leonard Schechtman
National Center for Toxicological Research, Rockville, Maryland

I. THE ICH PROCESS

The International Conference on Harmonisation of Technical Requirements for Registration of Pharmaceuticals for Human Use (ICH) is a project of the regulatory authorities of Europe, Japan, and the United States and of experts from the pharmaceutical industry in these three regions. The objectives are to establish internationally harmonized recommendations (ICH guidances) for the studies conducted to assure the quality, safety, and efficacy of pharmaceuticals. The six cosponsors of ICH are: (1) the Commission of the European Union (EU); (2) the European Federation of Pharmaceutical Industries Associations (EFPIA); (3) the Ministry of Health and Welfare, Japan (MHW); (4) the Japan Pharmaceutical

Manufacturers Association (JPMA); (5) the U.S. Food and Drug Administration (FDA); (6) the Pharmaceutical Research and Manufacturers of America (PhRMA). The International Federation of Pharmaceutical Manufacturers Associations (IFPMA) serves as a coordinating institution for the pharmaceutical industry and provides the ICH secretariat. The World Health Organization, the European Free Trade Area, and Canada have an observer status in the ICH process. A steering committee with individuals from the represented parties of the three regions oversees and guides the harmonization process. In a first step, the ICH steering committee, in close cooperation with experts from the three regions, selects topics for which there is a realistic chance that a guidance acceptable in all three regions could be developed. Each regulatory party agrees to accept completed topical guidances as regulatory policy in their respective region. Thereafter, expert working groups (EWGs) are established to develop a guidance(s) on their topic. These EWGs consist of representatives of the six (three regulatory and three industrial) sponsoring parties. These experts are responsible for communication, consultation, scientific contact, and acceptance of decisions in their respective regions. The EWG creates a draft guidance (step 1) which, after several further steps of consultation (Fig. 1), is established as an ICH guidance to provide the necessary direction in the EU, Japan, and the United States (step 5). Once ICH guidances have been finalized and published, they are officially adopted in the three ICH regions and put into practice (Fig. 1). Detailed information on the ICH process, the decision-making process, and on guidance texts are available in the proceed-

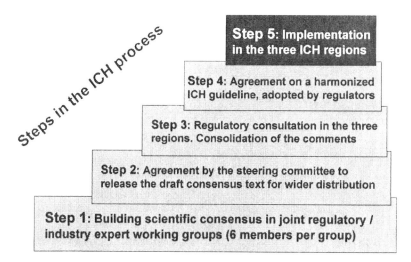

FIG. 1. The ICH guidance development process

ings of the four ICH conferences (1–4). Guidances are also available from the internet at http://www.ifpma.org/ich1.html.

Genotoxicity and carcinogenicity testing for regulatory approval or registration of a product is generally conducted according to guidances issued by national or international authorities. The new ICH guidances in these areas of product safety testing for pharmaceuticals now provide an internationally accepted approach to pharmaceutical development and approval. These guidances are complemented by the existing guidelines issued by OECD. This article addresses some specifics of the ICH guidances on genotoxicity and carcinogenicity, provides background information on their development, and gives reasoning why the ICH EWGs have chosen the type of experimental approach described in the guidances.

II. ICH GENOTOXICITY GUIDANCES

Genotoxicity was selected as an ICH topic at the first ICH conference, which held in November 1991 in Brussels. After the nomination of the genotoxicity experts for the six parties, the first meeting of the EWG took place in Tokyo, Japan in December 1992. Existing regulations or routine operating processes in the three ICH regions served as the basis for harmonization work. Hence, the ICH genotoxicity guidances are an evolution of the pre-existing guidances in Japan and the EU and a crystallization of the FDA genotoxicity postulates. As an initial approach to harmonization, the ICH genotoxicity EWG identified areas of disharmony and differences in regulation with respect to genotoxicity in the three regions. It was decided to prioritize the guidance effort by generating the first of two guidances, namely, the guidance on "Specific Aspects of Regulatory Genotoxicity Tests for Pharmaceuticals" (ICH S2A, where "S" stands for safety, "2" for the topic, genotoxicity). This guidance was finalized by the EWG in July 1995. Because these specific aspects addressed issues as they related to existing guidances (with respect to pharmaceuticals), the introductory paragraph in this first ICH genotoxicity guidance still mentions the old (and now outdated) regulatory requirements in the EU and Japan and procedures still followed by certain centers within the US FDA. As a consequence of further negotiations and additional collaborative studies, the second ICH guidance on "A Standard Battery for Genotoxicity Testing of Pharmaceuticals" (ICH S2B) was finalized in July 1997. Both ICH guidances should be read and applied together. They replace previously existing genotoxicity guidances for pharmaceuticals in the three ICH regions. A comprehensive presentation and review of the ICH genotoxicity guidances, the evolutionary process of achieving harmonization, and a discussion of the applications of those genotoxicity guidances as components of other safety, quality, and multidisciplinary ICH guidances has been published (5).

Important aspects of ICH guidances on genotoxicity are (a) specific recommendations for valid genotoxicity tests, (b) the formulation of testing strategies, and (c) guidance for interpretation of test results. In the same period in which the ICH genotoxicity guidances have been developed, OECD updated a number of their genotoxicity test guidelines (6). Both processes influenced each other and resulted in a number of similar recommendations. The OECD guidelines describe individual test methodologies and do not specifically address testing strategies and overall assessment issues. Additionally, the ICH guidances contain pharmaceutical-related recommendations, which differ in some respects from the recently revised (21 July 1997) OECD guidelines.

Figures 2 and 3 depict the major points of recommendation for in vitro and in vivo genotoxicity tests derived from both ICH genotoxicity guidances. Reasoning for some of these recommendations is discussed below.

A. The Standard Battery of Genotoxicity Tests

The standard battery of genotoxicity testing of pharmaceuticals, which is given in the ICH guidance S2B, is as follows:

1. A test for gene mutation in bacteria.
2. An in vitro test with cytogenetic evaluation of chromosomal damage with mammalian cells or an in vitro mouse lymphoma *tk* assay (MLA)
3. An in vivo test for chromosomal damage using rodent hematopoietic cells

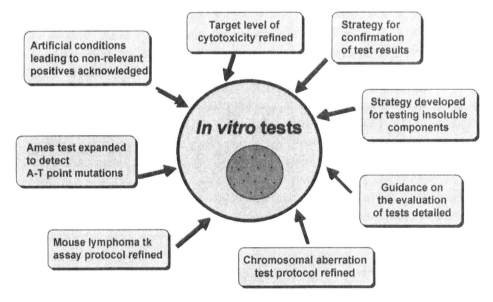

FIG. 2. Issues of conduct and interpretation of in vitro genotoxicity tests addressed in ICH guidances

ICH Background and Practice

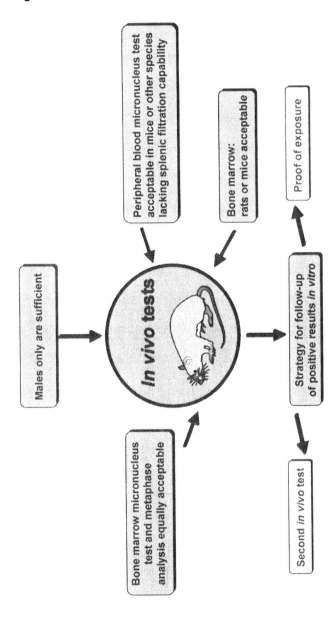

FIG. 3. Issues of conduct and interpretation of in vivo genotoxicity tests addressed in ICH guidances

As the guidance indicates, negative results of such a test battery are usually sufficient to demonstrate absence of genotoxic activity with a high level of confidence. Deficiencies of this battery to detect genotoxins may be attributable to a number of possible factors, such as (a) the limitations associated with the particular target cell or tissue or with the test itself, (b) inordinate toxic effects resulting from drug treatment, (c) the constraints imposed by the specific limitations of the genetic endpoint being monitored, (d) target tissue accessibility to drug exposure and, (e) metabolic limitations. In recognition of such possibilities, the guidance provides for those instances when the standard genotoxicity test battery may warrant modification, allowing for justifiable alternative approaches. Examples include (a) the use of alternative validated in vitro and/or in vivo tests, (b) protocol modifications, (c) the use of modified metabolic activation conditions, and (d) monitoring genetic impairment directly in the target organs of tumor burden. Thus, although the ICH guidance provides for a standard three-test genotoxicity battery that is designed to identify the majority of potential genotoxins examined, other such tests may prove to be more appropriate in a limited number of situations. Nevertheless, as structured, the standard three-test battery serves as the primary screen for assessing genotoxic potential, and any alternative tests are to serve as additions to that battery for the purpose of ancillary analyses.

Table 1 gives an overview of the types of genetic lesions that can be detected in the various standard screening assays for genotoxicity in relation to those lesion that are considered important for the carcinogenesis process. Further justification for the combination of assays comprising the standard battery is discussed below.

1. Reasoning for the Bacterial Reverse Mutation Assay

The bacterial reverse mutation test (BRM), often referred to as the Ames test, is the genotoxicity test with the largest database and is probably the most widely used test worldwide. Although it is ultimately desirable to make predictions for mammalian/human cells in genetic toxicology, the BRM has been shown to detect relevant genetic changes and the majority of genotoxic rodent carcinogens. The test is included in most regional and international guidances as basic for evaluating most product categories. Testing of pharmaceuticals is no exception to this general rule. Consequently, the BRM was adopted as part of the standard battery of genotoxicity tests.

The selection of standard *Salmonella typhimurium* or *Escherichia coli* strains in the BRM, according to some previously existing guidelines, revealed two deficiencies: (a) their insensitivity to radiomimetics because of the inability to detect specific changes in adenine-thymine base pairs, and (b) the insensitivity towards cross-linking agents. The first of these deficiencies was addressed by including among the standard set of strains used in bacterial mutation assays those that will detect point mutations at A-T sites. Such strains are *S. typhimu-*

TABLE 1. Genetic Changes Observed in Tumor-Related Genes and in Standard Selection Targets for Mutation Screening

	Detected in Standard Mutation or Chromosomal Aberration Tests					Involved in Carcinogenesis	
	BRM	Hprt	MLA (tk)	In vitro CA	In vivo MN	Oncogene	Suppressor gene
Point mutations	Yes	Yes	Yes			Yes	Yes
Large chromosomal deletions		Yes (?)	Yes	Yes	Yes*		Yes
Small chromosomal deletions	Yes	Yes	Yes		Yes*		
Chromosomal translocations			Yes	Yes	Yes*	Yes	Yes
Mitotic recombination			Yes				Yes
Chromosome loss			Yes	Yes†	Yes		Yes

*Indirectly, by mechanistic association with micronucleus formation.
†With protocol modifications.

rium TA102, which detects such mutations within multiple copies of *hisG* genes or *E. coli* WP2 *uvrA*, which detects these mutations in the *trpE* gene, or the same *E. coli* strain possessing the plasmid pKM101. Because radiomimetics are selective for A-T sites, these strains detect most radiomimetics. Damage induced by cross-linking agents may be detected using repair proficient strains. As cross-linking agents are usually good clastogens, and a test for clastogenic events is part of the standard battery of pharmaceuticals, there is no need to include specific strains to detect such agents in the BRM.

2. Reasoning for the Mouse Lymphoma *tk* Assay

The necessity for a gene mutation test with mammalian cells in a standard battery for genotoxicity testing of pharmaceuticals was a major issue of discussion in the ICH EWG. The protocol for mammalian cell gene mutation assays as an aggregate was discussed by Aaron et al (7), and the respective OECD guideline was updated (6). However, among the assays described in this one guideline, the MLA has a unique place because of its apparent ability to detect several important types of genetic damage, the full range of which may not be detectable in other standard assays (8–11). Furthermore, there are unique aspects of the MLA, and the various methods by which it can be conducted, that warrant a separate OECD MLA guideline. It is noteworthy that certain types of damage detected by the MLA are important in oncogenes and tumor suppressor genes (Table 1).

In comparison, the *HPRT* (hypoxanthine guanine phosphoribosyl transferase) assays using various cell lines detect only a limited spectrum of such genetic changes (Table 1). From Table 1, it appears also that it may be sufficient for the in vitro component of the standard genotoxicity test battery to consist of the BRM plus a single assay with mammalian cells, that is, either the in vitro *CA* (A chromosomal aberration test) or the MLA. This is in agreement with regulatory experience (12). It is accepted that the MLA detects a wide range of genetic effects and that the genetic damage manifest is expressed in living cells. However, for the use in a standard battery, ICH guidances specify that an in vitro CA and the MLA can be used interchangeably. In addition to the considerations given in Table 1, this conclusion was supported by the following studies:

1. A reinvestigation of nine previously designated mouse lymphoma *tk* uniquely positive rodent carcinogens reported by the NTP revealed that a majority of these compounds are clastogenic in vitro (structural aberrations or polyploidy) when tested according to updated protocol principles for the chromosomal aberration test (13).

2. Fifty-nine percent (20/34) of in vitro clastogens were detected in the MLA through the use of the standard 3 to 4 hours of treatment with test compound in the presence and absence of an exogenous metabolic activation system (14, 15).

ICH Background and Practice

3. Among the 41% (14/34) of compounds that were not detected, it was noted that they often induced chromosomal changes (chromosomal aberrations, polyploidy, or both) only when they were not removed from the treatment medium after the more routinely used short 3- to 6-hour treatment period (16, 17). Out of 15 such chemicals, including nucleoside analogues and spindle poisons, 11 compounds produced positive results in the MLA with a continuous treatment period of 24 hours in the absence of metabolic activation using the microwell methodology. These data led to the adoption of this latter treatment modification in the ICH S2B guidance.

Utilization of the ICH-recommended modified protocols for the MLA and the in vitro CA have thus provided essentially the same level of safety information within the ICH standard battery. Further protocol and data interpretation guidance can be found in the communications of two MLA workshops that were held under ICH auspicies (18, 19). Regarding the protocol detailed for the MLA in the ICH S2B guidance, there is a continuous effort to address the following issues (5): (a) the stipulation for a continuous treatment period of 24 hours may affect the specificity of the MLA for nongenotoxic compounds; (b) because the majority of the data included in the EWG analysis of the MLA were provided for the microwell version, it was not clear whether the same conclusions could be drawn for the agar plate version. The recommendation for the use of the microwell method that appeared in the ICH S2B guidance document, which was based on limited comparative data made available to the EWG at the time of guidance finalization, may need to be re-evaluated through an ICH maintenance process. Because it is presumed that both methods give valid results, either methodology, microwell or agar plate, should be acceptable.

Other mammalian cell gene mutation assays, such as those that monitor mutation at the HPRT locus, in addition to or as an alternative to the MLA, are advocated under ICH guidance in situations in which the BRM is not considered appropriate. Such situations can occur, for example, if the test compound is an antibacterial agent or interacts with the mammalian cell replication system.

3. Reasoning for the In Vitro Chromosomal Aberration Test (In Vitro CA)

A cytogenetic evaluation of chromosomal damage in vitro has traditionally been part of the screening for genotoxicity in all of the ICH regions. Its inclusion in the newly designed ICH standard battery was put into perspective when comparing its usefulness with the MLA (see above). The protocol for the in vitro CA has been redesigned in detail (20), and the recommendations in ICH guidances are in agreement with the recent OECD guideline update (6). However, two protocol aspects are still a matter of discussion: (a) the levels and method(s) of quantification of test compound-induced cytotoxicity for a valid test, and (b) the necessity to test compounds into the insoluble concentration range. The current recommendations, including those in ICH guidances and OECD guidelines in both areas, are thought by some to account for some positives of questionable relevance (12, 21, 22). Aid for interpretation of such results is given in the ICH S2A "Guidance on Specific Aspects of Regulatory Genotoxicity Tests." Further discussion regarding this topic can be found below in "Regulatory Interpretation of Genotoxicity Test Results." Alternatives to the in vitro chromosomal aberration assay that may be capable of providing comparable information, for example, the in vitro micronucleus test, are being evaluated (23–27).

4. Reasoning for the In Vivo Test Using Rodent Hematopoietic Cells

The ICH guidance S2B recommends the inclusion of an in vivo test with rodent hematopoietic cells in the standard battery. Historically, hematopoietic cells have been used for in vivo analysis of chromosomal damage, and a large database of study outcomes has been accumulated. In vivo tests such as the bone marrow micronucleus test can be relatively insensitive (see below, "'Site of Contact' Evaluation of Genotoxicity," more than 80% of human carcinogens are positive in this assay (28). Compared to in vitro tests, in vivo tests such as the bone marrow micronucleus test appear to respond only to a subset of genotoxins, that is, those in vitro genotoxins that can reach the target cells in this tissue at sufficient concentrations to induce chromosomal damage. Nevertheless, the ICH guidance S2A reinforces the importance of an in vivo test component in the standard battery because of such additional factors as absorption, distribution, metabolism, and excretion, which are only partially manifest in in vitro tests but are relevant to human use. This reasoning, together with the fact that in vivo clastogenesis is highly predictive of carcinogenesis (28–30), argues strongly for the inclusion of such an in vivo assay in the core test battery. For the standard battery, a test using rodent hematopoietic cells satisfies this requisite because the bone marrow is a well-perfused tissue, and levels of drug-related materials in bone marrow are often similar to those observed in blood or plasma. This was shown by direct comparisons of drug levels in the two compartments for a large series of different pharmaceuticals (31). The ICH guidances do not give any preference for the evaluation of micronuclei over the analysis of chromosomal aberrations using hematopoietic cells. However, it is acknowledged that relative to the analysis of chromosomal aberrations in bone marrow cells, the analysis of micronuclei in erythrocytes is more expedient and convenient and that the latter is sensitive to the effects of some aneugenic compounds. Additionally, the presence of micronucleated cells can be assessed in the peripheral blood of mice (32, 33) and rats (34), which offers additional approaches for monitoring induction of micronuclei. Should the sensitivity and specificity of the peripheral blood micronucleus test prove to be comparable to that of the bone marrow micronucleus test, using the former could provide some advantages over the latter. Some of these include (a) ease of sample collection, (b) the ability to examine multiple samples harvested from a given treated animal, (c) the ability to determine the attainment of

ICH Background and Practice

steady-state levels of micronucleated cell induction (35), thereby optimizing for the time of bone marrow harvesting, and (d) avoiding the need to sacrifice treated animals unless bone marrow confirmation of the peripheral blood response is necessary.

One promising application of the in vivo micronucleus assay and a potentially significant advantage is its integration into the general toxicology screen (30, 35–39). The endpoint is compatible with longer term (e.g., 14-, 28-, and 90-day) general toxicology testing (which uses repeated exposures over the duration of the study) and can be monitored over the course of such studies. The demonstration of attaining steady-state levels of micronucleus induction (35) in bone marrow after repeated treatments at lower doses that are comparable to the peak levels achieved after acute exposure to higher doses supports the practicability of such an approach. In addition, the use of peripheral blood (37) makes it feasible to monitor for the temporal manifestation of micronuclei during the course of longer term toxicology studies performed in rodents. Some advantages offered by the concomitant assessment of such genotoxic effects over the course of general toxicology studies include (a) examining endpoints such as micronucleus induction in animals undergoing repeated exposure, (b) precluding the need to conduct an independent test for chromosome damage, and (c) limiting the unnecessary use of additional animals.

5. Other Test Models in Genetic Toxicology

Regulatory genotoxicity testing is focused on widely used, sufficiently standardized, and well-characterized test methods. Such test methods constitute the standard battery of genotoxicity tests for pharmaceuticals according to ICH guidances. However, genetic toxicology offers a wide variety of test methods in addition to those named for the standard battery of tests. Because most of these methods are not very widely used and not sufficiently standardized and characterized, such tests may be used to give supplementary information but, under normal circumstances, are not used to replace the standard methods. Specific circumstances in which additional tests may yield valuable information are described throughout the text and in the notes to the ICH guidances. One area in which such tests are important is the follow-up evaluation of in vitro positive results in in vivo test approaches in addition to the test for chromosomal damage in rodent bone marrow.

From time to time, consideration is given in the scientific community to implementing new test procedures that necessarily involve an appraisal of their status of use, degree of standardization, and potential for regulatory acceptance. One such effort is the International Workshop on Genotoxicity Test Procedures (IWGTP; Washington, DC/US, March 1999). In this workshop, the following four test methods were reviewed and discussed: (1) the in vitro micronucleus test; (2) the Comet assay, either in vitro or in vivo; (3) in vivo transgenic mutation assays; and (4) test systems to detect photochemical genotoxicity. It is understood that these methods either provide more efficient use of resources for genetic toxicology testing, for example, the in vitro micronucleus test, or could fill information gaps in regulatory decision making in many cases.

B. Regulatory Interpretation of Genotoxicity Test Results

Experience with testing, data collection for regulatory submission, and regulatory review largely influenced the recommendations given in both ICH genotoxicity guidances. In this regard, ICH guidances should facilitate appropriate interpretation of genotoxicity test results. Some issues of regulatory interpretation are discussed below, on the basis of experience with submissions to the Federal Institute for Drugs and Medical Devices (BfArM) of Germany.

1. Experience with Submissions for New Pharmaceuticals

Table 2 gives an overview of the number of tests that have been submitted in standard battery genotoxicity test categories for 335 new pharmaceutical entities within the 8 years from 1990–1997. "New" in this context means "active ingredient not previously available on the German pharmaceutical market." In the period analyzed, mutagenicity testing of pharmaceuticals was performed in accordance with the EU guideline (40) still valid at the time, encompassing four tests from the following categories:

- Test for gene mutation in bacteria

TABLE 2. Results of Standard Battery Genotoxicity Tests for New Pharmaceuticals

System	No. of compounds tested	No. of compounds with positive test results	% Compounds with positive test results
Bacterial reverse mutation	298	23	7.7%
MLA *tk* assay	104	28	26.9%
HPRT test	162	6	3.7%
In vitro CA	266	77	28.9%
In vivo bone marrow	283	19	6.7%

Data are given for 335 new pharmaceutical entities submitted for registration to the BfArM between 1991 and 1997.
Abbreviations: MLA, mouse lymphoma assay; HPRT, hypoxanthine guanine phosphoribosyl transferase assay.

ICH Background and Practice

- Test for gene mutation in a eukaryotic system
- Test for chromosomal aberrations in mammalian cells in vitro
- An in vivo test for genetic damage

In practice, a standard test battery with the following four tests was usually submitted:

- The *Salmonella* reverse mutation test
- HPRT mutation in CHO or V79 cells or a mouse lymphoma *tk* assay
- Chromosomal aberrations in a cell line or human lymphocytes
- A bone marrow test for micronuclei or chromosomal aberrations

Between 80% and 90% of the 335 new compounds have been tested in the four standard battery test categories. There are instances, however, when compounds were not tested for genotoxicity at all or were evaluated with use of a reduced battery. Some reasons for this include: (a) compounds such as peptides may a priori have been judged as devoid of genotoxic activity, (b) some compounds with a known genotoxicity activity profile have been subjected to a reduced test battery, (c) some antibacterial compounds have not been tested in the bacterial reverse mutation assay, (d) other genotoxicity tests, such as yeast assays or *UDS* (unscheduled DNA synthesis) tests, have been submitted instead of systems comprising the four standard categories. Table 2 includes data on the number of new active ingredients positive in one of the test systems that constitute the standard battery of genotoxicity tests, according to the prevailing EU guideline in place at the time. Twenty three of 335 new compounds yielded positive findings in the BRM test; 28 compounds produced positive results in the MLA, six were positive for HPRT mutations, 77 compounds produced chromosomal aberrations in mammalian cells in vitro (in vitro CA), and 19 compounds were positive for chromosomal damage in vivo in bone marrow cells. It is obvious that the frequency of positives exhibited by the MLA and in vitro CA systems (26.9% and 28.9%, respectively) exceeds that of the other standard battery systems by about a factor of 4. Figure 4 gives the percentage of in vitro clastogenic compounds among the 335 new pharmaceutical entities on a year-by-year basis. It appears that there is a gradual increase of in vitro clastogens in the period reviewed that culminates in more than 40% in 1997. The reasons for this increase are unclear (12).

Among these new pharmaceutical enitities, many compounds that were positive in the BRM specifically or some other in vivo test were judged to fullfil criteria of relevant genotoxic risk for humans (12). However, a detailed analysis of the available data for the MLA and the in vitro CA study results revealed some problematic aspects of relevance interpretation for a number of positives in these systems. These aspects include (a) genotoxicity that was only observed at high levels of cytotoxicity, and (b) evidence for threshold mechanisms of genotoxicity,

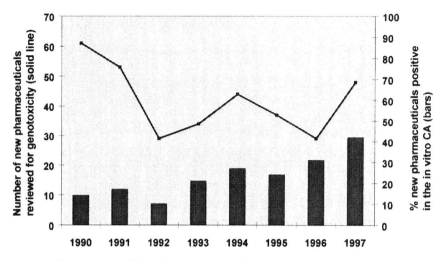

FIG. 4. Percentage of in vitro clastogenic new pharmaceutical compounds among submissions to the BfArM

two aspects of which were generally not observed/discussed in the context of interpretation of positive results from the BRM and the in vivo test (12).

2. Problems Associated with High Levels of Cytotoxicity

Cytotoxicity has been addressed by a number of authors as a phenomenon that may be (causally) associated with the induction of chromosomal damage. Vock et al (41) described cytotoxicity-related mechanisms of DNA double-strand break induction in vitro with several different agents. The association between a number of cytotoxicity parameters and chromosomal aberrations in vitro was analyzed in detail by a group at Merck Research Laboratories. Those researches concluded that there is a strong association between these two events for a number of compounds that they regard as non-DNA reactive. These compounds were designated as "thresholdable" in vitro clastogens (22, 42). Some of the 355 diverse compounds and chemical classes that were submitted for registration as new pharmaceuticals and analyzed by BfArM (12) seem to exhibit a similar relationship between these two events. A common feature for compounds that were positive for chromosomal aberrations in vitro but negative in the other assays of the standard ICH genotoxicity test battery was the association of clastogenicity and conspicuous cytotoxicity. The mechanisms of clastogenesis for some compound groups such as alkylating agents or nucleoside/-tide analogues are understood, but for the majority of in vitro clastogens, a cytotoxicity-based mechanism accounting for the clastogenic effects could not be established. It is difficult to verify in vivo the relationship between cytotoxicity and clastogenicity because chemical concentrations comparable to those achievable in vitro often cannot be achieved in the desired target tissues in vivo.

Two pharmaceutical compound groups, that is, angiotensin-converting enzyme (ACE) inhibitors and angiotensin II receptor antagonists, which include a high number of in vitro clastogenic members among the regulatory submissions for these groups, have been examined for possible events associated with the cytogenetic effects observed (12). Compared with other pharmaceuticals/chemicals that are also highly cytotoxic but not clastogenic, the ACE inhibitors and angiotensin II receptor antagonists did not show any specific features in association with their cytotoxicity, which could explain the high incidence of in vitro clastogens within these pharmacological groups. Whether these events are reciprocally related in a "cause-and-effect" manner or are simply manifest in consort remains to be determined.

It is therefore concluded that other class-specific, but as yet unidentified pharmacologically based properties, may render these compounds prone to clastogenicity. According to Galloway (22), DNA synthesis inhibition may be an effect that is very common to clastogens that are only positive at high levels of cytotoxicity. In an analysis of the in vitro genotoxic potential of antihistamines, Snyder (43) suggests that "the apparent genotoxicity of antihistamines and possibly many other pharmaceuticals derives from a hitherto unappreciated propensity of these drugs for stabilized intercalative DNA binding." On the basis of such proposals, several potential candidate mechanisms (not necessarily limited to cytotoxicity) could serve as the basis for the in vitro genotoxic effects associated with pharmaceuticals. For risk assessment, the question prevails whether the association of cytotoxicity and clastogenesis in vitro is suggestive of some type of a "threshold" and, hence, whether such results are relevant when concentrations encountered in vivo are much lower. In most cases, a determination of the validity of this association vis-à-vis threshold effects is not feasible because of the following reasons: (a) methods for determination of cytotoxicity are diverse and of varying reliability; (b) concentrations used in the definitive test are often based on cytotoxicity information derived from a pretest (and the pretest may use a methodology that differs from that of the definitive test); (c) often, the spacing of concentrations investigated does not allow us to identify exactly the genesis of genotoxic effects as it may relate to cytotoxicity; (d) in the same compound class, cytotoxic clastogens as well as cytotoxic nonclastogens can be identified. Therefore, information on alternative mechanisms involved in chromosome breakage could facilitate an assessment of the basis of positive findings. The ICH guidance S2A states that "at very low survival levels in mammalian cells, mechanisms other than direct genotoxicity per se can lead to 'positive' results that are related to cytotoxicity and not genotoxicity (e.g. events associated with apoptosis, endonuclease release from lysosomes, etc.)"

3. Thresholds in Genotoxicity

The concept that both nongenotoxic and genotoxic chemicals may act via diverse mechanisms appears to be gaining wider acceptance. One case in point is the view that some chemicals that induce DNA damage through nondirect mechanisms may exhibit thresholds. Some of these mechanisms are addressed in contributions to a special issue of *Mutation Research* (12, 21). Table 3 summarizes some mechanisms of genotoxicity that may involve nonlinear dose-effect relationships with thresholds (21, 44). Among the various chemical groups considered are compounds that interact with the mammalian, bacterial, or viral replication machinery, the clinical importance of which is apparent.

A prime example of such agents are nucleoside/nucleotide analogues, whose clinical importance is often related to anticancer treatment or control and suppression of viral infections such as HIV or hepatitis. This group of compounds is inherently genotoxic, but their mechanism of action is certainly different from that of other classes of genotoxic carcinogens. For example, all such compounds submitted for registration to BfArM were negative in the bacterial mutation test (12). Nucleoside/-tide analogues, when designed for use as drugs, usually require phosphorylation before they can be incorporated into the DNA, where they can then cause mutation or blockage of DNA synthesis. For antiviral agents, such as those 11 compounds submitted to the BfArM with positive in vitro CA data (12), the viral phosphorylase enzyme is usually many times more efficient than its mammalian counterpart. Consequently, incorporation of analogues into viral DNA occurs at concentrations significantly (orders of magnitude) below those needed to achieve incorporation into mammalian DNA. Hence,

TABLE 3. Some Indirect Mechanisms/Conditions for Genotoxicity That May Involve Nonlinear Dose Effects and/or Thresholds (see also 36)

- Changes in pH and osmolality
- Precipitation
- High levels of cytotoxicity
- Interaction with the replication process
  - Imbalance of DNA precursors
  - DNA synthesis inhibition
  - Inhibition of DNA repair enzymes
  - Inhibition of ribonucleotide reductase
  - Topoisomerase/gyrase inhibition
- Spindle interaction
- Inhibition of apoptosis
- Generation of active oxygen species
- Bacterial-specific metabolism (e.g., nitroreduction)

ICH Background and Practice

it is often possible to establish an acceptable safety margin for an antiviral nucleoside analogue. Scientific justification for the clinical use of such agents has involved individual risk-benefit analyses regarding their genotoxic and carcinogenic potential and has, at times, included threshold arguments. However, other extenuating factors may also be operational such as the conditions of use, the exposed population, and the processes followed by a given regulatory region, among others.

A specific example of such a situation involves nucleoside analogues that are intended for anticancer treatment (e.g., 5-fluorouracil [FU]). Such compounds act by incorporation after phosphorylation by mammalian enzymes, and much higher concentrations are often required than for effects on viruses. Thus, therapeutic concentrations can approach those capable of producing genotoxicity in mammalian cell test systems, thereby affording a relatively narrow safety margin. Whereas this may be acceptable for an anticancer drug (because of the benefits to be gained), the basis for use of such agents for other non-life-threatening conditions, such as psoriasis (21), appears to differ for different regulatory bodies.

C. The Future of Regulatory Genotoxicity Testing

The ICH guidances are derived from a continued development of pre-existing regulatory approaches in the three ICH regions. They describe a harmonized "status quo," but genetic toxicology is a rapidly developing field and other test methods will sooner or later replace the existing methods. Below, three areas of genotoxicity testing are addressed in which new methods may call for changes in guidance recommendations. Within ICH, a maintenance process has been established that will address updates and revisions of guidances when there is a recognized regulatory need and sufficient information is available upon which to take action.

1. Simplification in Hazard Identification

The search for new pharmaceutical principles usually involves the generation of hundreds or thousands of new compounds in a very short time. The techniques of combinatorial chemistry can even provide hundreds of chemicals in small quantities every week. Toxicity to the genetic apparatus is considered in most companies to be an important aspect of product safety. Consequently, there is a demand for simple and relatively rapid screening methods in genetic toxicology. Approaches such as the Ames II system (45), bacterial assays based on colorimetric changes (46), or the in vitro micronucleus with automated analysis (27, 47, 48) are under active consideration for high throughput screening. Such high throughput methods may ultimately prove to yield results as reliable as conventional genotoxicity tests but require less resources and in a shorter time frame. The in vitro micronucleus test using mammalian cell lines or human lymphocytes, for which an OECD guideline is being generated, offers the possibility to investigate chromosomal damage in vitro in a much shorter time than the traditional analysis of chromosomal aberrations. The incorporation of such methods into standard batteries of genotoxicity testing for agents such as pharmaceuticals is a potential option for the future and could lead to further simplification of the hazard identification process.

2. "Site of Contact" Evaluation of Genotoxicity

The reliability of the in vivo test is directly connected with the ability to measure genotoxic effects at important sites of metabolism and contact of reactive metabolites. In this regard, despite their demonstrated responsiveness and predicitivity of potentially genotoxic/carcinogenic agents, hematopoietic cells in the bone marrow may not be ideal targets for genetic toxicology testing despite the fact that the tissue is generally well perfused. It is known for some compounds that their ultimate carcinogenic metabolite(s) is only very short-lived and, therefore, is preferentially active, that is, genotoxic, carcinogenic, or both, only at the site where it is formed.

Dimethylnitrosamine, 2-nitropropane, and 2,4 dinitrotoluene are examples of such agents that yielded negative results in the bone marrow but are detected when liver cells are analyzed for genotoxic effects in vivo (49, 50). However, target tissue systems other than bone marrow, which may use more predictive target cells such as the liver unscheduled DNA synthesis (UDS) test, the Comet assay, ^{32}P post-labeling, or mutation detection in transgenes, suffer certain limitations. For example, they offer only an indirect measurement of mutations or chromosomal damage, or, as in the case of transgenes, there are not enough data on the comparability of mutation frequencies in transgenes (neutral reporter genes) and cell-own (coding) genes (51, 52). If scientifically justified, such systems under ICH guidance may be considered as optional in vivo test approaches to provide additional data if in vitro tests are positive and the rodent hematopoietic test does not allow a conclusive assessment.

Regarding the further in vivo assessment of genotoxicity beyond in vivo cytogenetic evaluation when positive in vitro test results are obtained, the ICH guidances acknowledge the paucity of "validated, widely used in vivo system[s] which measures gene mutation." They go on to state that a large database exists for the liver UDS assay. A review of the literature shows that a combination of the liver UDS test and the bone marrow micronucleus test will detect most genotoxic carcinogens with few false positive results (49). Still, false negative results with this combination of assays are known, for example, for very unstable genotoxic compounds and certain aromatic amines (49). Therefore, the guidances recommend that further in vivo testing should not be restricted to liver UDS tests, as other assays may be more appropriate (e.g., ^{32}P post-labeling, DNA strand-breakage assays, and others), depending on the compound in question. Although the guidances caution that it is important to recognize that for all of these in vivo endpoints, "their relationship to mutation is not precisely known," it could be argued that to the extent that that relationship is understood, the mutation rate cannot be directly inferred from the measurement of these endpoints.

3. Genetic Changes in Tumor-Related Genes

The determination of the likelihood of carcinogenic effects by damage to the genetic apparatus is one of the main objectives of genetic toxicology testing (see introductory paragraph to the ICH S2B guidance). Over the last several years, investigations into the mechanisms of tumor formation and the genes involved in the carcinogenesis process have generated much evidence in support of the importance of changes in tumor suppressor genes such as *p53* and in oncogenes such as *ras* (53, 54) Consequently, it seems reasonable to incorporate investigations into changes in such important genes into the standard testing program.

For screening purposes, this seems not to be necessary because the mutational characteristics of each chemical are, in principle, conserved across different genes. The pattern of mutations in terms of type and sequence specificity for a specific mutagen that can be found in the *hprt* gene, the *lacI* gene, or any tumor-related gene are largely identical (55). A mere substitution of conventional target genes by cancer-related genes will, therefore, in all likelihood, not give a significantly different genotoxicity profile of a compound or even improve the predictivity of potential carcinogenicity of a chemical.

Another important point is the known high diversity of different DNA lesions involved in carcinogenicity, which should be detected when screening for genotoxic carcinogens. No single cancer-related gene nor carcinogenic pathway appears suitable as an overall indicator whose induction could serve as the consummate marker for risk analysis. Thus, for screening, it makes less sense to look for unique genetic alterations in specific oncogenes than to seek detection of the broadest possible spectrum of DNA lesions. An assay such as the MLA, which offers the advantage of detecting a wide variety of genetic alterations, will sufficiently cover a substantial part of the lesions involved in carcinogenesis.

Investigations into mutations in tumor-related genes, although not necessary for screening purposes, may be useful for the interpretation of tumor findings in a carcinogenicity study. Notwithstanding the fact that such analyses are not currently a part of the ICH recommendations, one promising aspect in chemical mutagenesis in this context is the awareness that mutagens may cause a characteristic and unique mutational spectrum (53), a finding that has given rise to the term "mutagen signature" or "mutagen fingerprint" in DNA. This capability can be used to investigate the etiology/pathogenesis of induced tumors in rodent bioassays (in conventional 2-year studies as well as the new medium-term transgenic models) (56). In the simplest case, mutational spectrum analysis of a tumor-related gene, such as *p53*, from a carcinogen-induced tumor will show a compound-specific pattern of mutations that is not found in the same tumor type from control animals but which correlates well with patterns seen in in vitro or in vivo genotoxicity models. Such a finding would provide direct evidence for a genotoxic mechanism in the development of that specific tumor type.

A logical further development of this approach is the application to human tumors. There are several well-characterized examples of this experimental concept. One such example is the G:C to T:A transversions in the p53 gene at codon 249 on the third nucleotide in liver cancer in association with dietary exposure to aflatoxin B1 (57). Another example is the CC:GG to TT:AA tandem transitions in skin cancers in association with UVB exposure (58).

It is clear that this experimental approach of tumor mutation analysis in cancer-related genes goes far beyond routine mutagenicity testing and that further experience and the resolution of several issues regarding its application are needed. One such issue is the selection of genes to be analyzed, that is, determining which is (are) the relevant gene(s) for neoplastic development of a specific tumor. Important alterations may be missed when focusing on only one gene or even one exon. Besides specificity of carcinogen-DNA interaction, other important factors, such as selection processes, and repair of endogenous DNA damage, among others can influence the final set of mutations observed in tumors. It may be possible that mutation spectra are insufficiently distinctive for identification of the specific causative agent. However, further progress in this rapidly developing field of science should provide a better understanding of the potential uses of this technique for risk assessment procedures.

III. ICH CARCINOGENICITY GUIDANCES

Within the ICH process, three carcinogenicity guidances have been developed that address different aspects of regulatory carcinogenicity testing and assessment in lifetime rodent bioassays and alternative models. The "Guidance on the need for carcinogenicity studies of pharmaceuticals" (S1A) was finalized by the ICH EWG in late 1995. The guidance "Testing for carcinogenicity of pharmaceuticals" (S1B) was finalized in July 1997. The guidance "Dose selection for carcinogenicity studies of pharmaceuticals" (S1C) was already agreed upon in October 1994 and an addendum "Addition of a limit dose and related notes" was issued in July 1997.

A. Dose Selection for Carcinogenicity

One issue of major concern for the carcinogenicity debate was the many positive results observed in traditional rodent bioassays. A review of the U.S. Physician's Desk Reference showed that more than half of pharmaceuticals tested for carcinogenicity are reported to elevate the incidence of some type of tumors in rodents (59). Whereas many of these pharmaceuticals are limited to use in treating life-threatening diseases, many are not in this class. It is believed that a significant proportion of tumor incidence manifest in rodent bioassays results from the exposure of the animals to doses of test agent that may not be relevant for human exposure situations, for example, doses much higher than would be used in practice.

Ideally, the criteria used for determining dose levels in rodent carcinogenicity testing should facilitate detection of all important and relevant carcinogens but avoid tumor responses caused by disruptions in homeostasis that are considered to be irrelevant for the human exposure situation. This conflict cannot be easily resolved. For instance, very high levels of exposure had to be used to demonstrate the rodent tumorigenicity of phenacetin (60). This compound is considered to be both carcinogenic and genotoxic in vivo, but the underlying mechanism(s) of its tumorigenicity remain to be identified. The ICH guidance "Dose selection for carcinogenicity studies of pharmaceuticals" and its subsequent addendum provide guidance for selection of reasonable and justifiable dose levels in carcinogenicity studies. Specifically, the guidance stipulates that the doses selected "should provide an exposure to the agent that (1) allow an adequate margin of safety over the human therapeutic exposure, (2) are tolerated without significant chronic physiological dysfunction and are comparable with good survival, (3) are guided by a comprehensive set of animal and human data that focus broadly on the properties of the agent and the suitability of the animal (4) and permit data interpretation in the context of clinical use." In this context, the ICH guidance offers alternatives to the use of the maximally tolerated dose (MTD) that has been traditionally used for high-dose selection in rodent bioassays. One such alternative is to base high-dose selection on a 25-fold ratio of rodent to human plasma *AUC* (area under curve) of parent compound or metabolites (61). This was based, among other data, on an evaluation of *IARC* (International Agency on Research on Cancer) Category 1 and 2A carcinogens (61). All of these compounds were detected at exposure levels of at most 25-fold human/rodent body surface area dose ratio or systemic exposure ratio. An addendum to the guidance specifies that for genotoxic or suspected genotoxic compounds, a limit dose lower than the MTD cannot be accepted for rodent carcinogenicity testing. This seems reasonable, as it is important to rigidly investigate whether such compounds are tumorigenic. However, at MTD levels, nongenotoxic mechanisms may play a greater role than envisaged in tumor formation, even for genotoxic compounds. In this situation, it is important to clarify the underlying mechanism of tumor formation when observed.

B. Models for Testing for Carcinogenicity of Pharmaceuticals

1. Traditional Animal Models

Traditionally, the carcinogenic potential of chemicals is tested in rodent lifetime bioassays through the use of rats and mice. New pharmaceutical entities have usually been tested in both of these rodent species, but under ICH auspices, the usefulness of this approach was re-examined. An analysis of several databases (62, 63) has suggested that carcinogenicity results derived from rats versus mice may carry different degrees of impact on the overall decision-making process. For example, regulatory action on pharmaceuticals in some regions can be more heavily weighted on the carcinogenic response observed in the rat than the mouse, although not all regulatory bodies respond similarly. The benefit derived from conducting a two (rodent) species carcinogenicity bioassay is the identification of transspecies carcinogens, which, by virtue of this property, may be of human significance. Strategic attempts are being made to reduce the size of the carcinogenicity bioassay, for example, by limiting the number of species to one, specifically, more often the rat. A single-species bioassay usually would be accompanied by additional studies, which could include, among others, genetic toxicology and in vitro neoplastic transformation assays, mechanistic studies, alternative, relevant, short-term toxicological in vivo and in vitro assays, pharmacology and pharmacokinetic studies, available human data, a second rodent bioassay in another rodent species (if desired). The philosophy in support of this paradigm is that together, these studies would constitute a weight-of-evidence approach from which human risk could be assessed. Some of these alternative approaches are specified in Table 4 and discussed below.

2. Alternative Models to Detect Tumorigenic Effects

Traditional lifetime bioassays for determination of carcinogenicity are very costly and time-consuming. It takes about 3 years from the initial set-up of a study to completion, when all pathology data are finally assessed and reviewed. Because the endpoint of a carcinogenicity study in rodents, that is, tumor formation, is of utmost importance for the assessment of the carcinogenic potential of genotoxins and other possibly carcinogenic compounds, researchers have been interested in designing more efficient animal tumor models. In this regard, the ICH guidance "Testing for carcinogenicity of pharmaceuticals" is establishing a basis and serving as a prototype for regulatory acceptance of alternative

TABLE 4. Considerations on Testing of Pharmaceuticals for Carcinogenicity

The basic carcinogenicity testing principle	• One long-term (lifetime) rodent carcinogenicity bioassay • Appropriate species (based on pharmacology, repeated-dose toxicology, metabolism, toxicokinetics, route of administration) • Plus one other study that supplements the rodent long-term bioassay
A choice of supplementary studies	• Short or medium-term in vivo rodent test systems with tumor endpoints • A long-term rodent carcinogenicity study in a second species
Available short- or medium-term in vivo rodent test systems	Examples include: • Initiation-promotion model (e.g., for hepatocarcinogenesis) • Transgenic mouse model ($p53^{+/-}$, TG.AC, TgrasH2, XPA$^{-/-}$) • Neonatal rodent tumorigenicity model Other systems may be preferable on a case-by-case basis
Additional mechanistic studies	• Cellular changes • Biochemical measurements • Considerations for additional genotoxicity testing • Modified protocols (for the long-term bioassay)

models to the standard rodent lifetime bioassay. This guidance still emphasizes the importance of at least one lifetime bioassay; however, a number of alternative and more rapid in vivo tumor models are specifically addressed in Note 3 to this guidance.

3. Transgenic Animal Models

Transgenic animals carrying a proto-oncogene or animals lacking one or both alleles of a tumor suppressor or DNA repair gene are expected to be more susceptible to carcinogens and develop tumors faster than "normal" animals. Several transgenic animal models have been proposed for use as alternatives to the traditional rodent bioassay. Those that are undergoing further detailed characterization in several collaborative studies are: (1) $p53^{+/-}$-knockout mice, (2) v-Ha-ras transgenic mice (TG.AC mice), (3) XPA$^{-/-}$-deficient mice, and (4) human c-Ha-ras transgenic mice (TgHras2). The types of genes that have been knocked out (tumor suppressor gene, DNA repair gene) or inserted (ras oncogene) in these mouse models are all considered to be of major importance for human carcinogenesis. Current evidence suggests that tumor formation in these models can be observed for a variety of carcinogens within a 6-month period, which is considerably faster than the duration of the traditional (lifetime) rodent bioassays (64).

Based on knowledge of the genes involved and the way the models are constructed, response characteristics such as those summarized in Table 5 for various categories of presumed carcinogenic action, that is, "purely initiating," "complete genotoxic," and "nongenotoxic" carcinogens are pending. For instance, the presumed mechanism of chemically induced tumors in the heterozygous $p53+/-$ model may involve a loss of the remaining functional allele or other effects on the $p53$ gene pathway and will preferentially detect genotoxic carcinogens. On the other hand, the TgH*ras*2 mouse, which carries five to six copies of the human c-Ha-ras gene with its own promotor, is likely to be able to detect both genotoxic and nongenotoxic carcinogens (65). It is understood that most, if not all, genotoxic carcinogens exert some kind of tumor-promoting activity, which implies that the concept of a "purely initiating" carcinogen may be more or less academic. However, it is important to recognize that a single treatment with an efficient initiating carcinogen is sufficient to elucidate a tumor response in the newborn mouse model.

For interpretation of tumor data derived from these models, detailed knowledge of their strengths and limitations is needed. For example, the $p53$ tumor suppressor gene does not seem to play a major role in liver tumor development, for example, $p53$ heterozygosity does not enhance the rate of development of diethylnitrosamine-induced hepatocellular adenoma and carcinoma (66). It should also be noted that whereas one of the primary advantages of such tumor models is the accelerated development of tumors, the tumor expressed results from activation of a specific target(s). Therefore, the model used needs to be chosen carefully with particular regard to the pharmaceutical (pharmaceutical class) being evaluated. The potential role of transgenic animal mutation model systems, both in

TABLE 5. Short- and Medium-Term Rodent In Vivo Systems for Tumorigenicity and Their Expected Tumor Outcome for Genotoxic (DNA-Damaging) and Nongenotoxic (Tumor-Growth-Promoting) Carcinogens

Test model (duration)	"Pure" (genotoxic) initiator	"Complete" (genotoxic) carcinogen	Nongenotoxic carcinogen
TG.AC mice (6 months)	Negative	Positive	Positive
Tg rasH2 mice (6 months)	Negative (?)	Positive	Positive
$p53^{+/-}$ mice (6 months)	Positive (?)	Positive	Negative
XPA$^{-/-}$ mice (6 months)	Positive (?)	Positive	Negative
Newborn mice (1 year)	Positive	Positive	Negative

ICH Background and Practice

mutagenesis research and in the regulatory decision-making process, has been reviewed previously (67). The possibility of their use in carcinogenic risk assessment has also been explored (68).

The ICH guidance "Testing for carcinogenicity of pharmaceuticals" (S1B), in its Note 3, has introduced the potential for regulatory use of transgenic animal models, such as those mentioned above for carcinogenicity assessment of pharmaceuticals. The reliability of the different transgenic tumor models is under investigation in several studies, including an international collaborative study coordinated by the International Life Science Institute (ILSI), Health and Environmental Science Institute (HESI), Alternatives to Carcinogenicity Testing Committee. Contrera and DeGeorge (64) have pointed out that, "Based on available information there is sufficient experience with some in vivo transgenic rodent carcinogenicity models to support their application as complementary second species studies in conjunction with a single 2-year rodent carcinogenicity study." They go on to state, "The use of the in vivo transgenic mouse model in place of a second 2-year mouse study will improve the assessment of carcinogenic risk by contributing insights into the mechanisms of tumorigenesis and potential human relevance not available from a standard 2-year bioassay."

Certain of the regulatory authorities in the ICH regions have established a system by which to provide scientific advice to the private sector regarding the use of alternative toxicological methods. Such guidance is also available to pharmaceutical companies that intend to use alternative models for carcinogenicity testing.

4. Other Alternative In Vivo Models

In addition to transgenic tumor models, the ICH guidance S1B mentions, in its Note 3, two specific tumor models that may be used to replace a traditional rodent bioassay, that is, initiation-promotion test methodlogies and the neonatal rodent tumorigenicity model (newborn mouse model). Neither approach to test for tumorigenicity is new, and their suitability has been assessed in ICH negotiations on the background of a large database (69, 70). The newborn mouse model seems to be suitable primarily for detection of clearly genotoxic compounds, that is, "pure initiators" and "complete carcinogens" (Table 5), because the limitation of treatment to two to three doses over the first 2 weeks of life would generally not facilitate detection of nongenotoxic mechanisms of tumorigenesis.

5. In Vitro Models to Detect Cell Transformation

Several in vitro assays for malignant cell transformation have been developed in the past, the majority of which use fibroblastic target cells. Among them, the Syrian hamster embryo cell transformation assay (SHE assay) makes use of early passage (most often primary, secondary, or tertiary) cultures of embryo cells, which, upon treatment with known carcinogens, are "transformed" into morphologically anomalous cells, recognized by their atypical clonal growth patterns into morphologically aberrant colonies (71–73). Other assays such as the BALB/3T3 (74) and the C3H10T½ (75) assays use established mouse cell lines. Aberrant cellular growth patterns atop confluent, contact-inhibited, monolayers of these cell lines are manifest as foci of transformed, misoriented, non-contact-inhibited cells.

Each of these in vitro transformation assay systems offers advantages and disadvantages as well as differences in their inherent capacity to metabolize (activate/detoxify) compounds to bioreactive metabolites or respond to metabolites generated by an exogenous metabolic supplement (76, 77). Transformed cell populations derived by expansion of transformed colonies or foci are tumorigenic in vivo, which is the basis for the use of such assays to predict carcinogenic potential of chemicals. These and other properties of transformation assay systems methodologies as well as surveys of their respective carcinogenic predicitivity have been reported (78–83).

Irrespective of the transformation system, however, the mechanism(s) of the transformation event has not been fully elucidated (83, 84). Furthermore, because the overwhelming majority of rodent and human tumors is of epithelial origin, it would appear that the predictivity of in vitro transformation systems (the preponderance of which rely on fibroblastic cell types) is independent of in vivo tumor type or tumor etiology. A successful malignant transformation of human primary cells of epithelial (embryonic kidney) and fibroblastic (foreskin) origin has been reported (85). In addition to loss of tumor suppressor gene function (P53 and RB) and gain of oncogenic activity (ras), expression of telomerase activity was found to be essential for creating tumorigenic human cells from normal ones (85).

Although in vitro cell transformation assays seem to be good predictors of animal carcinogenicity (80, 82, 86, 87), their added contribution in assessing risk remains obscure. One possible advantage of cell transformation assays vis-à-vis genetic toxicology assays as screening systems for carcinogens may be related to the apparent ability of the former to detect nongenotoxic carcinogens (80). However, it is not clear whether the supposed mechanism(s) of nongenotoxic carcinogenesis in rodents and in cell transformation assays in vitro are the same. The ICH carcinogenicity guidance S1C makes reference to the use of in vitro cell transformation assays for prediction of carcinogenicity "at the compound selection stage," and thereby encourages the use of such assays along with the routinely used standard genetic toxicology screening tests.

C. The Future of Regulatory Carcinogenicity Testing

1. Integrated Approaches to Evaluate Genotoxicity and Carcinogenicity

The preponderance of available scientific information suggests that cancer is a process in which genetic changes are inevitably involved (88). The testing of

ICH Background and Practice

compounds for their genotoxic potential mainly addresses safety concerns regarding a carcinogenic potential by induction of irreversible changes in the genetic material of cells as the initial step in the carcinogenesis process. The new ICH guidances on genotoxicity and carcinogenicity represent a more integrated approach to these fields than has previously existed. For instance, it is acknowledged in the ICH guidance that "unequivocally genotoxic compounds, in the absence of other data, are presumed to be trans-species carcinogens, implying a hazard to humans. Such compounds need not be subjected to long-term carcinogenicity studies." To allow for this type of situation, appropriate product labeling has been put into practice to reflect the presumed carcinogenicity on the basis of positive genotoxicity, when other (e.g., carcinogenicity) results are lacking. Both the ICH genotoxicity and carcinogenicity guidances seek additional genotoxicity data "for compounds that were negative in the standard test battery (for genotoxicity) but which have shown effects in a carcinogenicity test with no clear evidence for an epigenetic mechanism." Such additional data may come from investigations into DNA adduct formation, mutation induction in transgenes, or changes in tumor-related genes. In addition, new transgenic tumor models may provide other options for evaluating potential carcinogenicity of compounds demonstrated to be genotoxic, such as the $p53^{+/-}$ knockout model (64).

2. Carcinogenicity Testing of Nongenotoxic Compounds

Mechanistic understanding of the action of human carcinogens, that is, IARC Category 1 carcinogens, suggests that such agents may be subdivided roughly into five carcinogenic groups: (1) genotoxic compounds, (2) fibers, (3) compounds resulting in hormonal imbalances, (4) immunosuppressants, (5) mitogens. Representatives of these groups, with the exception of carcinogenic fibers, are also found among pharmaceuticals (89). The question may be posed whether future experience and a more detailed epidemiological analysis of human carcinogenesis will add to the list of carcinogens others that may produce tumors by completely different modes of action. For the present, however, it seems reasonable to concentrate mainly on the detection of possible human pharmaceutical carcinogens that operate through the above-named mechanisms. It is understood that the majority of genotoxic carcinogens are identifiable by the genotoxicity test battery plus short-term tumor models such as the $p53^{+/-}$ knockout mouse (see previous sections). Conversely, the identification of relevant nongenotoxic carcinogens seems to necessitate the exploitation of alternative approaches.

The current database for the rodent bioassay includes results for each of the above-specified mechanisms. Yet, the database includes compounds whose tumorigenicity in rodents may not be relevant for human exposure situations. Induction of liver tumors in rats through peroxisome proliferation, induction of mammary tumors in rodents by enhanced secretion of prolactin, induction of liver tumors in rodents by effects on the cytochrome p450 system (barbiturates) are examples of tumor responses in rodents from exposure to pharmaceuticals that are believed not to be relevant for the patient situation. Information on all of these probable mechanisms can be derived from toxicological studies other than the rodent bioassay. Further investigations into existing and evolving alternative predictive models may ultimately provide the desired knowledge to supplement an abridged version of the rodent bioassay. It is expected that such an integrated approach should, in most instances, accurately predict tumor outcome. When that situation is achieved, the conduct of rodent bioassays in two species on new compounds in well-defined pharmacological classes of action may, at times, be considered superfluous. Monroe's conclusion that "if a pharmaceutical is not (i) genotoxic, (ii) an immunosuppressant, (iii) hormonal, or (iv) a chronic irritant at human exposure levels, it is unlikely to pose a carcinogenic hazard to humans" may be seen as an extreme position (90).

The ICH guidances emphasize the importance of using all corroborative information when designing the rodent bioassay experiment for carcinogenicity assessment. The future, however, could witness the replacement of the rodent bioassay completely by an assessment of the "likelihood of tumorigenesis" using "alternative" information from pharmacological and toxicological studies plus various short-term tumor models without any forfeiture of safety. In this regard, transgenic short-term tumor models capable of detecting both genotoxic effects and relevant nongenotoxic mechanisms would be most advantageous. Current evidence suggests that the human c-Ha-*ras* transgenic mouse model fulfills some of these conditions, because it may not respond to nongenotoxic mouse-only carcinogens (65). Further pursuit of such model systems may, hopefully, abbreviate the effort, time, and expense, as well as limit the use of animals, associated with current approaches to assess carcinogenic potential. In fact, recent developments already indicate that short-term transgenic tumor models could ultimately replace the traditional rodent bioassays for assessing both genotoxic and nongenotoxic carcinogenesis (64, 65, 91, 92). As explained by Contrera and De-George (64), there are several potential and promising applications of alternative assay systems for assessing carcinogenicity. Some of these are to serve as (a) surrogates for the 2-year rodent carcinogenicity study performed in a second species; (b) complementary and confirmatory tests for equivocal carcinogenicity results derived from the rodent bioassay; (c) prescreens for carcinogenic potential before conducting comprehensive carcinogenicity studies; (d) an option to repeating an inadequate, unacceptable, or uninformative lifetime rodent bioassay; (e) methods for determining the carcinogenic activity of newly introduced genotoxic impurities or degradants present in the final marketed product that may not have been present at the time that the rodent carcinogenicity bioassay was performed. With these encouraging prospects, it is not unreasonable to predict that the future of regulatory decision-making for pharmaceuticals will rely heavily on alternative methodologies and mechanistic investigations on issues of of genotoxic and nongenotoxic carcinogenesis, and that this reliance could supplant the current dependence on guidance-conforming performance of rodent carcinogenicity bioassays.

REFERENCES

1. PF D'Arcy, DWG Harron. Proceedings of the First International Conference on Harmonisation (ICH). Antrim, N. Ireland: Greystone Books Ltd, 1992.
2. PF D'Arcy, DWG Harron. Proceedings of the Second International Conference on Harmonisation (ICH). Antrim, N. Ireland: Greystone Books Ltd, 1994.
3. PF D'Arcy, DWG Harron. Proceedings of the Third International Conference on Harmonisation (ICH), Antrim, N. Ireland: Greystone Books Ltd, 1996.
4. PF D'Arcy, DWG Harron. Proceedings of the Fourth International Conference on Harmonisation (ICH). Antrim, N. Ireland: Greystone Books Ltd, 1998.
5. L Müller, Y Kikuchi, G Probst, L Schechtman, H Shimada, T Sofuni, D Tweats. ICH-harmonised guidances on genotoxicity testing of pharmaceuticals: Evolution, reasoning and impact. Mutat Res 436:195–225, 1999.
6. OECD: Ninth Addendum to the OECD Guidelines for the Testing of Chemicals. Published by OECD, Paris, February, 1998.
7. CS Aaron, G Bolcsfoldi, H-R Glatt, M Moore, Y Nishi, L Stankowski, J Theiss, E Thompson. Mammalian cell gene mutation assays working group report. Mutat Res 312:235–239, 1994.
8. J Hozier, J Sawyer, D Clive, M Moore. Chromosome 11 aberrations in small colony L5178Y TK$^{-/-}$ mutants early in their clonal history. Ann N Y Acad Sci 107:423–425, 1985.
9. J Hozier, J Sawyer, D Clive, M Moore. Cytogenetic distinction between the TK$^+$ and TK$^-$ chromosomes in the L5178Y TK$^{+/-}$-3.7.2C cell line. Mutat Res 105:451–456, 1982.
10. ML Applegate, MM Moore, CB Broder, A Burrell, G Juhn, KL Kasweck, P-F Lin, A Wadhams, JC Hozier. Molecular dissection of mutations at the heterozygous thymidine kinase locus in mouse lymphoma cells. Proc Natl Acad Sci U S A 87:51–55, 1990.
11. RD Combes, H Stopper, WJ Caspary. The use of L5178Y mouse lymphoma cells to assess the mutagenic, clastogenic and aneugenic properties of chemicals. Mutagenesis 10:403–408, 1995.
12. L Müller, P Kasper. Human biological relevance and the use of threshold-arguments in regulatory genotoxicity assessment: Experience with pharmaceuticals. Mutat Res, 1999, in press.
13. A Matsuoka, K Yamakage, H Kusakabe, S Wakuri, M Asakura, T Noguchi, T Sugiyama, H Shimada, S Nakayama, Y Kasahara, Y Takahashi, KF Miura, M Hatanaka, M Ishidate Jr, T Morita, K Watanabe, K Hara, K Odawara, N Tanaka, M Hayashi, T Sofuni. Re-evaluation of chromosomal aberration induction on nine mouse lymphoma assay "unique positive" NTP carcinogens. Mutat Res 369:243–252, 1996.
14. T Sofuni, M Honma, M Hayashi, H Shimada, N Tanaka, S Wakuri, T Awogi, K Yamamoto, Y Nishi, M Nakadate. Detection of *in vitro* clastogens and spindle poi-

sons by the mouse lymphoma assay using the microwell method: Interim report of an international collaborative study. Mutagenesis 11:349–355, 1996.
15. M Honma, M Hayashi, H Shimada, N Tanaka, S Wakuri, T Awogi, KI Yamamoto, N-U Kodani, Y Nishi, M Nakadate, T Sofuni. Evaluation of the mouse lymphoma *tk* assay (microwell method) as an alternative to the *in vitro* chromosomal aberration test. Mutagenesis 14:5–22, 1999.
16. M Honma, L-S Zhang, H Sakamoto, M Ozaki, K Takeshita, M Momose, M Hayashi, T Sofuni. The need for long-term treatment in the mouse lymphoma assay. Mutagenesis 14:23–29, 1999.
17. H Shimada. Sensitivity and specificity of the extended exposure protocol of the MLA. In: PF D'Arcy, DWG Harron, eds. Proceedings of the Fourth International Conference on Harmonisation (ICH). Antrim, N. Ireland: Greystone Books Ltd, 1998, pp 259–261.
18. T Sofuni, P Wilcox, H Shimada, J Clements, M Honma, D Clive, M Green, V Thybaud, RHC San, BM Elliot, L Müller. Mouse lymphoma workshop: Victoria, British Columbia, Canada, March 27, 1996. Protocol issues regarding the use of the microwell method of the mouse lymphoma assay. Environ Mol Mutagen 29:434–438, 1997.
19. D Clive, G Bolcsfoldi, J Clements, J Cole, M Honma, J Majeska, M Moore, L Müller, B Myhr, T Oberly, M-C Oudelhkim, C Rudd, H Shimada, T Sofuni, V Thybaud, P Wilcox. Consensus agreement regarding protocol issues discussed during the mouse lymphoma workshop: Portland, Oregon, May 7, 1994. Environ Mol Mutagen 25:165–168, 1995.
20. SM Galloway, MJ Aardema, M Ishidate Jr, JL Ivett, DJ Kirkland, T Morita, P Mosesso, T Sofuni. Report from working group on *in vitro* tests for chromosomal aberrations. Mutat Res 312:241–261, 1994.
21. DJ Kirkland, L Müller. Interpretation of the biological relevance of genotoxicity test results: The importance of thresholds. Mutat Res, 1999, in press.
22. SM Galloway, JE Miller, MJ Armstrong, CL Bean, TR Skopek, WW Nichols. DNA synthesis inhibition as an indirect mechanism of chromosome aberrations: Comparison of DNA-reactive and non-DNA-reactive clastogens. Mutat Res 400:169–186, 1998.
23. M Kirsch-Volders, ed. The CB *in vitro* micronucleus assay in human lymphocytes. Mutat Res Special Issue 392, 1–210, 1997.
24. M Kirsch-Volders, A Elhajouji, E Cundari, P van Hummelen. The *in vitro* micronucleus test: A multi-endpoint assay to detect simultaneously mitotic delay, apoptosis, chromosome breakage, chromosome loss and non-disjunction. Mutat Res 392:19–30, 1997.
25. B Miller, S Albertini, F Locher, V Thybaud, E Lorge. Comparative evaluation of the *in vitro* micronucleus test and the *in vitro* chromosome aberration test: Industrial experience. Mutat Res 392:45–59, 1997.
26. B Miller, F Pötter-Locher, A Seelbach, H Stopper, D Utesch, S Madle. Evaluation of the *in vitro* micronucleus test as an alternative to the *in vitro* chromosomal aberration assay: Position of the GUM working group on the *in vitro* micronucleus test. Mutat Res 410:81–116, 1998.

ICH Background and Practice

27. D Roman, F Locher, W Suter, A Collier, M Bobadilla. Evaluation of a new procedure for the flow cytometric analysis of *in vitro*, chemically induced micronuclei in V79 cells. Environ Mol Mutagen 32:387–396, 1998.
28. T Morita, N Asano, T Awogi, YF Sasaki, S-I Sato, H Shimada, S Sutou, T Suzuki, A Wakata, T Sofuni, M Hayashi. Evaluation of the rodent micronucleus assay in the screening of IARC carcinogens (Groups 1, 2A and 2B). The summary report of the 6th collaborative study by CSGMT/JEMS/MMS. Mutat Res 389:3–122, 1997.
29. RJ Preston, W Au, MA Bender, JG Brewen, AV Carrano, JA Heddle, AF McFee, S Wolff, JS Wassom. Mammalian in vivo and in vitro cytogenetic assays: A report of the U.S. EPA's Gene-Tox Program. Mutat Res 87:143–188, 1981.
30. KH Mavournin, DH Blakey, MC Cimino, MF Salamone, J Heddle. The in vivo micronucleus assay in mammalian bone marrow and peripheral blood: A report of the U.S. Environmental Protection Agency Gene-Tox Program. Mutat Res 239:29–80, 1990.
31. GS Probst. Validation of target tissue exposure for in vivo tests. In: PF D'Arcy, DWG Harron, eds. Proceedings of the Second International Conference on Harmonisation (ICH). Antrim, N. Ireland: Greystoke Books Ltd, 1994, pp 249–252.
32. JT MacGregor, R Schlegel, NW Choy, CM Wehr. Micronuclei in circulating erythrocytes: A rapid screen for chromosomal damage during routine toxicity testing in mice. In: AW Hayes, RC Schnell, TS Miya, eds. Developments in the Science and Practice of Toxicology. Amsterdam: Elsevier, 1983, pp 555–558.
33. The Collaborative Study Group for the Micronucleus Test, CSGMT. Micronucleus test with mouse peripheral blood erythrocytes by acridine orange supravital staining: The summary report of the 5th collaborative study by CSGMT/JEMS:MMS. Mutat Res 278:83–98, 1992
34. A Wakata, Y Miyamae, S Sato, T Suzuki, T Morita, N Asano, T Awogi, K Kondo, M Hayashi. Evaluation of the rat micronucleus test with bone marrow and peripheral blood: Summary of the 9th collaborative study by CSGMT/JEMS:MMS. Environ Mol Mutagen 32:84–100, 1998
35. JT MacGregor, CM Wehr, PR Henika, MD Shelby. The in vivo erythrocyte micronucleus test: Measurement at steady state increases assay efficiency and permits integration with toxicity studies. Fund Appl Toxicol 14:513–522, 1990.
36. S Hamada, S Sutou, T Morita, A Wakata, S Asanami, S Hosoya, S Ozawa, K Kondo, M Nakajima, H Shimada, K Osawa, Y Kondo, N Asano, S Sato, H Tamura, N Yajima, R Marshall, C Moore, DH Blakey, JL Weaver, DK Torous, R Proudlock, S Ito, C Namiki, M Hayashi. Evaluation of the rodent long-term micronucleus assay: Summary of the 13th collaborative study by CSGMT/JEMS-MMS. Environ Mol Mutagen, 1999, submitted.
37. JA Heddle, MC Cimino, M Hayashi, F Romagna, MD Shelby, JD Tucker, PH Vanparys, JT MacGregor. Micronuclei as an index of cytogenetic damage: Past, present, and future. A summary report of the micronucleus workshop at the 22nd annual meeting of the Environmental Mutagen Society, April, 1991. Environ Mol Mutagen 18:277–291, 1991.
38. JT MacGregor, JD Tucker, DA Eastmond, AJ Wyrobeck. Integration of cytogenetic assays with toxicology studies. Environ Mol Mutagen 25:328–337, 1995.

39. ML Garriott, JD Brunny, DEF King, JW Parton, LS Schwier. The in vivo rat micronucleus test: Integration with a 14-day study. Mutat Res 342:71–76, 1995.
40. Anon. Testing of Medicinal Products for their Mutagenic Potential. The Rules Governing Medicinal Products in the European Union. Vol 3B, 1998 edition, pp 45–50.
41. EH Vock, WK Lutz, P Hormes, HD Hoffmann, S Vamvakas. Discrimination between genotoxicity and cytotoxicity in the induction of DNA double-strand breaks in cells treated with etoposide, melphalan, cisplatin, potassium cyanide, Triton X-100 and γ-irradiation. Mutat Res 413:83–94, 1998.
42. CA Hilliard, ML Armstrong, CI Bradt, RB Hill, SK Greenwood, SM Galloway. Chromosome aberrations *in vitro* related to cytotoxicity of non-mutagenic chemicals and metabolic poisons. Environ Mol Mutagen 31:316–326, 1998.
43. RD Snyder. A review and investigation into the mechanistic basis of the genotoxicity of antihistamines. Mutat Res 411:235–248, 1998.
44. D Scott, SM Galloway, RR Marshall, M Ishidate Jr, D Brusick, J Ashby, BC Myhr. Genotoxicity under extreme culture conditions. A report from ICPEMC task group 9. Mutat Res 257:147–204, 1991.
45. P Gee, CH Sommers, AS Melick, XM Gidrol, MD Todd, RB Burris, ME Nelson, RC Klemm, E Zeiger. Comparison of responses of base-specific Salmonella tester strains with the traditional strains for identifying mutagens: The results of a validation study. Mutat Res 412:115–130, 1998.
46. NF Cariello, S Narayanan, P Kwanyuen, H Muth, WM Casey. A novel bacterial reversion and forward mutation assay based on green fluorescent protein. Mutat Res 414:95–105, 1998.
47. SD Dertinger, DK Torous, KR Tometsko. Simple and reliable enumeration of micronucleated reticulocytes with a single-laser flow cytometer. Mutat Res 371:283–292, 1996.
48. SD Dertinger, DK Torous, KR Tometsko. Flow cytometric analysis of micronucleated reticulocytes in mouse bone marrow. Mutat Res 390:257–262, 1997.
49. DJ Tweats. Follow-up of *in vitro* positive results. In: PF D'Arcy, DWG Harron, eds. Proceedings of the Second International Conference on Harmonisation (ICH). Antrim, N. Ireland: Greystoke Books Ltd, 1994, pp 240–244.
50. S Madle, SW Dean, U Andrae, G Brambilla, B Burlinson, DJ Doolittle, C Furihata, T Hertner, CA McQueen, H Mori. Recommendations for the performance of UDS tests *in vitro* and *in vivo*. Mutat Res 312:263–285, 1994.
51. JHM van Delft, A Bergmans, FJ van Dam, AD Tates, L Howard, DJ Winton, RA Baan. Gene mutation assays in λ*lacZ* transgenic mice: Comparison of *lacZ* with endogenous genes in splenocytes and small intestinal epithelium. Mutat Res 415:85–96, 1998.
52. MG Manjanatha, SD Shelton, A Aidoo, LE Lyn-Cook, DA Casciano. Comparison of in vivo mutagenesis in the endogenous *hprt* gene and the *lacI* transgene of Big Blue® rats treated with 7,12-dimethylbenz(a)anthracene. Mutat Res 401:165–178, 1998.
53. MS Greenblatt, WP Bennett, M Hollstein, CC Harris. Mutations in the *p53* tumor suppressor gene: Clues to cancer etiology and molecular pathogenesis. Cancer Res 54:4855–4878, 1994.
54. LA Stanley. Molecular aspects of chemical carcinogenesis: The roles of oncogenes and tumor suppressor genes. Toxicology 96:173–194, 1995

55. H Okonogi, T Ushijima, XB Zhang, JA Heddle, T Suzuki, T Sofuni, JS Felton, JD Tucker, T Sugimura, M Nagao. Agreement of mutational characteristics of heterocyclic amines in lacI of the Big Blue® mouse with tumor related genes in rodents. Carcinogenesis 18:745–748, 1997.
56. JK Dunnick, RA Herbert, JF Hardisty, J Seely, EM Furedi-Machacek, JF Foley, S Stasiewicz, G Lacks, JE French. Phenolphthalein rapidly induces malignant hematopoietic tumors and loss of heterozygosity in the $p53$ wild type allele in heterozygous $p53$ deficient (+/–) mice. Toxicol Pathol 25:533–540, 1997.
57. H-M Chen, C-N Ong. Mutations of the $p53$ tumor suppressor gene and ras oncogenes in aflatoxin hepatocarcinogenesis. Mutat Res 366:23–44, 1996.
58. HN Ananthaswamy, SM Loughlin, P Cox, RL Evans, SE Ullrich, ML Kripke. Sunlight and skin cancer: Inhibition of $p53$ mutations in UV-irradiated mouse skin by sunscreens. Nat Med 3:510–514, 1997.
59. TS Davies, AM Monro. Marketed human pharmaceuticals reported to be tumorigenic in rodents. J Am Coll Toxicol 14:90–107, 1995.
60. Anon. Phenacetin (Group 2A) and analgesic mixtures containing phenacetin (Groups 1). IARC Monograph Suppl 7: 310–312, 1987.
61. JF Contrera. How was the relative systemic exposure ratio derived? In: PF D'Arcy, DWG Harron, eds. Proceedings of the Second International Conference on Harmonisation (ICH). Antrim, N. Ireland: Greystoke Books Ltd, 1994, pp 265–268.
62. AM Monro. Testing for carcinogenic potential. Rapporteur's report. In: PF D'Arcy, DWG Harron, eds. Proceedings of the Third International Conference on Harmonisation (ICH). Antrim, N. Ireland: Greystoke Books Ltd, 1996, pp 261–268.
63. JPJ van Oosterhout, JW van der Laan, EJ de Waal, K Olejniczak, M Hilgenfeld, V Schmidt, R Bass. The utility of two rodent species in carcinogenic risk assessment of pharmaceuticals in Europe. Regul Toxicol Pharmacol 25:6–17, 1997.
64. JF Contrera, JJ DeGeorge. *In vivo* transgenic bioassays and assessment of the carcinogenic potential of pharmaceuticals. Environ Health Perspec 106:71–80, 1998.
65. S Yamamoto, K Urano, H Koizumi, S Wakana, K Hioki, K Mitsumori, Y Kurokawa, Y Hayashi, T Nomura. Validation of transgenic mice carrying the human prototype c-Ha-ras gene as a bioassay model for rapid carcinogenicity testing. Environ Health Perspec 106 (suppl 1):57–69, 1998.
66. CJ Kemp. Hepatocarcinogenesis in $p53$-deficient mice. Mol Carcinog 12:132–136, 1995.
67. JT MacGregor. Transgenic animal models for mutagenesis studies: Role in mutagenesis research and regulatory testing. Environ Mol Mutagen 32:106–109, 1998.
68. P Schmezer, C Eckert, UM Liegibel, RG Klein, H Bartsch. Use of transgenic mutation test systems in risk assessment of carcinogens. Arch Toxicol 20:321–330, 1998.
69. K Mitsumori. Evaluation of new models. I. Initiation and promotion models and the H*ras*2 mouse model. In: PF D'Arcy, DWG Harron, eds. Proceedings of the Fourth International Conference on Harmonisation (ICH). Antrim, N. Ireland: Greystone Books Ltd, 1998, pp 263–272.
70. JS MacDonald. Evaluation of new models II. In: PF D'Arcy, DWG Harron, eds. Proceedings of the Fourth International Conference on Harmonisation (ICH). Antrim, N. Ireland: Greystone Books Ltd, 1998, pp 272–277.
71. Y Berwald, L Sachs. In vitro transformation of normal cells to tumor cells by carcinogenic hydrocarbons. J Natl Cancer Inst 35:641–661, 1965.

72. RJ Pienta, JA Poiley, WB Lebhertz III. Morphological transformation of early passage golden Syrian hamster embryo cells derived from cryopreserved primary cultures as a reliable in vitro bioassay for identifying diverse carcinogens. Int J Cancer 19:642–655, 1977.
73. RJ Isfort, GA Kerckaert, DB Cody, J Carter, KE Driscoll, RA LeBoeuf. Isolation and biological characterization of morphological transformation-sensitive Syrian hamster embryo cells. Carcinogenesis 17:997–1005, 1996.
74. T Kakunaga. A quantitative system for assay of malignant transformation by chemical carcinogens using a clone derived from BALB/3T3. Int J Cancer 12:463–473, 1973.
75. CA Reznikoff, JS Bertram, DS Brankow, C Heidelberger. Quantitative and qualitative studies of chemical transformation of cloned C3H mouse embryo cells sensitive to postconfluence inhibition of cell division. Cancer Res 33:3239–3249, 1973.
76. LM Schechtman, RE Kouri. Control of benzo(a) pyrene-induced mammalian cell cytotoxicity, mutagenesis and transformation by exogenous enzyme fractions. In: D Scott, BA Bridges, FH Sobels, eds. Progress In Genetic Toxicology, Proceedings of the Second International Conference on Environmental Mutagens, New York: Elsevier/North Holland, 1977, pp 307–316.
77. LM Schechtman. Metabolic activation of procarcinogens by subcellular enzyme fractions in the C3H 10T1/2 and BALB/c 3T3 cell transformation systems. In: T Kakunaga, H Yamasaki, eds. Transformation Assay of Established cell lines: Mechanisms and application. IARC Scientific Publications No. 67, International Agency for Research on Cancer, Lyon, 1985, pp 137–162.
78. VC Dunkel, SHH Swierniga, RL Brillinger, JPW Gilman, ER Nestman. Recommended protocols based on a survey of current practice in genotoxicity testing laboratories: III. Cell transformation in C3H/10T1/2 mouse embryo cells, BALB/c 3T3 mouse fibroblasts and Syrian hamster embryo cell cultures. Mutat Res 246:285–300, 1991.
79. EJ Matthews, JW Spalding, RW Tennant. Transformation of BALB/c 3T3 cells: V. Transformation responses of 168 chemicals compared with mutagenicity in Salmonella and carcinogenicity in rodent bioassays. Environ Health Perspec 101(suppl 2):347–482, 1993.
80. H Yamasaki, ed. Use of the Syrian hamster embryo (SHE) cell transformation assay for predicting the carcinogenic potential of chemicals. Mutat Res Special Issue 356, 1996.
81. LM Schechtman. BALB/c 3T3 cell transformation: Protocols, problems and improvements. In: T Kakunaga, H Yamasaki, eds. Transformation assay of established cell lines: Mechanisms and application. IARC Scientific Publications No. 67, International Agency for Research on Cancer, Lyon, 1985, pp 165–184.
82. C Heidelberger, AE Freeman, RJ Pienta, A Sivak, JS Bertram, BC Casto, VC Dunkel, MW Francis, T Kakunaga, JB Little, LM Schechtman. Cell transformation by chemical agents—a review and analysis of the literature. A report of the US Environmental Protection Agency Gene-Tox Program. Mutat Res 114:283–385, 1983.
83. T Kakunaga, H Yamasaki. Transformation assay of established cell lines: Mechanisms and application. IARC Scientific Publications No. 67, International Agency for Research on Cancer, Lyon, 1–225, 1985.

84. N Mishra, V Dunkel, M Mehlman. Advances in Modern Environmental Toxicology, Vol 1; Mammalian Cell Transformation by Chemical Carcinogens. Princeton Junction, NJ: Senate Press Inc, 1980.
85. WC Hahn, CM Counter, AS Lundberg, RL Beijersbergen, MW Brooks, RA Weinberg. Creation of human tumor cells with defined genetic elements. Nature 400:464–468, 1999.
86. T Kuroki, K Sasaki. Relationship between cell transformation and in vivo carcinogenesis based on available data on the effects of chemicals. In: T Kakunaga, H Yamasaki, eds. Transformation assay of established cell lines: Mechanisms and application, IARC Scientific Publications No. 67, International Agency for Research on Cancer, Lyon, 1985, pp 93–118.
87. JS Bertram. Neoplastic transformation in cell cultures: In vitro/in vivo correlations. In: T Kakunaga, H Yamasaki, eds. Transformation Assay of Established Cell Lines: Mechanisms and Application. IARC Scientific Publications No. 67, International Agency for Research on Cancer, Lyon, 1985, pp 77–91.
88. AG Knudson, Jr. Genetic influences in human tumors. In: FF Becker, ed. Cancer, A Comprehensive Treatise 1, Etiology: Chemical and Physical Carcinogenesis. New York: Plenum Press, 1982, pp 73–88.
89. Y Kurokawa. Guidelines for dose selection: Issues in the assessment of carcinogenic potential of therapeutic agents: What other types of endpoints are acceptable. In: PF D'Arcy, DWG Harron, eds. Proceedings of the Second International Conference on Harmonisation (ICH). Antrim, N. Ireland: Greystone Books Ltd, 1994, pp 262–264.
90. A Monro. How useful are chronic (life-span) toxicology studies in rodents in identifying pharmaceuticals that pose a carcinogenic risk to humans? Adverse Drug Reac Toxicol Rev 12:5–34, 1993.
91. RW Tennant, J Spalding, JE French. Evaluation of transgenic mouse bioassays for identifying carcinogens and noncarcinogens. Mutat Res 365:119–127, 1996.
92. RW Tennant, JE French, JW Spalding. Identifying chemical carcinogens and assessing potential risk in short-term bioassays using transgenic mouse models. Environ Health Perpect 103:942–950, 1995.

10

New OECD Genetic Toxicology Guidelines and Interpretation of Results

Michael C. Cimino
U.S. Environmental Protection Agency, Washington, D.C.

I. INTRODUCTION

The United States Environmental Protection Agency (USEPA) and other U.S. regulatory agencies historically have used mutagenicity information as part of a weight-of-evidence approach to evaluate the potential for human carcinogenicity. In addition, agencies consider heritable mutation a regulatory endpoint. In the interest of easing the testing burden faced by industry, the USEPA has developed a series of harmonized genetic toxicology guidelines under the Office of Prevention, Pollution and Toxic Substances (OPPTS) for use in the evaluation of toxic substances and pesticides. This provides a single set of guidelines that minimizes variation in the testing procedures needed to meet data requirements under regulatory legislation, such as the Toxic Substances Control Act (TSCA) and the Federal Insecticide, Fungicide and Rodenticide Act (FIFRA). Further-

---

This manuscript has been reviewed by the Office of Prevention, Pesticides and Toxic Substances, U.S. Environmental Protection Agency, and approved for publication. Mention of trade names or commercial products does not constitute endorsement or recommendation for use.

more, to ease the burden on industries with international markets, for many years it has been USEPA policy to accept data produced in accordance with guidelines of the Organisation for Economic Co-operation and Development (OECD), an international organization with its own genetic toxicology guidelines. The harmonization within the USEPA thus extends to OECD: for purposes of regulatory submissions, USEPA requirements are fully satisfied by the OECD guidelines.

For most toxic substances and pesticides that are regulated under TSCA and FIFRA, the testing scheme for mutagenicity is a three-tier system. Chemicals that are negative in all first-tier tests generally require no further testing for mutagenicity. However, where exposure data, structure-activity relationships (SAR), or other factors warrant, even these chemicals may be subject to a data review, subsequent testing in a cancer bioassay and, possibly, testing for heritable mutagenicity.

II. THE OECD

The OECD develops and coordinates environmental health and safety activities on an international level. These activities include (a) harmonizing chemical testing and hazard assessment procedures; (b) harmonizing classification and labeling; (c) developing principles for Good Laboratory Practices (GLPs); (d) cooperating in the investigation of existing chemicals; and (e) sharing and exploring possible cooperative activities on risk management.

The OECD is not a supranational organization, like the United Nations, but rather a forum or center for discussion, wherein the governments of member nations express their points of view, share their experiences, and search for agreement. When appropriate, member countries can formulate an accord into a formal OECD Council Act, which is agreed upon at the highest level of the OECD. There are two types of such Council Acts: (1) the Decision of Council, which is legally binding, and (2) the Council recommendation, which is a strong recommendation that does not have as great a force as the Decision of Council.

Currently the OECD comprises 29 member countries. Most are European, but three are North American (Canada, Mexico, and the United States), and four are Asian or Pacific (Japan, Australia, New Zealand, and Korea). Additional countries are also affiliated informally with the OECD: (a) "partners in transition" (the Republic of Slovakia); (b) other nonmember countries (Brazil, Argentina, Chile, and Malta); (c) Center for European Economies in Transition, and; (d) Dynamic Asian Economies in Transition (Hong Kong, Malaysia, Singapore, Taiwan, and Thailand).

III. THE OECD GUIDELINES

As part of its program begun in 1971 to evaluate the safety of chemicals, the OECD developed "Guidelines for the Testing of Chemicals." This collection of the most relevant internationally agreed upon testing methods is to be used by government, industry, and independent laboratories when testing the safety of new and existing chemicals as well as chemical preparations such as pesticides, pharmaceuticals, and food additives. These guidelines cover tests for physical-chemical properties, human health effects, environmental effects, and degradation and accumulation in the environment.

The first OECD test guidelines were adopted in 1981. They have been periodically updated to keep pace with the state of the science. In 1997, OECD completed an update of seven guidelines in genetic toxicology. Additionally, new guidelines are developed based on changing needs identified by the OECD member countries. Furthermore, OECD-wide networks of National Coordinators and national experts provide input from scientists in government, industry, and academia.

A vital component of using OECD guidelines is the commitment by member countries to the Mutual Acceptance of Data (MAD) generated by studies properly conducted in accordance with the relevant OECD guidelines for assessing chemical safety. In recognition of the importance of "concerted action...amongst OECD Member...countries to protect man and his environment from exposure to hazardous chemicals," the importance of "international production and trade in chemicals," and the importance of minimizing testing costs while generating valid data, an OECD Decision of Council (adopted May 12, 1981) states that data generated in any member country in accordance with OECD guidelines and GLPs will be accepted in other member countries (1). It is the policy of the United States Environmental Protection Agency (USEPA) to conform to MAD and to accept data generated in accordance with OECD guidelines.

Because the OECD is not a government entity, copies of the OECD health effects guidelines must be purchased (2). Exceptions are drafts of new or updated guidelines, which are available on the Internet (www.oecd.org/ehs/test/testlist.htm). Acquisition of old guidelines and of adopted updates requires purchase of the entire guideline set, available in paper or CD-ROM form, and currently including the 9th Addendum (published in March 1998). Information on purchase is available on the OECD website, www.oecd.org.

Table 1 lists all 15 of the OECD genetic toxicology guidelines currently in force, including the seven recently updated ones (indicated by the date "21july1997," the official OECD adoption date) as well as the older guidelines that were not revised. Note that two old OECD guidelines (471 and 472) are now condensed into one and no longer exist as separate entities. Note also that two new guidelines remain in draft stage and are not included in published form.

TABLE 1. OECD and USEPA Guidelines

OECD#	OECD name	Date	Tier	OPPTS#	OPPTS name	40 CFR#	Canada	EU	Comments
Gene mutations									
471 & 472	Bacterial Reverse Mutation Test	21july97 [NEW]	1	870.5100	Bacterial reverse mutation test	798.5100 & .5265	4.5.4	5.4.1	Joint OECD 471 & 472; HAPS
471	Genetic Toxicology: Salmonella typhimurium, Reverse Mutation Assay	26may83		none	Salmonella typhimurium reverse mutation assay	798.5265	4.5.4	5.4.1	Combined into OECD 471 & 472; no longer available
472	Genetic toxicology: Escherichia coli, Reverse Mutation Assay	26may83		none	Escherichia coli WP2 and WP2 uvrA reverse mutation assays	798.5100	4.5.4	5.4.1	Combined into OECD 471 & 472; no longer available
none				870.5140	Gene mutation in Aspergillus nidulans	798.5140	4.5.4	5.4.1	
none				870.5250	Gene mutation in Neurospora crassa	798.5250	4.5.4	5.4.1	
480	Saccharomyces cerevisiae, Gene Mutation Assay	23oct86		none		none	4.5.4	5.4.1	
477	Sex-linked Recessive Lethal Assay in Drosophila melanogaster	4apr84		870.5275	Sex-linked recessive lethal assay in Drosophila melanogaster	798.5275	4.5.8	5.4.3	

OECD Guidelines

								draft may94
none: draft	Somatic Mutation and Recombination Tests (SMART) in Drosophila melanogaster		none		none	4.5.8		
476	In vitro Mammalian Cell Gene Mutation Test	21july97	870.5300	1	In vitro mammalian cell gene mutation test	798.5300	4.5.5	5.4.1	HAPS
484	Mouse Spot Test	23oct86	none		none	none	4.5.5	5.4.1
none			870.5195	3	Mouse biochemical specific locus test	798.5195	4.5.8	5.4.1
none			870.5200	3	Mouse visible specific locus test	798.5200	4.5.8	5.4.1
Chromosome mutations								
473	In vitro Mammalian Chromosome Aberration Test	21july97 [NEW]	870.5375	1	In vitro mammalian chromosome aberration test	798.5375	4.5.6	5.4.1
474	Mammalian Erythrocyte Micronucleus Test	21july97 [NEW]	870.5390	1	Mammalian erythrocyte micronucleus test	798.5395	4.5.7	5.4.2	HAPS
475	Mammalian Bone Marrow Chromosome Aberration Test	21july97 [NEW]	870.5385	1	Mammalian bone marrow chromosome aberration test	798.5385	4.5.7	5.4.2	HAPS
483	Mammalian Spermatogonial Chromosome Aberration Test	21july97 [NEW]	870.5380	2	Mammalian spermatogonial chromosome aberration test	none	4.5.8	5.4.2	HAPS
478	Rodent Dominant Lethal Assay	4apr84	870.5450	2	Rodent dominant lethal assay	798.5450	4.5.8	5.4.3
none			870.5460	3	Rodent heritable translocation assays	798.5460	4.5.8	5.4.2
none			none		Heritable translocation test in drosophila melanogaster	798.5955		

(Continued)

TABLE 1. (Continued)

OECD#	OECD name	Date	Tier	OPPTS#	OPPTS name	40 CFR#	Canada	EU	Comments
485	Mouse Heritable Translocation Assay	23oct86	3	none		none	4.5.8	5.4.3	
DNA effects									
none				870.5500	Bacterial DNA damage or repair assays	798.5500	4.5.8	5.4.2	
481	Saccharomyces cerevisiae, Mitotic Recombination Assay	23oct86		870.5575	Mitotic gene conversion in Saccharomyces cerevisiae	798.5575	4.5.8	5.4.2	
482	DNA Damage and Repair, Unscheduled DNA Synthesis in Mammalian Cells in vitro	23oct86		870.5550	Unscheduled DNA synthesis in mammalian cells in culture	798.5550	4.5.8	5.4.2	
479	In vitro Sister Chromatid Exchange Assay in Mammalian Cells	23oct86		870.5900	In vitro sister chromatid exchange assay	798.5900	4.5.6	5.4.1	
486	Unscheduled DNA Synthesis (UDS) Test with Mammalian Liver Cells in vivo	21july97 [NEW]		none		none	4.5.8	5.4.2	
none				none	Detection of DNA damage by alkaline elution in mammalian testicular cells in vivo	none	4.5.8	5.4.2	
none				none	Unscheduled DNA synthesis in mammalian testicular cells in vivo				

OECD#	OECD name	date	tier	OPPTS#	OPPTS name	guidance	40 CFR	EU
none				870.5915	*In vivo* sister chromatid exchange assay		798.5915	4.5.7 5.4.2
	Morphological cell transformation			none	Morphological transformation of cells in culture		795.285	4.5.8
none: draft	In vitro Syrian Hamster Embryo (SHE) Cell Transformation Assay	draft mar96		none			none	4.58

The first column (OECD#) gives the number of the OECD guideline. The second column contains the OECD guideline name. The third column (date) is the date that the current version of the guideline was adopted by the OECD. The fourth column (tier) indicates in which of the three tiers of the USEPA mutagenicity test scheme the test is used. The fifth column (OPPTS#) contains the new OPPTS number, that is, that of the harmonized "870" series. The sixth column contains the new, harmonized OPPTS name. The seventh column (40 CFR#) provides the old OPPT number as it appears in the FR and in 40 CFR 798. The eight and ninth columns provide the applicable guidance numbers associated with testing for Canada and the European Union, respectively. Note that two old guidelines (OECD numbers 471 and 472, equivalent to OPPTS 798.5265 and 798.5100) are now condensed into one and no longer exist as separate tests, and that two OECD guidelines are still in draft stage and thus not included in published form.

Abbreviation: HAPS, test rule for hazardous air pollutants.

A. The USEPA/OPPTS Harmonized Guidelines

The USEPA's OPPTS developed a series of harmonized guidelines in genetic toxicology and other health effects endpoints for use in the testing of pesticides and toxic substances and for generating the test data that must be submitted to the Agency for review under Federal regulations. These guidelines were developed through a process of harmonization that combined the testing guidance and requirements that existed in the Office of Pollution Prevention and Toxics (OPPT; "Toxics"[1]), which appeared in the Code of Federal Regulations (40 CFR Part 798 - Health Effects Testing Guidelines), with the testing guidance used in the Office of Pesticide Programs (OPP; "Pesticides"[1]), which appeared in publications of the National Technical Information Service (NTIS). The purpose of harmonizing these documents into a single set of OPPTS guidelines was to minimize variations among the testing procedures necessary to meet the data requirements of the USEPA under TSCA and FIFRA. Key to this harmonization between Toxics and Pesticides is that it extends to harmonization with the current OECD guidelines, including the most recently updated versions. The OPPTS 870 harmonized genetic toxicity guidelines are identical to the equivalent OECD guidelines, except for minor format differences necessary to meet requirements of the U.S. Government Federal Register (FR) Office.

In August 1998, OPPTS issued an FR notice announcing the availability of these harmonized guidelines (63 FR 41845, August 5, 1998). These harmonized Series 870 guidelines supersede the older OPPT 798 guidelines published in the CFR. The 798 guidelines remain in force for existing testing actions only. As soon as the last study that cites 798 guidelines(s) is received, they will be retired from the CFR and copies placed in the public docket for the record.

Table 1 lists the 18 OPPTS 870 guidelines in genetic toxicology that are harmonized between Toxics and Pesticides. The 870 guidelines are available from the U.S. Government Printing Office, Washington, DC 20402; on the Federal Bulletin Board; or by phone for disks or paper copies. In addition, the guideline documents can be obtained at no charge from the USEPA website www.epa.gov/docs/OPPTS_Harmonized/870_Health_Effects_Test_Guidelines/Drafts).

The 870 guidelines become TSCA test standards when cited under a new subpart in 40 CFR 799. The test standards are the same as the 870 guidelines except for minor modifications required to render them legally enforceable when

---

[1]The terms "Toxics" and "Pesticides," with initial capitals, are used to distinguish between the Office of Pollution Prevention and Toxics (OPPT) and the Office of Pesticide Programs (OPP). Each of these suboffices, headed by a director, is part of OPPTS, headed by an assistant administrator. When the words "toxic(s)" and "pesticide(s)" are used without initial capitals, they refer to chemical categories, not USEPA offices.

they are cited as TSCA Section 4 Test Rules (primarily the inclusion of "should"/"shall" language in appropriate locations). Four guidelines have already been modified for use in the HAPS (Hazardous Air Pollutants) Test Rule (62 FR 43820, August 15, 1997). The four guidelines are those equivalent to OECD guideline numbers 471/472, 474, 475, 476.

B. USEPA Use of OECD Guidelines

For purposes of regulatory submissions, USEPA requirements are fully met by the OECD guidelines. However, specific departures from the guidelines are permitted on a case-by-case basis with appropriate justification and the concurrence of the relevant Office before initiation of the study. In addition, for certain chemicals, the USEPA may require that specific testing conditions be met (e.g., species, cell line, activation conditions, route of administration). Proposed departures from the published guidelines always should be discussed with the Agency before committing time and resources to testing. The OECD guidelines apply to submissions for regulatory purposes to Toxics and Pesticides, but they may not apply to testing conducted for other offices (e.g., Water, Air, and Research and Development), as the latter studies usually are not conducted for regulatory purposes (see section VII below).

The old separation by both OPPTS and OECD of prokaryote gene mutation tests into separate tests in *Salmonella* and *Escherichia* has yielded to the recent decision by OECD to combine the two into a single microbial assay, a decision that USEPA follows. Thus, the old USEPA *Salmonella* (40 CFR 798.5265) and *Escherichia* (40 CFR 798.5100) guidelines no longer exist as separate entities; they are now combined into one new bacterial reverse gene mutation test guideline (OPPTS 870.5100).

A cell transformation guideline was developed for a specific test rule (cresols: CFR Section 795.285; Table 1). Presently, no corresponding OECD guideline exists, although OECD is actively considering its development. The potential for use of the various cell transformation assays has not been widely exploited because until recently, there was insufficient validation of these systems for general use (3).

The OECD also is considering drafting a guideline for transgenic animals. Transgenic assays are potentially of great value in providing information at lower cost and in less time than traditional long-term bioassays; in addition, they may provide tissue-specific information that may be relevant to cancer risk assessments (see, for example [3]). However, transgenic systems are not yet routine because adequate validation of the system has not occurred and standardized guidelines do not exist.

One old OPPT "798" guideline—the *Drosophila* heritable translocation assay (40 CFR 798.5955)—was excluded from the OPPTS harmonization effort for lack of use. Two other test guidelines were developed and used for a specific Toxics test rule chemical but were not harmonized, although it is likely that they will be used in future test rules. These two assays are the *in vivo* unscheduled DNA synthesis (UDS) and the *in vivo* alkaline elution assays, both in testicular cells.

Finally, four OECD guidelines have no USEPA equivalents. One is newly published, the *in vivo* UDS assay in somatic (liver) cells (OECD guideline 486).

The current OPPTS testing scheme for mutagenicity has been published and discussed at length by Dearfield et al (4; see also 3, 5). This test scheme is presented in Figure 1, with the corresponding OECD and USEPA guideline numbers added in appropriate locations. Details of the various assays included in the OPPTS mutagenicity testing scheme are discussed below.

This testing scheme has been in force for almost 9 years. The USEPA (and other U.S. regulatory agencies) does not change existing assessment methods until the advantages of the innovations have been firmly established. The primary reason behind this stability is the amount of effort involved in changing testing schemes, guidelines, regulations, or statutes. Other factors affecting change are the continued value of old tests; modern trends toward patenting/licensing new technologies, which increase the cost and difficulty of validating them; litigation concerns; and international regulatory requirements. Nevertheless, the potential of these new systems to clarify mechanisms and sites of action and gene expression, to provide greater sensitivity and specificity, to reduce time and cost, and to offer alternatives to whole-animal systems must be encouraged.

IV. NEW METHODS

A recently established mechanism exists for introducing new test methods into the evaluation scheme, not only for USEPA but also for other regulatory agencies. As a result of the 1993 National Institutes of Health Revitalization Act (NIHRA, [6]), the National Institutes of Environmental Health Sciences (NIEHS) has established the Interagency Coordinating Committee on the Validation of Alternative Methods (ICCVAM). The functions of ICCVAM (7) are to (a) communicate criteria and procedures that agencies should use in considering new methods; (b) encourage development of new methods that will provide improved assessment of potential toxicity; (c) provide guidance for validating and evaluating new methods; (d) contribute to the increased likelihood of regulatory acceptance of scientifically valid new methods at both national and international levels; (e) encourage the use of such new methods, and; (f) encourage alternatives to animal testing that demonstrate reduction, refinement, and replacement (the "3 Rs") of whole-animal methods.

The USEPA's Offices of Research and Development (ORD) and Water (OW) actively are using or investigating several such new methodologies. In

OECD Guidelines

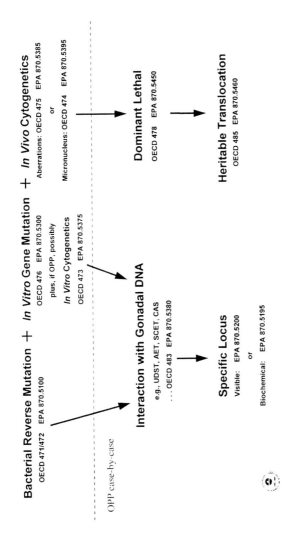

FIG. 1. Mutagenicity testing scheme for the U.S. Environmental Protection Agency Office of Prevention, Pesticides and Toxic Substances (USEPA OPPTS). This three-tier scheme applies to existing chemicals (e.g., test rules) and for pesticide (OPP) registration. Test guideline numbers for OECD and OPPTS guidelines (as OECD and EPA, respectively) are indicated where appropriate. *Abbreviations: UDST,* unscheduled DNA synthesis in testicular cells *in vivo; AET,* alkaline elution in testicular cells *in vivo; SCET,* sister chromatid exchange in testicular cells *in vivo; CAS,* chromosomal aberrations in spermatogonial cells *in vivo.* The guideline numbers under "Interaction with Gonadal DNA" apply to the CAS test. (Modified from Ref. 5.).

Toxics and Pesticides, data on alternate systems rarely have been required or submitted, but when available, such data contribute to a weight-of-evidence determination. Submitters are encouraged to send available data from alternative methodologies along with data from routine tests, so that Agency reviewers can become familiar with the use, evaluation, and relevance of the new methods.

Ultimately, acceptance of a new method remains the decision of each individual agency according to its specific regulatory mandates. If a new test method is accepted by Federal agencies, appropriate follow-up actions involve implementation, training, and use of the methods in the risk assessment process. The recently formed NTP (National Toxicology Program) Interagency Center for the Evaluation of Alternative Toxicological Methods (NICEATM) works with the ICCVAM to carry out these activities. Sponsors of new methods are strongly encouraged to contact the NICEATM before submission of test methods for guidance on the information needed and the process for its submission and review. The ICCVAM/NICEATM Web site is http://iccvam.niehs.nih.gov.

V. USEPA/OPPT

TSCA (8) provides the USEPA with the authority "to regulate commerce and protect human health and the environment by requiring testing and necessary use restrictions on certain chemical substances . . ." TSCA supplements other existing laws, such as FIFRA (9), the Clean Air Act (10), the Clean Water Act (11), and the Occupational Safety and Health Act (12). It was designed to fill the gap in the government's authority to test and regulate chemicals. Under TSCA, the term "chemical" encompasses a wide variety of organic and inorganic substances manufactured or imported for industrial uses, such as dyes, pigments, lubricant additives, chemical intermediates, synthetic fibers, structural polymers, coatings—essentially any commercial chemical except those used as drugs, food additives, cosmetics, pesticides, and certain other uses. TSCA is administered by OPPT, under the Assistant Administrator of OPPTS.

TSCA has two main regulatory features: 1. acquisition of sufficient information by the USEPA to identify and evaluate the potential hazards from chemical substances; and 2. regulation of the production, use, distribution, and disposal of such substances when necessary.

Under TSCA, the USEPA historically has used mutagenicity information as part of a weight-of-evidence approach to discern the potential for human carcinogenicity, as detailed in the Agency's Carcinogen Risk Assessment Guidelines (13). However, this is not the sole use for mutagenicity data. Through its Mutagenicity Risk Assessment Guidelines [14], the USEPA also considers heritable mutation as a regulatory endpoint. These two applications of mutagenicity data are very closely interlinked.

A. Existing Chemicals

Section 4(a) of TSCA gives the USEPA the authority to require the testing of existing chemicals if unreasonable risk to health or the environment is suspected. To require such testing, the USEPA must find that: (a) the chemical may present an unreasonable risk or a significant potential for exposure, (b) insufficient data are available to perform a reasoned risk assessment; and (c) testing is necessary to generate such data (and such testing is not already underway). A testing requirement is promulgated in a Test Rule, which must: (a) identify the substance to be tested and the tests to be conducted; (b) provide (or reference) guidelines for performance of the tests; and (c) specify a reasonable period for completion of testing.

There are several mutagenicity endpoints of concern, including point (gene) mutations (i.e., submicroscopic changes in the base sequence of DNA) and chromosomal mutations (i.e., structural and numerical chromosome aberrations). Structural chromosome aberrations include deletions, duplications, insertions, inversions, and translocations of parts of chromosomes, whereas numerical chromosomal aberrations are gains or losses of whole chromosomes (i.e., aneuploidy, trisomy, monosomy) or sets of chromosomes (e.g., haploidy, polyploidy). The OPPTS test scheme (Fig. 1) identifies those agents that induce gene mutations or structural chromosome aberrations. Until there is a properly validated and routinely available procedure for testing for numerical chromosome aberrations, the USEPA will evaluate agents that may induce numerical chromosome aberrations on a case-by-case basis. However, OPPTS believes that the existing battery provides sufficient information about chemical activity to allow a reasonable assessment of potential mutagenicity hazard.

The OPPTS mutagenicity test scheme (Fig. 1) is a three-tiered system. The first tier comprises a battery of three tests: the bacterial gene mutation assay (OECD guideline 471/472; OPPTS guideline 870.5100), an *in vitro* assay for gene mutation (OECD 476; OPPTS 870.5375) and an *in vivo* assay for chromosomal effects, which may be either an assay for chromosome aberrations in bone marrow (OECD 475; OPPTS 870.5385) or an assay for micronucleus induction in erythrocytes (OECD 474; OPPTS 870.5395). All four tests are covered by the recent updates to the OECD guidelines.

The bacterial gene mutation assay is a reverse mutation assay that uses several specific tester strains of the prokaryotes *Salmonella typhimurium* and *Escherichia coli* for the detection of point mutations. Because the genetics of each tester strain have been well defined, it is possible to identify the specific type of genetic effect induced by agents that are active in this system (i.e., base substitutions, frameshifts). This assay also is both easy to perform and used routinely. Consequently it is well validated and has an extensive database of tested chemicals. It is very useful for detecting the intrinsic mutagenicity of many classes of biologically active chemicals, especially ones that appear to act by way of an electrophilic mechanism. Although disagreement exists regarding the assay's precise ability to predict chemical carcinogenicity accurately, it nonetheless provides useful information in predicting the carcinogenic, as well as the mutagenic, potential of chemical agents.

Despite the proven utility of the bacterial reverse mutation test, it is not the only assay the USEPA uses to assess gene mutation. In conducting a hazard evaluation of a chemical, the USEPA places greater weight on tests conducted in eukaryotes than in prokaryotes, and in mammalian than in submammalian species (14). Major differences between mammalian and bacterial cells, such as membrane structures, DNA repair capabilities, and the organization and complexity of mammalian genomes, suggest that it is necessary to include a mammalian system in the battery. Furthermore, there are chemicals that give negative results in bacteria but are mutagenic when tested for gene mutations in mammalian cell culture. Finally, the deliberations of the 1987 Williamsburg Workshop (15), which evaluated the databases of the USEPA Gene-Tox Program (16) and of NTP studies (17) for the use of short-term tests as predictors of carcinogenicity, show that the L5178Y mouse lymphoma cell assay may be sensitive to a different subset of chemicals than the bacterial and the *in vitro* cytogenetics assays (4). Therefore, OPPTS concludes that the combination of the bacterial assay with a mammalian cell culture assay for gene mutation provides more information than that obtained from the bacterial assay alone. This additional information may provide a better idea of the mechanism of mutagenic activity, a refinement of possible mutagenicity concern, and a basis for further testing.

Because chromosome abnormalities in structure and number are involved in the induction or expression of cancer, testing for the effect of a chemical on chromosomes is a necessary adjunct to testing for gene mutations. For determining effects on chromosomes, the assay of choice is an *in vivo* cytogenetics assay. This may either be metaphase analysis for structural chromosome aberrations or a micronucleus assay, both usually conducted in rodents. These tests have been performed routinely for many years and each has a substantial database of tested chemicals. Although these tests are usually conducted in bone marrow, other organ or tissue sites may be considered (e.g., liver, lymphocytes, spleen), particularly if knowledge about the test chemical provides support for such selection. The micronucleus assay frequently is evaluated in peripheral blood erythrocytes. Because the Mutagenicity Risk Assessment Guidelines place greater weight on results from *in vivo* tests than from *in vitro* tests (14), cytogenetics should be performed *in vivo* to allow for such factors as whole-animal metabolism, repair capabilities, pharmacokinetic factors (e.g., biological half-life, absorption, distribution, excretion), and target specificities.

Under the provisions of TSCA Section 4, the USEPA may require a cancer bioassay immediately. However, criteria of "may present an unreasonable risk," such as structure-activity relationships (SAR), chemical production or release amount, and mutagenicity testing results are used when the USEPA makes a policy decision to trigger carcinogenicity testing. In such cases, the decision to trigger a cancer bioassay from short-term test results would occur as follows:

1. An automatic trigger to a bioassay depends on a minimum of two positive responses, at least one of which must be in an *in vivo* assay. Thus, a positive response in all three first-tier tests, or a positive in the bacterial assay and the *in vivo* assay for chromosomal effects, or a positive response in the *in vitro* mammalian gene mutation assay and the *in vivo* assay for chromosomal effects, would lead directly to a 2-year bioassay.

2. Any other combination of responses, including a single positive response in one assay, or positive responses in both of the *in vitro* tests (bacterial and mammalian cells in culture) but not in the *in vivo* assay for chromosomal effects, results in a "data review." This review, which occurs before a decision is made to require further testing, considers all available information, including other test results, SAR, production volume, and exposure figures.

Two tests that were in the former test scheme for Toxics have been removed (4). An *in vitro* cytogenetics assay (OECD 473; OPPTS 870.5375) no longer serves as trigger to a cancer bioassay and is no longer part of the test scheme, although it may contribute to the weight-of-evidence for chemical hazard. Additionally, the *Drosophila* sex-linked recessive lethal assay (OECD 477; OPPTS 870.5275) no longer serves as a second-tier gene mutation trigger to a cancer bioassay.

The Toxics first tier mutagenicity test scheme of Section 4 is nearly identical to the Pesticides initial battery (see below). However, whereas Pesticides specifically designates the L5178Y/TK+/– mouse lymphoma assay as the assay of choice for *in vitro* gene mutation, Toxics does not so designate, preferring to make a decision about use of a particular assay at the time of promulgation of a Section 4 test rule. However, Toxics agrees that, generally, chemicals are best tested in either the L5178Y mouse lymphoma or CHO AS52 assays.

It is anticipated that no further testing would be required for the majority of chemicals that are negative in all three first-tier tests. However, where exposure data, SAR, or other factors indicate it is warranted, even such first-tier-negative agents may be subject to a data review and subsequent testing in a cancer bioassay.

B. Further Actions Beyond Testing in the Initial Tier

Once intrinsic mutagenicity has been identified, the test scheme requires assay(s) that assess the ability of the chemical to interact with DNA in the gonad. This may apply regardless of whether a cancer bioassay has been triggered. Evaluations for effects in the gonad *in vivo* comprise the second tier of the test scheme (Fig.1).

Currently, there are no practical assays for *in vivo* evaluation of potential gene mutagenicity in the mammalian gonad. As mentioned in the previous section, the *Drosophila* sex-linked recessive lethal assay (OECD 477; OPPTS 870.5275) no longer serves as a second-tier trigger to a bioassay. It also no longer serves in second tier evaluation of *in vivo* heritable gene mutation. The assays currently used for the second tier assess interaction with gonadal DNA are, for example, the *in vivo* unscheduled DNA synthesis (UDS), alkaline elution (AE), sister chromatid exchange (SCE), or chromosomal aberration (OECD 483, a new guideline; OPPTS 870.5380) assays in testicular tissues, or the rodent dominant lethal assay (OECD 478; OPPTS 870.5450). Results in testicular UDS and AE assays correlate well with the specific locus assay, based on review of the USEPA Gene-Tox database (16, 18). This scheme, already used on one occasion in the Test Rule process, provides a needed alternative to an immediate trigger to a specific locus assay from a positive response in an *in vitro* gene mutation assay (bacterial or mammalian cells in culture), a requirement that was deemed to be too costly without evidence of *in vivo* mammalian gonadal effect.

At present, gene mutagens that are positive in second-tier gonadal assay(s) may be tested, after a data review, in either the third-tier biochemical or visible specific locus assay (OPPTS 870.5195 and 870.5200, respectively; there are no OECD equivalents). The choice between the two is left to the sponsor. Correspondingly, chromosome mutagens that are positive in second-tier assay(s) may be tested in the third-tier heritable translocation assay (OECD 485; OPPTS 870.5460). Although existing data indicate that chemicals positive in the heritable translocation assay are also positive in the specific locus assay, there is no assurance that all gonadal chromosome mutagens also will be gonadal gene mutagens. Evidence based on testing in somatic cells suggests that such agents may exist (e.g., acrylates [19]).

The need for additional testing to support a quantitative risk assessment depends on both available mutagenicity data and other relevant considerations, such as human exposure levels, chemical use patterns, and release to the environment. Although quantitative mutagenicity risk assessments have been conducted rarely, previous efforts to perform quantitative risk assessment will be re-examined to identify and apply appropriate methods to quantify the risk associated with the chemical under consideration (20; see also references in [4]).

C. New Chemicals

Section 5 of TSCA requires that manufacturers and importers of "new" chemicals submit a Premanufacturing Notification (PMN) to the USEPA 90 days before beginning the manufacture or import of the new substance. By "new," TSCA means those chemicals not appearing on the Inventory of Existing Commercial Chemicals, which was originally compiled in 1977 and is continuously updated.

OECD Guidelines

The manufacturer may proceed with production of the new chemical after this 90-day notification period, unless notified otherwise by the USEPA.

Section 5 does not require that submitters conduct toxicity testing before submission of a PMN, although it does mandate submission of any such data that is either in their possession or readily obtainable. Because there is no statutory requirement for up-front testing, OPPT has developed techniques for mutagenicity hazard assessment that can be used when there are little or no test data on the substance itself. The overall approach involves (a) evaluation of available toxicity data on the PMN chemical, if any; (b) evaluation of test data available either on substances analogous to the PMN chemical, on key potential metabolites of the PMN chemical, or on analogues of those metabolites; and (c) use of knowledge and judgment of scientific assessors in the interpretation and integration of the information developed in the course of the assessment.

Mutagenicity data generally are used for three purposes in new chemical evaluation: (1) as part of exposure-based testing; (2) as part of the weight-of-evidence that a chemical may be a potential carcinogen; and (3) to assess the potential of the PMN chemical to induce heritable genetic effects.

For new chemicals that meet specified criteria for potential volume and exposure, the USEPA requires mutagenicity testing with a two-test battery comprising the bacterial and *in vivo* mouse micronucleus assays (Fig.2). Both tests are covered by new OECD guidelines. In supporting a concern for the potential carcinogenicity of a PMN chemical, the USEPA generally cites data on analogue(s) known to be carcinogenic (i.e., possessing demonstrated tumor-forming ability in one or more animal studies). In such instances, mutagenicity data on the PMN chemical or on analogue(s) of the chemical are used to support the case for potential carcinogenicity. When no analogue of the PMN chemical has been tested for carcinogenicity, mutagenicity data alone generally are not considered sufficient to support a concern for potential carcinogenicity. Regulatory action under Section 5 is seldom taken on the basis of mutagenicity data alone, especially on the basis of *in vitro* mutagenicity data.

Although manufacturers generally must submit data from the bacterial and micronucleus assays when the USEPA requires mutagenicity testing for Section 5, for certain chemical classes (e.g., acrylates, vinyl sulfones), there is a requirement for *in vitro* gene mutation data in mammalian cells, specifically the mouse lymphoma test.

Secondly, where an appropriate carcinogenic analogue has been tested in the same assay(s) as those required for the PMN chemical, this agent is generally included in the mutagenicity test as an additional concurrent positive control. In some cases, when genotoxic activity of the analogue chemical in short-term tests was not known, the USEPA has required the simultaneous testing of both the PMN chemical and the analogue. In this latter case, positive results for both the PMN chemical and the analogue are used to support the weight-of-evidence conclusion

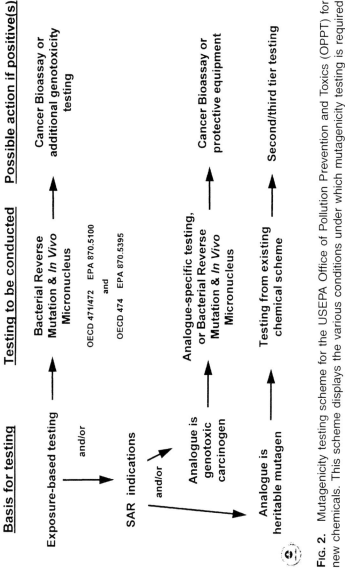

FIG. 2. Mutagenicity testing scheme for the USEPA Office of Pollution Prevention and Toxics (OPPT) for new chemicals. This scheme displays the various conditions under which mutagenicity testing is required under Section 5 of the Toxic Substances Control Act, and the testing required under each condition. Test guideline numbers for OECD and OPPTS guidelines (as OECD and EPA, respectively) are indicated where appropriate. *Abbreviation: SAR*, structure-activity relationships. (Modified from Ref.3.)

that the PMN chemical may be a carcinogen. In such instances, a 2-year bioassay on the PMN chemical or the use of protective equipment to limit exposure generally is required.

Negative mutagenicity results for the PMN chemical, in the face of positive mutagenicity results for the carcinogenic analogue, are taken as an indication that the PMN chemical is probably noncarcinogenic, or in cases where the analogy may have been uncertain, that the analogue was not appropriate. In either case, concern for potential carcinogenicity is lessened as a result of the mutagenicity data, and a 2-year bioassay generally is not considered necessary. In those instances where a carcinogenic analogue is inactive in short-term mutagenicity tests, negative results for the PMN chemical do not alleviate concern for potential carcinogenicity.

A requirement to conduct a long-term cancer bioassay on a chemical usually is treated by the regulated industry as equivalent to a ban on its production/import. Chemicals subjected to the requirement for a long-term cancer bioassay often are withdrawn by submitters because the high cost of such testing cannot be justified. The USEPA believes that judicious and reasonable use of short-term testing in the Section 5 process reduces the number of chemicals subject to a long-term bioassay.

Finally, where data on an appropriate analogue indicate a potential for heritable mutagenicity, without a concurrent concern for potential carcinogenicity, testing similar to that for existing chemicals (the three-tier system of Fig.1) may be required. Given the nature of PMN assessment and the limitations in the size of the database of chemicals tested for heritable genetic effects, concern for a chemical's ability to induce heritable gene or chromosomel aberrations is rarely supportable under Section 5.

VI. USEPA/OPP

FIFRA (9), Section 3 (Registration of Pesticides) permits the USEPA to publish guidelines specifying the kinds of information required to support the registration of a pesticide. Part 158 of the Code of Federal Regulations (40 CFR - Protection of Environment) specifies the types and minimum amounts of data required to make regulatory judgments about the risks of pesticide products in the registration process. FIFRA is administered by the OPP, under the Assistant Administrator of OPPTS.

Current guidance states that mutagenicity testing is required for all general use patterns of pesticides (including terrestrial, aquatic, greenhouse, forestry, domestic outdoor and indoor for both food crop and nonfood uses). The OPP's Pesticide Assessment Guidelines, Subdivision F, Hazard Evaluation: Human and Domestic Animals provides guidance on implementation of Part 158 requirements.

Mutagenicity test results are considered by OPP in decisions about the carcinogenicity of the pesticide. The technical grade of the active ingredient is the substance to be used for testing. If a pesticide has been tested for carcinogenicity, available mutagenicity data will be used along with carcinogenicity test results as part of the weight-of-evidence for its classification (13). When carcinogenicity testing is conditionally required in accordance with the Part 158 Toxicology Data Requirements, evidence of chemical mutagenicity may provide the basis to require a carcinogenicity study for that chemical.

The tests included in the Pesticides initial battery are the same as those in the first tier of the Toxics scheme (Fig. 1): (a) bacterial reverse mutation assay; (b) mammalian cells in culture forward gene mutation assay which allow's detection of point mutations as well as large deletions and chromosome rearrangements, such as in mouse lymphoma L5178Y cells at the thymidine kinase (*tk*) locus maximizing assay conditions for small colony expression and detection; or Chinese hamster ovary (CHO) AS52 cells at the xanthine-guanine phosphoribosyl transferase (*xprt*) locus; or Chinese hamster lung fibroblast (CHL) V79 cells at the hypoxanthine-guanine phosphoribosyl transferase (*hgprt*) locus, accompanied by an appropriate *in vitro* test for clastogenicity (OECD 473; OPPTS 870.5375); and (c) *in vivo* assay for chromosomal effects using either chromosome aberration or micronucleus analysis. All tests are covered by new OECD guidelines.

The preference in the Pesticides battery for the L5178Y/TK+/– mouse lymphoma assay generated much comment during the Subdivision F Agency and public examination period (4). In general, the key consideration in this decision was the ability of the mouse lymphoma assay to maximize information on the genotoxicity of the chemical. Recent advances in the understanding of the types of genetic events detectable by mammalian cell assays indicate that two assays, the mouse lymphoma and the CHO AS52 assays, may detect chemicals capable of inducing both point mutations and chromosomal events. By using a combination of molecular analysis and banded karyotype analysis, research in the mouse lymphoma assay has shown that *tk* mutants appear to include presumed point mutations (no visible alteration in karyotype or Southern blot pattern), total *tk* gene deletions, mitotic nondisjunction, translocation, homologous mitotic recombination, and gene conversion. An assay capable of detecting the above mutation events provides more information and is thus preferred. Pesticides considers the mouse lymphoma assay to be better characterized than the CHO AS52 assay with regard to the types of genetic damage detected.

The Office of Pesticide Programs uses the test guidelines of the OPPTS harmonized 870 series, which are in turn harmonized to the corresponding OECD guidelines (see above).

This test battery was designed for chemicals of unknown genotoxic potential, to satisfy the minimum regulatory requirement for initial mutagenicity testing. By using this battery, Pesticides accomplishes two things: (1) it provides a defined set of mutagenicity tests to be performed on chemicals to be submitted for registration, and (2) it ensures the generation of a body of data for decisions on the need for further testing and the degree of concern about the potential mutagenicity of the test agent. However, alternative tests may be proposed if such testing is based on knowledge specific to the test chemical. In such cases, either the submitter or Pesticides may propose discussion of these points before testing is initiated.

In addition to this battery, other provisions of Part 158 of the CFR are designed to enhance the information submitted to the OPP for registration. If other tests for endpoints that may be predictive of mutagenicity are performed in addition to the initial battery, these results must also be submitted to Pesticides. Also, a Data Call-In (DCI) requires submission of a reference list of all studies/papers known to the submitter that concern the mutagenicity of the chemical. Submission of other relevant data is encouraged (e.g., metabolism, distribution studies, reproductive studies, which in many cases are already required by Part 158, Toxicology Data Requirements). This additional information may provide better insight into the significance and interpretation of mutagenicity test results, and facilitates OPP's effort to provide a timely and accurate assessment of the chemical.

In the event of equivocal or discordant results among the tests initially submitted to Pesticides, confirmatory testing or other relevant information may be required. This would provide information to clarify the potential genotoxic hazard of the chemical. For example, additional *in vivo* cytogenetics testing may be required to address such concerns as target tissue/organ or species specificity, differences in metabolism or distribution, or SAR alerts.

A. Further Actions Beyond Testing in the Initial Tier

As stated earlier, results from the initial battery and any confirmatory testing are reviewed along with all other available, relevant information before decisions on subsequent regulatory action are made. If no mutagenicity hazard is identified from the available information, including the testing in the initial tier, further action is usually unnecessary. However, if available information suggests a mutagenicity hazard, a decision to take further action may be made to determine if the chemical may induce heritable mutation in mammals (14). As is the case with Toxics, when Pesticides evaluates a chemical for its potential to induce heritable mutations, all available data are considered, including mutagenicity data, exposure data, SAR considerations, mechanism of action, pharmacokinetics and metabolism, reproductive effects, target organ specificity, subchronic and chronic effects, and the ability of the chemical to reach the germ cell and interact with gonadal DNA. If further testing is necessary, that testing would involve the same second- and third-tier tests, and quantitative risk assessment would be conducted if appropriate.

VII. OTHER USEPA OFFICES

The main statutory mandates for mutagenicity testing at the USEPA reside in TSCA and FIFRA in the Toxics and Pesticides programs, and have been discussed in Sections V and VI above. Toxicity testing used in other Offices, such as the Offices of Air and Radiation (OAR), Solid Waste and Emergency Response (OSWER), Research and Development, (ORD), and Water (OW), generally are not part of regulatory submissions, and utilization of OECD guidelines may not be required for such nonregulatory cases. However, in cases in which one of these Offices calls upon the regulatory authority of TSCA or FIFRA to require testing, these OPPTS/OECD harmonized guidelines generally should be accepted. An example of this is OAR's HAPs rule, mentioned earlier. On the other hand, the USEPA's ORD and OW are actively using or investigating several new methodologies for which there are no current OECD or OPPTS guidelines.

VIII. OTHER U.S. REGULATORY AGENCIES

Although this chapter has focused on the OECD guidelines used by the United States Environmental Protection Agency, other federal regulatory agencies apply similar testing schemes to evaluate chemicals for mutagenicity.

The United States Food and Drug Administration (FDA) requires that manufacturers provide mutagenicity data for food additives, cosmetics, and human and animal drugs that they wish to market. Evaluation of mutagenicity data is a key component in the assessment by FDA of potential carcinogenicity, as well as for determination of the potential to induce heritable damage to the gene pool. For purposes of this assessment, several centers of the FDA have independently arrived at mutagenicity testing schemes that closely resemble each other and that of the USEPA OPPTS.

For example, the Center for Food Safety and Applied Nutrition (CFSAN), in its draft Redbook (Toxicological Principles for the Safety Assessment of Direct Food Additives and Color Additives Used in Food), requires three tests: (1) bacterial gene mutation, (2) *in vitro* mouse lymphoma assay, and (3) *in vivo* cytogenetics. This scheme is identical to that used by Pesticides of the USEPA. The Center for Veterinary Medicine (CVM) requires four tests: (1) bacterial gene mutation, (2) *in vitro* mammalian gene mutation assay, (3) *in vitro* cytogenetics assay, and (4) in vivo cytogenetics. The Center for Devices and Radiological Health (CDRH) requires three *in vitro* tests in three areas of genetic damage.

The Center for Drug Evaluation and Research (CDER) also requires three tests: (1) bacterial gene mutation, (2) *in vitro* mammalian cell assay, either the

OECD Guidelines

mouse lymphoma assay or a cytogenetics assay, and (3) *in vivo* cytogenetics. This is the test battery proposed by the Fourth International Conference of Harmonization (ICH4) for genotoxicity testing of pharmaceuticals, discussed elsewhere in this volume.

Food and Drug Administration scientists actively participated as U.S. experts during the drafting and updating of the current OECD guidelines. The tests used by the CFSAN, CVM, and CDER are covered by the new OECD guidelines. The OECD guidelines are generally accepted by the FDA centers, although the ICH has proposed some modifications, especially to the mouse lymphoma assay.

IX. SUMMARY

The OECD has developed Guidelines for the Testing of Chemicals as part of its international program to evaluate the safety of chemicals. A vital component of the utilization of the OECD guidelines is the commitment by member countries to the Mutual Acceptance of Data generated by studies properly conducted in accordance with the relevant OECD guidelines, a commitment to which the USEPA subscribes. For purposes of regulatory submissions, USEPA requirements are fully satisfied by the OECD guidelines. However, specific departures from the guidelines may be permitted or required on a case-by-case basis with prior justification and concurrences.

The USEPA also has a series of harmonized OPPTS genetic toxicology guidelines for use in the testing of toxic substances and pesticides. This provides the Agency with a single set of guidelines to minimize variations among the testing procedures necessary to meet the data requirements under the laws TSCA and FIFRA. This harmonization extends to the OECD guidelines.

The USEPA historically has used mutagenicity information as part of a weight-of-evidence approach to evaluate potential human carcinogenicity. Additionally, the USEPA considers heritable mutation as a regulatory endpoint. For existing chemicals and pesticides, the OPPTS mutagenicity test scheme is a 3-tier system. The first tier of the Toxics scheme comprises three tests: the bacterial gene mutation assay, an *in vitro* assay for gene mutation, and an *in vivo* assay for chromosomal effects. Pesticides specifies the mouse lymphoma cell assay for *in vitro* gene mutation.

For the majority of chemicals that are negative in all first-tier tests, no further testing is required. However, where exposure data, SAR, or other factors warrant, even these chemicals may be subject to additional testing.

TSCA does not require that submitters conduct toxicity testing for new chemicals. However, for new chemicals that meet exposure criteria, the USEPA requires two mutagenicity tests: the bacterial and the mouse micronucleus assays. In some instances, mutagenicity data on the PMN chemical or on analogue(s) of the chemical lend support to the case for potential carcinogenicity. For certain chemical classes, other specific tests may be required.

If intrinsic mutagenicity is identified, the test scheme requires assay(s) that assess the ability of the chemical to interact with DNA in the gonad. This may apply regardless of whether a cancer bioassay has been triggered. Evaluations for effects in the gonad *in vivo* comprise a second tier of testing. Mutagens positive in second-tier assay(s) may be tested in third-tier tests (either a specific locus or a heritable translocation assay). If warranted by the test results, quantitative risk assessment will be performed.

Toxicity testing used in USEPA Offices other than Toxics and Pesticides generally is not part of a regulatory submission, and utilization of OECD guidelines may not be required in such cases. However, when one of these Offices calls upon the regulatory authority of Toxics or Pesticides, the OPPTS/OECD harmonized guidelines generally are accepted.

Other federal regulatory agencies, such as the FDA, apply testing schemes similar to that of the USEPA to evaluate chemicals for mutagenicity, thus providing a level of consistency across agencies.

ADDENDUM

Since the completion of this chapter in early 1999, the USEPA has developed a modification of the testing scheme for the Chemical Right-to-Know (RtK) project for High Production Volume (HPV) chemicals. The HPV project is the product of the initiative set forth on Earth Day of 1998 by Vice President Al Gore. This initiative requires that information be gathered and widely disseminated to the public as well as to scientists, policymakers, and industry on chemicals that are produced in high volume and for which there is likely significant human exposure. The project has identified approximately 2,800 chemicals that require toxicology profiles for acute, reproductive, and genetic toxicity. In view of the large number of chemicals involved and the potential requirement for extensive testing to obtain needed data, whole-animal testing could require large numbers of animals, if such testing were conducted according to the normal procedures established for these endpoints. In response to public comment and extensive deliberations between the Agency, industry, public interest groups, and the public, given the nature of HPV as a screening program, and because of the very large number of chemicals involved, the testing scheme required for genetic toxicity has been modified for purposes of this program only. Detailed information about the HPV program is available on the USEPA website at http://www.epa.gov/chemrtk/volchall.htm.

Therefore, to minimize animal use, *in vitro* cytogenetic testing may be used in place of *in vivo* testing. This *in vitro* testing may be either the *in vitro* chromosome aberration assay (OECD 473), or the *in vitro* micronucleus test. The latter does not have an established guideline available, either for USEPA or for OECD.

It is emphasized that this modification pertains only to the HPV program because of its unique nature as a screening program for a very large number of chemicals. It does not supersede or replace the testing schemes presented in the body of this chapter.

REFERENCES

1. OECD. Decision of Council, May 12 1981. Document C(81)30(Final) in OECD 1998, op. cit., 1981.
2. OECD. OECD Guidelines for the Testing Of Chemicals. OECD, Paris, France, 1998.
3. JT MacGregor, BS Shane, J Spalding, J Huff. Carcinogenicity. In: U.S. Congress, Office of Technology Assessment. In Screening and Testing Chemicals in Commerce, OTA-BP-ENV-166 Washington, DC: Office of Technology Assessment, 1995, pp 11–28.
4. KL Dearfield, AE Auletta, MC Cimino, MM Moore. Considerations in the U.S. Environmental Protection Agency's testing approach for mutagenicity. Mutat Res 258:259–283, 1991.
5. AE Auletta, KL Dearfield, MC Cimino. Mutagenicity test schemes and guidelines: US EPA Office of Pollution Prevention and Toxics and Office of Pesticide Programs. Environ Mol Mutagen 21:38–45, 1993.
6. USC (United States Code). NIH/National Institutes of Health Revitalization Act. Public Law 103-43. 42 U.S.C. 1301, 1993.
7. NIEHS (National Institutes of Environmental Health Sciences). Validation and Regulatory Acceptance of Toxicological Test Methods: A Report of the ad hoc Interagency Coordinating Committee on the Validation of Alternative Methods. NIH Publication No. 97-3981. NIEHS, Research Triangle Park, NC, 1997.
8. USC (United States Code). Toxic Substances Control Act. 12 U.S.C. 2601, 1976.
9. USC (United States Code). Federal Insecticide, Fungicide and Rodenticide Act. 7 U.S.C. 135, 1972.
10. USC (United States Code). Clean Air Act. 42 U.S.C. 7401, 1970b.
11. USC (United States Code). Clean Water Act. 33 U.S.C. 121, 1977.
12. USC (United States Code). Occupational Safety and Health Act. 29 U.S.C. 651, 1970a.
13. USEPA. Guidelines for carcinogen risk assessment. Fed Register 51:33992–34003, 1986a.
14. USEPA. Guidelines for mutagenicity risk assessment. Fed Register 51:34006–34012, 1986b.
15. A Auletta, J Ashby. Workshop on the relationship between short-term test information and carcinogenicity: Williamsburg, Virginia, January 20–23, 1987. Environ Mol Mutagen 11:135–245, 1988.
16. NLM (National Library of Medicine). National Institutes of Health, U.S. Environmental Protection Agency, Office of Prevention, Pesticides and Toxic Substances—

GENE-TOX Agent Registry, online computer database—TOXNET file. U.S. Department Health and Human Services, Bethesda, MD 20894, 1999.
17. RW Tennant, BH Margolin, MD Shelby, E Zeiger, JK Haseman, J Spalding, W Caspary, M Resnick, S Stasiewicz, B Anderson, R Minor. Prediction of chemical carcinogenicity in rodents from *in vitro* genetic toxicity assays. Science 236:933–941, 1987.
18. KS Bentley, AM Sariff, MC Cimino, AE Auletta. Assessing the risk of heritable gene mutation in mammals: *Drosophila* sex-linked recessive lethal test and tests measuring DNA damage and repair in mammalian germ cells. Environ Mol Mutagen 23:3–11, 1994.
19. MM Moore, K Harrington-Brock, CL Doerr, KL Dearfield. Differential mutant quantitation at the mouse lymphoma *tk* and CHO *hgprt* loci. Mutagenesis 4:394–403, 1989.
20. VL Dellarco, LR Rhomberg. Quantification of germ-cell risk associated with the induction of heritable translocations. Banbury Report 34:397–411, 1990.

11

Dose–Response Relationships in Chemical Carcinogenesis and Cancer Risk Assessment

Werner K. Lutz
University of Würzburg, Würzburg, Germany

I. INTRODUCTION

A. Tumor Incidence in Animal Bioassays

In a standard test for carcinogenicity of a chemical with 50 animals per dose group, an increase in tumor incidence must be about 10% to become statistically significant. For humans, a 10% lifetime cancer risk (1 in 10) would not be acceptable, unless the exposure is voluntary, such as by cigarette smoking. For most human cancer risk assessments, therefore, extrapolation to lower doses is necessary.

Fitting mathematical models into the observed dose-incidence data points from animal bioassay points can hardly give meaningful results if the extrapolation has to span many orders of magnitude. It is unlikely that the function that best fits the dose response in the high percent range of tumor incidence can correctly describe the shape of the curve at much lower levels (e.g., at 0.001%, i.e., 1 in 100,000). The mechanisms that drive the dose response are expected to differ between the various dose ranges. Therefore, shapes of dose-response relationships in chemical carcinogenesis must be discussed with regard to the underlying mechanisms.

In Table 1, the raw data on tumor induction in the rat by vinyl chloride (VC) (1), formaldehyde (FA) (2), and 2,3,7,8-tetrachlorodibenzo[p]dioxin (TCDD) (3) are given. Vinyl chloride shows a dose-proportional increase up to 500 ppm, followed by a saturation (a supralinear shape) at higher exposure concentrations. Formaldehyde had no effect at the lowest dose level tested; a strongly sublinear (convex) shape was seen at a higher dose. With TCDD, the spontaneous liver tumor incidence of the control group disappeared at the lowest dose level(s), and an increase at the top dose was seen only in females. For cancer risk assessment in humans, the following questions have to be answered:

What cancer incidence is expected at 1 ppm VC, a concentration limit that could be followed at most workplaces? In view of the proportionality seen between 50 and 250 ppm, it appears acceptable to extrapolate linearly down to 1 ppm.

TABLE 1. Tumor Incidence in Rats After Treatment with Vinyl Chloride, Formaldehyde and 2,3,7,8-Tetrachlorodibenzo[p]dioxin (TCDD)

Vinyl chloride*	Exposure concentration (ppm)	Angiosarcoma in the liver
	0	0/60
	50	1/59
	250	4/59
	500	7/59
	2,500	13/59
	6,000	13/60
	10,000	9/61
Formaldehyde†	Exposure concentration (ppm)	Nasal tumors
	0	0/156
	2.0	0/159
	5.6	2/153
	14.3	94/140
TCDD‡	Daily dose (µg/kg)	Liver tumors
Males	0	2/85
	0.001	0/50
	0.01	0/50
	0.1	1/50
Females	0	1/86
	0.001	0/50
	0.01	2/50
	0.1	11/49

*By inhalation; 4 h/day, 5 days/wk, for 12 months; observation for 31 months (Ref. 1).
†By inhalation; 6 h/day, 5 days/wk, for 2 years (Ref. 2).
‡p.o.; for 2 years (Ref. 3).

Dose-Response Relationships

This would result in an estimate of 1 additional tumor case in 3000, an incidence that might be considered tolerable for a workplace.

With FA, the situation is more complex. What cancer incidence is expected at 0.1 ppm, a concentration that can be reached indoors? Linear extrapolation from the 5.6 ppm incidence would result in a tumor incidence of 1 in 765, after adjustment to a 24-hour daily exposure. This would hardly be acceptable for an environmental exposure of a large population, including children.

With TCDD, the situation is intriguing. The observed decrease of the spontaneous liver tumor incidence is obviously not statistically significant, but it occurred in both males and females. What mechanism could drive the observed J-shaped dose response? Could low dose levels even be beneficial with respect to cancer incidence?

B. The Process of Carcinogenesis

To understand the shapes of the dose-response relationships shown, the best approach is to first analyze the process of carcinogenesis and then to investigate the modes of action of the carcinogens.

Figure 1 (4) shows schematically the widely accepted understanding of carcinogenesis as a multistage process based on the accumulation of a number of mutations in critical genes involved in the control of cell division, cell death, and metastatic potential (5). Mutations can be induced by carcinogen-derived DNA

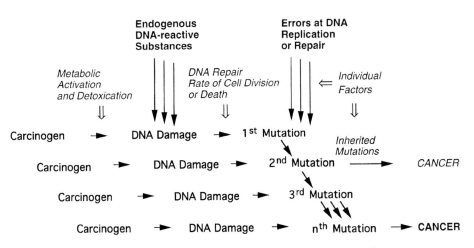

FIG. 1. Schematic representation of the multimutational process of carcinogenesis. Individual factors (*in italics*) modulate the rate of the process or affect the number of steps required for malignant transformation. (From Ref. 4, with permission of Elsevier Science.)

damage but also by damage associated with misrepair, DNA replication, and chromosome segregation. The most direct type of DNA damage is a chemical interaction of the chemical with the DNA molecule, for instance, by covalent binding to form carcinogen-DNA adducts.

C. Background DNA Damage

Mutations and DNA damage are not only the result of exposure to exogenous carcinogens. Endogenous DNA-damaging agents result in a background level of damage (6–8). Table 2 gives some examples. The endogenous part can be subdivided into chemical and biochemical processes, and it includes the consequences of oxygen stress. The exogenous sources have been subdivided according to the question of avoidability rather than type. Numerous exogenous sources can hardly be eliminated. Physical carcinogens, such as UV and ionizing radiation, have for a long time been accepted to result in a background DNA damage. Accordingly, exposure standards at the workplace have been set in relation to a background dose. For chemical carcinogens, this has not been done. The idea prevailed that chemically induced DNA damage could be avoided if it were possible to exclude exposure to exogenous carcinogens. On this basis, exposure standards for exogenous genotoxic carcinogens were often set to 0 (i.e., the "Delaney clause" for carcinogens in foods). Meanwhile, it has become increasingly clear that a zero tolerance is unrealistic. New guidelines for setting exposure limits are required. The knowledge of the background DNA damage observed under various conditions could help set a benchmark.

A total background damage on the order of 1 modification in 100,000 DNA nucleotides has been suggested recently (9). Depurinations, oxidative lesions, alkyl phosphotriesters, and cyclic adducts from unsaturated aldehydes predominated. On the basis of this background DNA damage, it can also be postulated that the process of carcinogenesis has a finite rate in the absence of exogenous DNA-damaging chemicals. A substantial fraction of the total cancer incidence could be explained on that basis (10, 11).

D. Modulation of the Process of Carcinogenesis

For a DNA-reactive chemical to induce a mutation, a number of conditions have to be met. Some are depicted in Figure 2. The first question is whether the phase I reaction of biotransformation results in the formation of a DNA-reactive intermediate. Secondly, does the reactive intermediate escape the various enzymatic and nonenzymatic detoxication processes? Thirdly, does it react with DNA or with another molecule? Although the reaction with water or other small molecules represents detoxification, the formation of protein adducts can be indirectly genotoxic, for instance by disturbing chromosome segregation or by cytotoxicity followed by regenerative hyperplasia.

TABLE 2. Sources of Background DNA Damage

Source	Type	Example
Endogenous		
Chemical	DNA instability	Depurination
Biochemical	Errors during DNA replication	Essential metal ions
	Errors during DNA repair	
	DNA-reactive chemicals	S-AdoMet
		Aldehyde forms of carbohydrates
Physiological	Oxygen stress-derived (ROS)	HO•, NO•, peroxides
		Lipid peroxidation products
Pathophysiological	Formation of carcinogens in vivo	Nitroso-compounds (NOC)
Exogenous		
Hardly avoidable	Radiation	UV, ionizing radiation
	Natural radioactive isotopes	^{222}Rn, ^{40}K
	Carcinogens in ambient air	PAH,* benzene
	Some therapeutic drugs	Tumor therapy
Avoidable in part	Natural dietary carcinogens	Estragole
	Carcinogens from food processing	Urethane
	Food pyrolysis products	Arylamines, PAH, NOC
	Exposure at the workplace	Vinyl chloride
	Carcinogens in ambient air	Passive smoking
Avoidable (in principle)	Active smoking	
	"Unnecessary" dietary, environmental, and work-related exposures	

*Polynuclear aromatic hydrocarbons

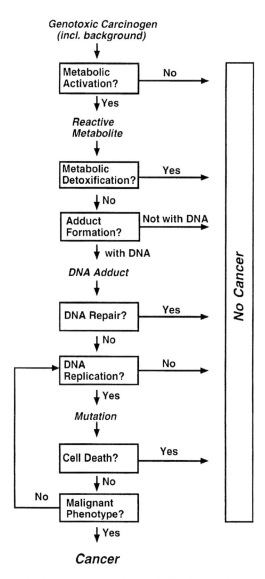

FIG. 2. Sequence of events that modulate the probability of cancer induction from exposure to endogenous or exogenous genotoxic carcinogens.

Dose-Response Relationships

Once a carcinogen-DNA adduct is formed, the question is whether it is repaired before the DNA is replicated. Figure 3 shows how a DNA methylation at the O^6-position of guanine can result in a permanent genetic change after two rounds of DNA replication. O^6-Methylguanine is known to pair with thymine instead of cytosine. If this mismatch is not repaired before the second replication, one of the four daughter cells will have undergone a GC to AT base pair substitution mutation. If the associated change in the amino acid sequence results in the loss of function, for instance of a tumor suppressor protein, one step in the process of malignant transformation could have been taken. A final question in the sequence shown is: does the mutation lead to the death of the cell? If not, the

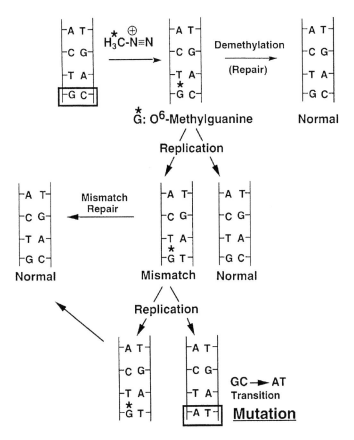

FIG. 3. Schematic representation of the fixation of a DNA damage as a permanent genetic change by two rounds of DNA replication. The example shows the methylation of guanine at the O^6 position, followed by mispairing with thymine, resulting in a GC to AT transition mutation.

cycle can be repeated until the necessary number of permanent changes is reached and the cell gains a fully malignant phenotype.

The sequence of events depicted in Figure 2 is probably not the only way to induce a permanent genetic change. Regulation of gene expression in the course of cell differentiation appears to be irreversible to some extent. It cannot be excluded, therefore, that tumor suppressor genes can be turned off without a change in the respective base sequence. At this time, however, the understanding that a DNA lesion is "fixed" as a mutation in the course of DNA replication is more conspicuous and is also supported by the evidence that nondividing cells rarely transform to malignancy. Other, so-called "epigenetic" mechanisms are not yet as well understood.

E. Susceptibility Factors

At all crossroads shown in Figure 2, numerous factors modulate the rate of the process (12). Genetic polymorphisms in tumor suppressor genes and proto-oncogenes can result in large differences in the susceptibility of individuals to develop cancer. As indicated in Figure 1, this could be caused by a reduction of the *number* of steps necessary for malignant transformation. On the other hand, susceptibility factors can modulate the *rate* of the process. This is expected for polymorphisms of DNA repair enzymes or of enzymes catalyzing metabolic activation and detoxication (13).

Modulation by lifestyle-dependent factors can also be important. For instance, exposure to one carcinogen can result in an increased potency of another carcinogen. Examples are the supra-additive (synergistic) cancer risks from smoking plus alcohol for cancer of the oral cavity, the larynx, and the esophagus, or from smoking plus radon or asbestos for lung cancer.

In conclusion, the potency of a carcinogen is not equal for all individuals. The consequence of this idea for low dose risk assessment will be discussed in chapter III.

II. DOSE-RESPONSE RELATIONSHIPS

A. Superposition of Dose-Response Relationships

The sequence of events shown in Figure 2 indicates that a chemical can increase tumor incidence in a number of ways. A few are listed in Table 3. Besides direct genotoxicity by interaction with the DNA molecule, a chemical can modulate the background process and increase the DNA damage indirectly. One chemical can be effective in more than one way, and each mode of action has its particular dose response. The overall dose-response relationship, therefore, is a superposition of a number of dose-response curves for the various effects of the chemical on the process of carcinogenesis.

Dose-Response Relationships

TABLE 3. Modes of Action of Carcinogens, Including the Increase of the Background Process of Carcinogenesis

Mode of action	Process or example
Interaction with DNA	DNA-adduct formation
Increasing the oxidative DNA damage	Peroxisome proliferation
Increasing the spontaneous mutation rate	Carcinogenic metal ions
Increasing the rate of metabolic activation	P450 inducers
Decreasing the rate of metabolic detoxication	Glutathione depletion
Delaying DNA repair (e.g., MGMT*)	O^6-benzylguanine
Disturbing chromosome segregation	Spindle poisons
Accelerating cell division	Growth factors, hormones
	Cytotoxicity
Affecting cell differentiation	Hormones
Slowing down cell death	"Tumor promoters"

One substance can influence several processes, with different dose-response relationships.
The dose-response relationship for cancer incidence results from a superposition of the various process-specific, dose-response relationships.

*O^6-methylguanine DNA methyltransferase.

B. From Carcinogen Dose to DNA Damage

It is helpful to divide the sequence of events into two parts, (1) the process that determines the level of DNA damage as a function of exposure dose, and (2) the relationship between the DNA lesions and permanent genetic changes, such as mutations. In Figure 4, general shapes for the respective dose-response relationships are shown (4).

All processes that govern the first part are expected to follow first order kinetics, as long as the enzymatic steps involved are far from being saturated. Proportionality is expected at the low-dose end. With increasing dose, enzymatic reactions can become saturated and introduce nonlinearities. If detoxication reactions or DNA repair processes are saturated with increasing dose, a sublinear deviation follows. Alternatively, if the activation of the carcinogen to the reactive intermediate is the rate-limiting step, a supralinear curve will be observed.

These aspects explain parts of the dose response observed with VC and FA shown in Table 1. Between 2 and 5.6 ppm FA, detoxication by glutathione conjugation becomes saturated, and the level of DNA-protein cross-linking increases overproportionately with dose (2). For VC, the formation of the DNA-reactive epoxide metabolite saturates above 500 ppm (1). Therefore, the level of DNA adducts cannot increase proportionally between 500 and 2500 ppm, and the in-

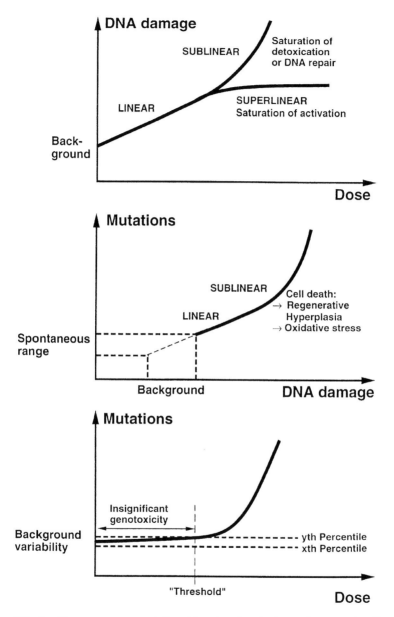

FIG. 4. Dose-response relationships in chemical carcinogenesis. **Top Chart:** DNA damage as a function of dose. **Center Chart:** mutations as a function of DNA damage. **Bottom Chart:** mutations as a function of dose, in a situation of high-dose toxicity associated with low genotoxic potency. (From Ref. 4, with permission of Elsevier Science.)

crease in tumor incidence levels off. A sublinear dose response as a result of saturation of DNA repair could be exemplified for methylating agents by the dose response for both the level of O^6-methylguanine (14) and tumor incidence (15).

In Figure 4, the dose-DNA damage curve does not originate from the origin (0,0) but from the background DNA damage introduced in section I C above. The slope of the "linear" part of the curve is equivalent to the DNA-damaging potency of the carcinogen at low dose. The lower the potency, the less important the increment over background.

For chronic exposure situations that result in a steady-state level of DNA damage, low-dose linearity depends on the assumption that the rate of DNA repair is proportional to the level of DNA damage. This might hold if the carcinogen-derived adducts were the only damage on the DNA. In view of the presence of background damage and its repair, the situation might be more complex.

C. From DNA Damage to Mutation: The Importance of the Rate of Cell Proliferation

The dose-response curve for mutations as a function of the DNA damage is shown in the middle part of Figure 4. Proportionality is postulated at the low-dose end, based on the process illustrated in Figure 3: a doubling of the level of DNA damage is expected to result in a doubling of the number of mutations. Again, the line does not originate from 0 but from the background DNA damage, for which a range has been drawn to account for individual variability.

The slope of the first part of the dose-response relationship depends on the potency of the induced DNA damage to induce a mutation, for instance, with respect to the probability of mispairing. At higher DNA damage levels, a sublinear deviation is postulated as a result of cytotoxicity. Regenerative processes result in an increase in the number of cell divisions per unit time and tissue volume by a premature recruitment of stem cells into the cell cycle. As a consequence, the time available for DNA repair is reduced and the probability of mutation increased (16). Furthermore, cell division is a cancer risk factor in view of mitotic recombinations that can lead to the loss of the functional copy of a tumor suppressor gene in a heterozygous situation. Finally, debris of the necrotic cells will attract macrophages. Their phagocytic activity is expected to increase the level of oxidative DNA damage, which will further increase the probability of mutations. All this results in an upward bending of the dose-response curve, as shown in Figure 4.

Toxicity is of particular relevance when interpreting tumor incidence data from animal experiments. The standard protocol requires the top dose to produce some sign of general toxicity. If this is associated with organ or cell toxicity, the respective tumor incidence data should not be incorporated in any assessment without further examination of the underlying mechanism (17).

These considerations directly concern the issue of FA. The steep slope in the dose response for FA-induced tumor induction between 5.6 and 14.3 ppm was associated with marked toxicity and tissue regeneration, as evidenced by cell proliferation indices in the nose (18). An extrapolation of the tumor incidence data for FA from the higher dose levels is expected to result in an overestimation of the cancer risk at low dose.

D. Sublinearity Versus Threshold

The increase in nasal tumors by FA exposure was observed only at dose levels that also resulted in marked toxicity in the target tissue. Nevertheless, FA can react with DNA and has a direct genotoxic component to its activity. At the low-dose end, the slope of the dose response is expected to have a positive value. This is illustrated in the lower part of Figure 4.

A *true* threshold is defined by slope 0 up to the threshold dose, at which point the curve continues with slope > 0. A distinction between a truly thresholded and a strongly sublinear dose response cannot be made on the basis of experimental data or statistical analyses alone. In practical terms, however, distinct nonlinearities could be handled as thresholds, and dose levels could be defined that have only an insignificant effect. The background process, including its variability within the population, can form the grounds to define the tolerable increment. For most carcinogens, a background exposure is unavoidable. Under these conditions, a substance-specific background can (and should) be taken into account.

The question of "how much is insignificant" cannot be answered on the basis of toxicological criteria alone. Factors such as the exposure situation, the general acceptance of the type of exposure, and the avoidability of the background will have to be included. For some carcinogens, including FA, a part of the background burden is endogenous and therefore truly unavoidable. The increment that gives rise to an insignificant increment, as shown in Figure 4, has to be discussed case by case. In this figure, no numerical value has been attributed to the percentiles of the distribution.

E. Dose-Response Relationships for Indirectly Genotoxic and Tumor-Promoting Agents

"Indirect genotoxicity" means that a carcinogen does not itself interact with the DNA molecule but modulates processes that result in increased DNA damage. This includes a number of mechanisms already addressed in Table 3. The dose response for this type of activity is often considered to be thresholded, in view of the idea that homeostatic control should take care of minor disturbances and that only high doses can result in a significant effect. This interpretation is unlikely to be valid for processes that have a background rate, for instance, for the formation of superoxide anion, hydrogen peroxide, and the extremely reactive hydroxyl radical. Although superoxide dismutase, catalase, and peroxidases control the generation of the hydroxyl radical at a level compatible with life, any increase in the formation of hydrogen peroxide will result in an increased formation of the hydroxyl radical. The slope of the dose-DNA damage curve could well be very small at low dose and bend up only when the detoxication reactions are saturated, but it starts in a linear manner. Therefore, the shapes shown in Figure 4 for genotoxic carcinogens could also be valid for most aspects of indirect genotoxicity.

The question of the dose response for tumor-promoting agents must be answered on the basis of the underlying molecular mechanism(s). "Tumor promotion" is related to the control of cell survival, proliferation, and death, growth advantage and clonal selection, and changes in differentiation. It might come into play particularly in cells that have already taken the first step(s) toward malignant transformation ("initiated" or premalignant cells).

The central question again is whether the chemical accelerates a background process of tumor promotion or whether it does something completely new. In the first case, a linear incremental shape at low dose might be appropriate; in the second situation, a true threshold could be envisaged. Growth factor-like activity (hormone-related carcinogenesis) could be an example for the first and cytotoxicity by particles above the solubility limit (saccharin) for the second. Only in the latter case can the dose response show a true threshold. In all other situations, linearity must be included in the discussion, unless additional activities can be postulated to modulate the process of carcinogenesis in a different manner. Mechanisms to explain putative J-shaped dose-response relationships will be discussed later in this chapter.

III. TIME-TO-TUMOR; SUSCEPTIBILITY DISTRIBUTIONS

A. What Effect Can One Molecule Have?

For all mechanisms that show linear low-dose response, no dose can be said to be without effect. To better understand what effect one single molecule of a DNA-reactive carcinogen could have, it is important to recall that carcinogenesis has a background rate. Represented on a time axis, every person would eventually manifest a tumor, even if we could avoid exposure to all exogenous carcinogens. The time of manifestation differs among individuals, according to the genetic and lifestyle-dependent susceptibility factors already discussed.

A theoretical example is shown in Figure 5. It could represent the situation for prostate cancer, which is diagnosed in about every second 80-year old man. Exposure to a prostate-specific carcinogen would result in an acceleration of the background process of carcinogenesis, so that the tumor would be manifested at

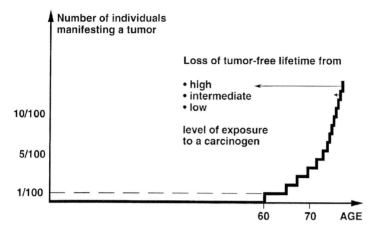

FIG. 5. Individual differences in time-to-tumor (e.g., prostate cancer) in a hypothetical group of 100 men. **Main Line:** background process of carcinogenesis. **Arrows:** acceleration of the process by exposure to a carcinogen, resulting in an earlier tumor manifestation, that is, a dose-dependent loss of tumor-free lifetime.

an earlier age. At high dose levels, most of the steps shown in Figure 1 would have a higher probability, so that the tumor could be manifested many years earlier. For a single low dose, on the other hand, only one step can be affected, the remaining steps being governed by the background process. Under such conditions, the tumor-free lifetime would be shortened only by a small period. For instance, for a tumor with a 4% background incidence within 75 years, and based on a multistage model with 6 stages, the loss of tumor-free lifetime was estimated to be on the order of 1 day, under conditions that would have resulted in an additional tumor incidence of 1 in 100,000 (19).

In terms of an interpretation of low-dose effects, it can be stated that even a minute dose of a DNA-reactive carcinogen can have an "effect." However, this effect can only result in a very small loss of the tumor-free lifespan and cannot be the induction of a tumor "out of the blue." At low dose, the background process determines the age at manifestation of a tumor.

B. Dose-Incidence Versus Time-Incidence Relationships

The connection between a *dose*-incidence relationship (Table 1; Fig. 4) and a *time*- (or *age*-)incidence relationship (Fig. 5) is schematically shown in Figure 6 for a control group and three dosed groups with 10 individuals each. A time-incidence relationship (upper part) provides information on all individuals and their time-to-tumor. In the control group (dose = 0), spontaneous carcinogenesis was assumed to result in a 30% tumor incidence within the mean lifespan (three

Dose-Response Relationships

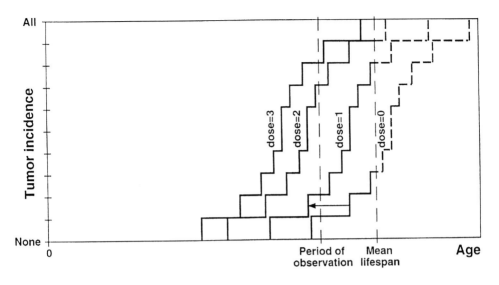

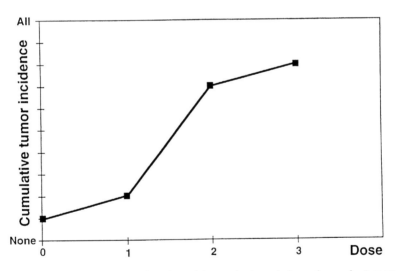

FIG. 6. Connection between time-(age-)dependent and dose-dependent representations for tumor incidence. **Top Chart:** schematic representation of the age-dependent tumor manifestation in four dose groups of 10 individuals. **Bottom Chart:** dose-response relationship for tumor incidence within the period of observation marked in the top chart.

tumors in the group of 10). With increasing dose, the curves are shifted to the left in some dose-dependent manner. At the top dose chosen, a 100% tumor incidence was assumed to be reached within the mean lifespan.

In both rodent bioassays and human cancer registries, the period of observation normally is not the mean lifespan but is a fixed period, for instance 2 years in rats, 65 or 75 years in humans. Based on the period of observation marked in the upper part of Figure 6, the incidence is 1, 2, 7, and 8 out of 10, at doses 0, 1, 2, and 3, respectively. Plotting these numbers as a function of dose, the familiar dose-response relationship results (lower part of Fig. 6). This representation provides information only about the given period of observation, and information on individual times-to-tumor is lost.

C. Who is Going to be Hit at the Lowest Dose?

When going from dose = 0 to dose = 1 in the upper part of Figure 6, one additional individual manifests the tumor within the period of observation (see the arrow). In view of the fact that the 10 individuals differ with respect to the rate of the background process of carcinogenesis, this also means that the newly recruited individual was the most susceptible to the combined action of the background plus the carcinogen-induced increment. The first step seen between dose 0 and 1 in the lower part of Figure 6 (the dose-incidence curve) was based on the most susceptible individual. Therefore, *the dose-response relationship is governed by the susceptibility distribution*.

Cancer risk assessment for low dose usually aims at estimating a number of additional cancer cases at a given exposure level (e.g., 1 in 100,000). However, among these 100,000 persons, the risk of being the victim at the given dose is not a matter of chance. The more we know about the mechanisms of carcinogenesis and about the individual modulation of the rate of the process, the smaller will be the stochastic part of the process. It will be most important to use the mechanistic information on the mode of carcinogenic action of a chemical to define those subpopulations or individuals who might be the most susceptible, that is, who runs the highest risk.

In the example of FA, nasal tumors in rats were seen only at dose levels that resulted in (a) saturation of detoxication reactions plus (b) toxic irritation and regeneration of the nasal epithelium. For FA-exposed humans, it would be reasonable to assume that the individuals at highest risk would be those who are either low in detoxication or who have pre-existing lesions in the respiratory tract. It is interesting to note a putative synergism of FA with wood dust from cancer epidemiology in chipboard production workers. A damaging effect of particles in the respiratory tract could have laid the ground for a particular susceptibility of this subpopulation for the additional toxicity of FA.

In terms of prevention of cancer risks from largely unavoidable sources, it might be most effective to identify susceptible subpopulations and to take measures to eliminate cocarcinogenic factors or add chemopreventive factors. This approach might also be appropriate for situations in which a general reduction of an exposure level might incur disproportionate costs.

IV. HYPOTHESES

A. J-Shaped Dose Response and Thresholds for DNA-Reactive Carcinogens?

Earlier in this chapter, it was shown that the rate of induction of mutations is proportional to both the level of DNA damage and the rate of cell division. Recent evidence now indicates that the two aspects are not independent of each other: DNA damage can result in a delay of the cell cycle (20, 21). An increase in the mutation rate by a carcinogen-induced DNA damage could be counteracted by a decrease in the rate of cell division. This effect is expected to be limited to a nontoxic dose range. At higher dose levels of a DNA-damaging carcinogen, cytotoxicity and regenerative hyperplasia might accelerate cell proliferation. Over the entire dose range, a J-shaped dose-response relationship could result for mutations. This is shown schematically in Figure 7, where a dose-linear increase in the level of DNA damage (upper part), multiplied with a J-shaped dose response for the rate of cell division (middle part), can result in a J-shaped dose response for mutation (lower part) if the decrease in the rate of cell division is larger than the increase in DNA damage.

Cell proliferation is important not only concerning the fixation of a DNA damage as a mutation but also with regard to the process of clonal expansion of premalignant cells. Using the two-mutation, clonal-expansion model of carcinogenesis, the requirements to produce a J-shaped dose-response curve from a dose-linear increase in the level of mutation and a J-shaped dose response for cell turnover were analyzed (22). The background values chosen for the model parameters resulted in a 10.5% "spontaneous" 2-year cumulative tumor incidence. Using this as a starting point, a decrease by 3%, 10%, and 30% in the rates of cell turnover resulted in a decrease in the spontaneous tumor incidence to 9.4%, 7.1%, and 3.0%. J-Shaped dose responses for the rates of cell turnover were modeled by parabolic curves, having their minimum at dose 1. Combination with linearly increasing mutation rates generated, under certain conditions, J-shaped dose-response curves for tumor incidence. One example is shown in Figure 8. The DNA-damaging potency of the carcinogen was assumed to result in a 30% increase in the spontaneous mutation rate at dose 1. The decrease in cell turnover at dose 1 was assumed to be 3%, 10%, or 30%. For the 10% and 30% decrease, the dose-linear increase in DNA damage was more than compensated for, result-

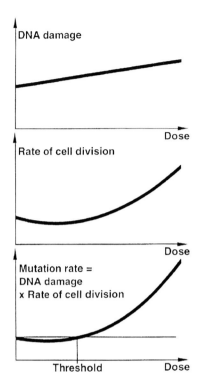

FIG. 7. Multiplicative combination of a linear dose-response relationship for DNA damage (**top chart**) with a J-shaped rate of cell proliferation (**center chart**) to result in a J-shaped dose response for mutations (**bottom chart**). (From Ref. 4, with permission of Elsevier Science.)

ing in a J-shaped dose response for tumor induction. "Threshold" doses of 0.8 and 1.6 dose units, respectively, could be deduced.

A decrease of the cell turnover by 10% or 30% could well be within physiological limits. For instance, treatment of rats with caffeic acid for 4 weeks resulted in a J-shaped dose response for the DNA replication in the forestomach, with an observed maximum decrease by as much as 46% (23).

B. Anticarcinogenic Effects of Carcinogens

The TCDD example shown in Table 1 is only one of a number of examples of anticarcinogenic effects of carcinogens. In an analysis of 218 bioassays for carcinogenicity performed by the National Toxicology Program, more than 90% of the tested chemicals showed at least one statistically significant decrease in site-specific tumor incidence (24). Random variability and reductions associated with

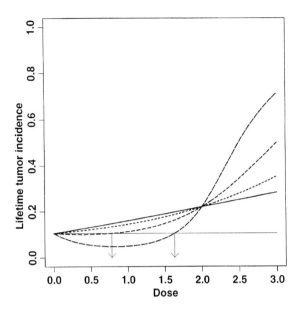

FIG. 8. Shapes of dose-response curves generated by superposition of a dose-linear increase in the rates of mutation with J-shaped dose-response curves for the rates of cell birth and death, in the two-stage clonal expansion model of carcinogenesis. The J-shape was generated by parabolic, quadratic functions intersecting the dose axis at dose 0 and 2 and reaching their minimum at dose 1. Percent decrease in the background rate of cell turnover at dose 1: - - - - 3%; – – – – 10%; —— — 30%). The full line represents the dose response obtained from the increase in the mutation rates alone (+30% at dose 1). Model parameters chosen for the background process: 1 million normal stem cells; background mutation rate of $\mu N = \mu P = 10^{-7}$ per day for both steps; birth and death rates for the premalignant cells $\beta = 0.5$, $\delta = 0.49$ per day. On this basis, the (spontaneous) cumulative "tumor incidence" over 2 years was 10.5%. (See text for discussion and Ref. 22 for details.)

reduced body weight can account for many of these decreases. For others, the mechanism by which reduced tumor incidence is achieved is yet to be determined, and aspects such as those in the preceding Section will have to be included.

For TCDD, J-shaped dose-response relationships were not only reported for liver tumors but also for altered hepatic foci (25) and for the rate of cell proliferation (26). A causal relationship between the three endpoints appears plausible. It also means that an inclusion of the high-dose data in a linearized extrapolation model is unlikely to give the best possible estimate of a low-dose cancer risk. A threshold model appears to be more appropriate.

C. Additivity Theory to be Revised?

J-Shaped dose responses for DNA-damaging carcinogens contradict the hypothesis that even the lowest doses of DNA-reactive carcinogens produce a damage increment to the background so that the cancer risk has to increase in an additive manner (27). The limitation in this argumentation is the understanding that all types of DNA damage add up with respect to their mutational consequences. In light of observations that DNA damage can affect the cell cycle, this hypothesis will have to be revised. Each type of DNA damage is expected to have its own specific consequence so that it cannot be considered to be just an increment to what is already there. This idea is supported by the finding that a number of DNA-reactive carcinogens generated J-shapes for the level of single-strand breaks in rat liver DNA (28).

V. CONCLUSIONS

The shape of a carcinogen dose-cancer incidence curve is the result of a superposition of dose-response relationships for various effects of the carcinogen on the process of carcinogenesis, including a modulation of the background process. In the dose range that results in observable increases of tumor incidence, nonlinearities as a result of saturation phenomena are the rule rather than the exception.

For the discussion of low-dose effects, the element of time is introduced. Based on the background rate of the process of carcinogenesis, a single low dose of a DNA-reactive carcinogen can have an effect on one or the other step of the multistage process, but this can only result in a shortening of the tumor-free lifetime and not in tumor formation "out of the blue."

Individual susceptibility is brought into the discussion as the basis of dose-response relationships. In low dose/low incidence situations, the chance for an individual to be hit depends on his/her genetic and lifestyle-derived susceptibility. For cancer prevention, identification of susceptible subpopulations might be more (cost-)effective than a general reduction of an exposure level.

Evidence showing that DNA damage can affect the cell cycle is introduced to challenge the additivity hypothesis for an incremental effect of carcinogen-induced DNA damage. Using the two-stage clonal expansion model, it is shown that a dose-linear increase in DNA damage can be counteracted effectively by reductions in the rate of cell turnover well within physiological limits. Therefore, linearity of the low-dose response curve for DNA-reactive carcinogens can be postulated for adduct formation but not for the induction of mutations or cancer.

REFERENCES

1. PJ Gehring, PG Watanabe, CN Park. Resolution of dose-response toxicity data for chemicals requiring metabolic activation: Example—vinyl chloride. Toxicol Appl Pharmacol 44:581–591, 1978.
2. JA Swenberg, CS Barrow, CJ Boreiko, HA Heck, RJ Levine, KT Morgan, TB Starr. Non-linear biological responses to formaldehyde and their implications for carcinogenic risk assessment. Carcinogenesis 4:945–952, 1983.
3. R Kociba. Rodent bioassays for assessing chronic toxicity and carcinogenic potential of TCDD. In: MA Gallo, RJ Scheuplein, K Van der Heijden, eds. Biological Basis for Risk Assessment of Dioxins and Related Compounds. Vol. 35, Banbury Report. Cold Spring Harbor: Cold Spring Harbor Laboratory Press, 1991, pp 3–11.
4. WK Lutz. Dose-response relationships in chemical carcinogenesis: Superposition of different mechanisms of action, resulting in linear-sublinear curves, practical thresholds, J-shapes. Mutat Res 405:117–124, 1998.
5. CC Harris. Chemical and physical carcinogenesis: Advances and perspectives for the 1990s. Cancer Res (suppl) 51:5023s–5044s, 1991.
6. BN Ames. Endogenous DNA damage as related to cancer and aging. Mutat Res 214:41–46, 1989.
7. LA Loeb. Endogenous carcinogenesis: Molecular oncology into the twenty-first century—Presidential address. Cancer Res 49:5489–5496, 1989.
8. WK Lutz. Endogenous genotoxic agents and processes as a basis of spontaneous carcinogenesis. Mutat Res 238:287–295, 1990.
9. RC Gupta, WK Lutz. Background DNA damage from endogenous and unavoidable exogenous carcinogens: A basis for spontaneous cancer incidence? Mutat Res 424:1–8, 1999.
10. WK Lutz, T Fekete. Endogenous and exogenous factors in carcinogenesis: Limits to cancer prevention. Int Arch Occup Envir Health 68:120–125, 1996.
11. M Otteneder, WK Lutz. Correlation of DNA damage with tumor incidence: Carcinogenic potency of DNA adducts. Mutat Res 424:237–248, 1999.
12. CC Harris. Interindividual variation among humans in carcinogen metabolism, DNA adduct formation and DNA repair. Carcinogenesis 10:1563–1566, 1989.
13. A D'Errico, E Taioli, X Chen, P Vineis. Genetic metabolic polymorphisms and the risk of cancer: A review of the literature. Biomarkers 1:149–173, 1996.
14. WK Lutz. Inducible repair of DNA methylated by carcinogens. Trends Pharmacol Sci 3:398–399, 1982.
15. L Zeise, R Wilson, EAC Crouch. Dose-response relationships for carcinogens: A review. Environ Health Perspect 73:259–308, 1987.
16. SM Cohen, LB Ellwein. Genetic errors, cell proliferation, and carcinogenesis. Cancer Res 51:6493–6505, 1991.
17. A Apostolou. Relevance of maximum tolerated dose to human carcinogenic risk. Regul Toxicol Pharmacol 11:68–80, 1990.
18. TM Monticello, JA Swenberg, EA Gross, JR Leininger, JS Kimbell, S Seilkop, TB Starr, JE Gibson, KT Morgan. Correlation of regional and nonlinear formaldehyde-

induced nasal cancer with proliferating populations of cells. Cancer Res 56:1012–1022, 1996.
19. WK Lutz, D Gaylor. Significance of DNA adducts at low dose: Shortening the time to spontaneous tumor occurrence. Regul Toxicol Pharmacol 23:29–34, 1996.
20. LH Hartwell, TA Weinert. Checkpoints: Controls that ensure the order of cell cycle events. Science 246:629–634, 1989.
21. SJ Elledge. Cell cycle checkpoints: Preventing an identity crisis. Science 274:1664–1671, 1996.
22. WK Lutz, A Kopp-Schneider. Threshold dose response for tumor induction by genotoxic carcinogens modeled via cell-cycle delay. Toxicol Sci 49:110–115, 1999.
23. U Lutz, S Lugli, A Bitsch, J Schlatter, WK Lutz. Dose response for the stimulation of cell division by caffeic acid in forestomach and kidney of the male F344 rat. Fund Appl Toxicol 39:131–137, 1997.
24. JK Haseman, FM Johnson. Analysis of National Toxicology Program rodent bioassay data for anticarcinogenic effects. Mutat Res 350:131–141, 1996.
25. HC Pitot, TL Goldsworthy, S Moran, W Kennan, HP Glauert, RR Maronpot, HA Campbell. A method to quantitate the relative initiating and promoting potencies of hepatocarcinogenic agents in their dose-response relationships to altered hepatic foci. Carcinogenesis 8:1491–1499, 1987.
26. RR Maronpot, JF Foley, K Takahashi, T Goldsworthy, G Clark, A Tritscher, C Portier, G Lucier. Dose response for TCDD promotion of hepatocarcinogenesis in rats initiated with DEN—histologic, biochemical, and cell proliferation endpoints. Environ Health Perspect 101:634–642, 1993.
27. D Krewski, DW Gaylor, WK Lutz. Additivity to background and linear extrapolation. In: S Olin, W Farland, C Park, L Rhomberg, R Scheuplein, T Starr, J Wilson, eds. Low-Dose Extrapolation of Cancer Risks: Issues and Perspectives. Washington, DC: ILSI/International Life Sciences Institute, 1995, pp 105–121.
28. KT Kitchin, JL Brown. Dose-response relationship for rat liver DNA damage caused by 49 rodent carcinogens. Toxicology 88:31–49, 1994.

12

Molecular Epidemiology and Biomarkers

Jia-Sheng Wang, Jonathan M. Links, John D. Groopman
Johns Hopkins University, Baltimore, Maryland

I. INTRODUCTION

Molecular epidemiology encompasses the use of biomarkers in epidemiological research through the incorporation of molecular, cellular, and other biochemical measurements into studies of etiology, prevention, and control of health risks encountered by human populations (1–3). Application of validated biomarkers to traditional epidemiological studies helps to delineate the continuum of events between an exposure and resulting disease; to identify smaller exposures to specific xenobiotics; to indicate earlier events in the natural history of diseases and reduce misclassification of dependent and independent variables; to enhance individual and group risk monitoring and assessments; and to reveal toxicologic mechanisms by which an exposure and a disease are related (3). A critical feature of molecular epidemiologic studies is the interdisciplinary collaboration between population, field scientists, and laboratory scientists from various disciplines, such as epidemiology, toxicology, molecular biology, genetics, immunology, biochemistry, pathology, and analytical chemistry. Because the analytic measurement of biomarkers are central to molecular epidemiologic studies, special attention to the collection, handling, and storage of biological specimens, as well as development and validation of analytical methods, is required (2). The use of biomarkers can also speed the process of risk assessment, because molecular epidemiologic studies many categories of human risk monitoring data (4).

Extrapolation of animal or other experimental data to humans is a traditional method of evaluating potential risk of disease from an exposure. Molecular epidemiology has the advantage of being directly targeted to human populations, in contrast to traditional epidemiology studies. These investigations have the potential to give an early warning by indicating preclinical effects of exposure and increased susceptibility. This should provide earlier evidence of associations between exposure and disease in population-based studies and signal opportunities to avert the related disease through timely intervention in specific individuals. Moreover, biomarker data depicting the distribution of preclinical changes and susceptibility factors in a population can improve the quantitative estimation of human risk from a given exposure (5). Over the past 15 years, the development and application of molecular biomarkers reflecting events from exposure through the formation of clinical disease have rapidly expanded our knowledge of the pathogenic mechanisms of human chronic disease, such as cancer and cardiovascular diseases, and have provided opportunities for development of improved tools for the treatment and prevention of these diseases.

II. OVERVIEW OF BIOMARKERS

As adapted from a report by the Committee on Biological Markers of the National Research Council (6), the development of disease that results from exposure to an environmental agent or other toxicant is multistage, starting with exposure, progressing to internal dose (e.g., deposited body dose), to biologically effective dose (e.g., dose at the site of toxic action), to early biological effect (e.g., at the subcellular level), to altered structure or function (e.g., subclinical changes), and finally, to frank, clinical disease. Any step in this process may be modified by host-susceptibility factors, including genetic traits and effect modifiers, such as diet or other environmental exposures. Therefore, biomarkers are indicators of events for physiological, cellular, subcellular, and molecular alterations in the multistage development of specific diseases (6).

Molecular biomarkers are typically used as indicators of exposure, effect, or susceptibility (6, 7). A biomarker of exposure refers to measurement of the specific agent of interest, its metabolite(s), or its specific interactive products in a body compartment or fluid, which indicates the presence (and magnitude) of current and past exposure. A biomarker of dose bears a quantitative/qualitative relationship with exposure. Such a biomarker may be an exogenous substance, an interactive product (e.g., between a xenobiotic compound and endogenous components), or other indicator. A biomarker of effect indicates the presence (and magnitude) of a biological response to exposure to an environmental agent. Such a biomarker may be an endogenous component, a measure of the functional capacity of the system, or an altered state recognized as impairment or disease. A biomarker of susceptibility is an indicator or a measure of an inherent or acquired ability of an organism to respond to the challenge of exposure to a specific xenobiotic substance or other toxicant. Such a biomarker may be the unusual presence or absence of an endogenous component or an abnormal functional response to an administered challenge. Molecular epidemiology and molecular dosimetry thus have great utility in addressing the relationships between exposure to environmental agents and development of clinical diseases and in identifying those individuals at high risk for the disease (1, 8).

The validation and application of molecular biomarkers for environmental agents should be based on specific knowledge of metabolism, interactive product formation, and general mechanisms of action (9, 10). Examples are studies on the relationships between tobacco smoking and lung cancer (11) and between aflatoxin exposure and liver cancer (12). A specific application of biomarker technology to human cancer is the study of the variation in response among individuals after exposures to tobacco products. For example, even in heavy tobacco smokers, less than 15% of these people develop lung cancer (13); thus, intrinsic susceptibility factors must affect the time course of disease development and eventual outcome. The identification of those at highest risk for developing cancers will be facilitated by biomarker studies. Extensive efforts have been made in the identification of these high-risk individuals through the use of various genetic and metabolic susceptibility markers, e.g., measurement of polymorphism of genotype and phenotype of various enzymes involved in phase I and phase II metabolic reactions of certain known carcinogens (14–17). This strategy has not yet proven to be broadly applicable to many other human diseases, although progress is being made for many types of cancers (2).

The validation of any biomarker-effect link requires parallel experimental and human studies (12). Ideally, an appropriate animal model is used to determine the associative or causal role of the biomarker in the disease or effect pathway, and to establish relations between dose and response. The putative biomarker can be validated in pilot human studies, where sensitivity, specificity, accuracy, and reliability parameters can be established. Data obtained in these studies can then be used to assess intra- or interindividual variability, background levels, relationship of the biomarker to external dose or to disease status, as well as feasibility for use in larger population-based studies. It is important to establish a connection between the biomarker and exposure, effect, or susceptibility. To fully interpret the information that the biomarker provides, prospective epidemiological studies may be necessary to demonstrate the role that the biomarker plays in the overall pathogenesis of the disease or effect. To date, few biomarkers have been rigorously validated using this entire process.

A. Biomarkers of Exposure

Although biomarkers of exposure can refer to any biomarker used to estimate current or past exposure to a specific environmental agent, the traditional definition of an exposure biomarker involves measurement of a xenobiotic, its metabolite, or its interactive products found in body tissue, fluids, and excreta, such as blood, urine, feces, or milk (18). These measures provide information about the actual concentration or internal dose of a specific agent that has been absorbed and distributed in the body. Measurement of the xenobiotic itself or its metabolites has been incorporated into a number of human epidemiological studies. For example, excretion of aflatoxin M_1, one of the major metabolites of aflatoxin B_1, has been used as a biomarker for the evaluation of human exposure to aflatoxin, and this marker was found to be associated with the risk of liver cancer (19, 20). Specific metabolites of one of the tobacco-specific nitrosamines, 4-(methylnitrosamino)-1-(3-pyridyl)-1-butanone (NNK), a potent chemical carcinogen, have been detected and quantified in the urine of smokers, but these metabolites were not found in the urine of nonsmokers (21). Intraindividual and interindividual variations in these metabolites of NNK in smokers' urine were noted, and this might be important in disease risk (22). Other examples include the measurement of blood and serum levels of heavy metals and pesticides (23, 24), such as DDE (1,1-dichloro-2,2-bis(p-chlorophenyl)-ethylene), the major metabolite of DDT (2,2-bis[p-chlorophenyl]-1,1,1-trichloroethane), which has been used as a biomarker in breast cancer studies in women (25, 26).

The metabolically activated ultimate forms of environmental carcinogens can covalently interact with cellular macromolecules such as DNA and proteins (27–30). These carcinogen-macromolecular adducts have an important role in human biomonitoring and molecular epidemiological studies (2, 9). They are specific biomarkers that provide a way to measure human exposure to these chemical carcinogens and provide information about specific dose to a carcinogen target site (DNA or protein). Moreover, it may be possible to establish a correlation between tumor incidence and exposure by measuring these adducts' level (5). Many different analytical methods have been developed to identify and measure carcinogen-macromolecular adducts, including enzyme-linked immunosorbent assay (ELISA), radioimmunoassay (RIA), and immunohistochemical staining assay (IHC), radiometric postlabeling methods, high performance liquid chromatography (HPLC), gas chromatography/mass spectrometry (GC/MS), liquid chromatography/mass spectrometry (LC/MS), and fluorescence spectroscopy (FS) (2, 8, 31, 32). These techniques have been used to measure composite and specific DNA adducts in cellular DNA isolated from peripheral lymphocytes, bladder, breast, lung, and colonic tissues, as well as excreted DNA adducts in urine. These methods are also used to measure hemoglobin (Hb) and albumin adducts in blood samples of people exposed to environmental carcinogens and

Molecular Epidemiology and Biomarkers

xenobiotics. In addition, these techniques have been applied in the clinical setting to examine carcinogen-macromolecular adducts of people undergoing chemotherapy with alkylating agents, in an attempt to associate adduct levels with clinical outcome (33, 34). Recently, these methods have also been applied to human clinical trials to validate various intervention tools for the assessment of chemopreventive agents in modulating various intermediate biomarkers (12).

In addition to the use of biomarkers for exogenous exposure, it is desirable to have a biomarker of endogenous oxidative damage, as many studies have found that endogenous oxidative DNA damage correlates with the formation of chronic degenerative diseases, including cancer (35). Among the many oxidatively damaged DNA bases formed, 8-oxo-2-deoxyguanosine is a lesion that can be sensitively measured. Several techniques, including HPLC-electrochemical detection (HPLC-EC), have been developed and applied to detect this damage product in biofluids and tissue samples from animals and humans (36). Immunoaffinity column methods have also been described for the analysis of oxidative damage products of nucleic acids excreted in urine (37). Quantitative analysis of these adducts in urine and tissues may eventually be used to assess the risk of disease from oxidative damage in people.

B. Biomarkers of Effect

A biomarker of effect has been defined by The International Programme on Chemical Safety as a measurable biochemical, physiological, behavioral, or other alteration within an organism that, depending on the magnitude, can be recognized as associated with an established or possible health impairment or disease (38). Although this broad definition covers the wide spectrum of functional alterations, in practice, biomarkers of effect represent changes at the subcellular level, particularly at the chromosomal and molecular levels, such as cytogenetic alterations and gene mutations (18).

Cytogenetic biomarkers currently applied in molecular epidemiological studies include chromosome aberrations (CAs), sister chromatid exchanges (SCEs), and micronuclei (MN). Chromosome aberrations are structural alterations, breaks, and rearrangements in chromosomes. Recently developed analytic methods using fluorescent in situ hybridization (FISH) and polymerase chain reaction (PCR) provide detection of new types of rearrangements and translocations of specific regions in certain chromosomes (39, 40). Exposure to ionizing radiation, alkylating cytostatics, tobacco smoking, benzene, and styrene has been found to induce CAs in humans (18). Sister chromatid exchange represent symmetrical exchanges of DNA segments between the sister chromatids of a duplicated metaphase chromosome. Tobacco smoking, alkylating cytostatics, and ethylene oxide can induce SCEs in human lymphocytes (41). Micronuclei are small additional nuclei observable in interphase cells. Increased MN frequencies in human lymphocytes have been found after exposure to ionizing radiation and formaldehyde (42, 43).

Biomarkers of gene mutations include somatic mutations in surrogate tissues and gene mutations in target tissues. The hypoxanthine phosphoribosyltransferase (HPRT) gene mutation and the glycophorin A (GPA) assay are two somatic gene mutation assays currently applied in molecular epidemiological studies for human risk monitoring (18). Gene mutations, particularly in critical target genes, are important biomarkers of biological effects, altered function, and preclinical disease (44, 45). Activation of protooncogenes and inactivation of tumor suppressor genes caused by mutations are critical genetic changes linked to eventual cancer formation. For example, the *ras* protooncogenes are targets for many genotoxic carcinogens. Activation of *ras* is an early event, possibly the initiating step, in the development of many chemical carcinogen-induced rodent tumors (46). The *ras* oncogene is also observed in different types of human tumors and at a higher frequency than any other oncogene. Both the activation of *ras* oncogenes and the inactivation of several suppressor genes, including *p53*, have been found in the development of human colon and lung cancers (47).

The tumor suppressor gene *p53*, the most commonly mutated gene detected in human cancers, has been used as a biomarker for molecular carcinogenesis, molecular epidemiology, and cancer risk assessment (48, 49). The number and type of mutations in this gene are not equally distributed but occur in specific hotspots that vary with tumor type (50). The differences in mutation patterns between tumors are consistent with different etiologies for the specific tumor types. One striking case in this research field is the studies on the relationship between aflatoxin exposure and development of human hepatocellular carcinoma (HCC), as summarized below.

The initial results from three independent studies of *p53* mutations in HCCs occurring in populations exposed to high levels of dietary aflatoxin showed high frequencies of G to T transversions, with clustering at codon 249 (51–53). On the other hand, studies of *p53* mutations in HCCs from Japan and other areas where there is little exposure to aflatoxin revealed no mutations at codon 249 (54). These studies provided the circumstantial, but as yet unproven, linkage between this signature mutation of aflatoxin exposure and the events detected in *p53* in liver tumors from China and Southern Africa.

Fujimoto et al (55) further examined the hypothesis that exposure to aflatoxin B_1 (AFB_1), either alone or coincident with other environmental carcinogens, might be associated with occurrence of allelic losses during the development of HCC in China. The HCC tissues were obtained from two different areas in China: Qidong, where exposure to hepatitis B virus (HBV) and AFB_1 is high, and Beijing, where exposure to HBV is high but to AFB_1 is low. They analyzed the tumors for mutations in the *p53* gene and loss of heterozygosity for the *p53*, *Rb*, and *APC* genes. The frequencies of mutation, loss, and aberration

Molecular Epidemiology and Biomarkers

(either mutation or loss) of the *p53* gene in 25 HCC specimens from Qidong were 60%, 58%, and 80%, respectively. The frequencies in nine HCC specimens from Beijing were 56%, 57%, and 78%; however, the frequency of a G to T transversion at codon 249 in HCCs from Qidong and Beijing were 52% and 0%, respectively. These data indicate that mutation or loss of heterozygosity in the *p53* gene, independent of the codon 249 mutation, plays a critical role in the development of HBV-associated HCCs in China. These results also show distinct differences in the pattern of allelic losses between HCCs in Qidong and Beijing, and suggest that AFB_1 and other environmental carcinogens may contribute to this difference.

The observation of the codon 249 mutation in *p53* with aflatoxin exposure is not limited only to China and Southern Africa. Senegal, a country where liver cancer incidence is one of the highest in the world, has high exposure to aflatoxins. Fifteen liver cancer samples were examined for mutation at codon 249 of the *p53* gene (56). Nontumoral DNA from the patients showed a wild type genotype. Mutation at codon 249 of the *p53* gene was detected in 10 of the 15 tumor tissues tested (67%). This frequency of mutation in codon 249 of the *p53* gene is the highest described to date in the literature, and these results confirm that there is an association between countries of high aflatoxin intake and a high frequency of mutation in codon 249 of the *p53* gene, and that HBV alone does not contribute to these base changes. Aguilar et al (57) re-examined the role of AFB_1 and *p53* mutations in HCCs and in normal liver samples from the United States, Thailand, and Qidong, where AFB_1 exposures are negligible, low, and high, respectively. The frequency of the AGG to AGT mutation at codon 249 paralleled the level of AFB_1 exposure, which also supports the hypothesis that this mycotoxin has a causative and probably early role in hepatocarcinogenesis.

Results from experimental studies have also linked aflatoxin as a causative agent in the described *p53* mutations (58). Previous work had shown that AFB_1 exposure causes almost exclusively G to T transversions in bacteria (59), and that aflatoxin-epoxide can bind to codon 249 to p53 in a plasmid in vitro, providing further indirect evidence for a putative role of aflatoxin exposure in *p53* mutagenesis (60). Aguilar et al (61) studied the mutagenesis of codons 247–250 of *p53* by rat liver microsome-activated AFB_1 in human HCC cells HepG2 using a RFLP/PCR genotypic strategy. They found that AFB_1 preferentially induced the transversion of G to T in the third position of codon 249; however, AFB_1 also induced G to T and C to A transversions into adjacent codons, although at lower frequencies. Cerutti et al (58) studied the mutability of codons 247–250 of *p53* with AFB_1 in human hepatocytes using a similar technique. Aflatoxin B_1, preferentially induced the transversion of G to T in the third position of codon 249, generating the same mutation found in a large fraction of HCCs from regions of the world with AFB_1-contaminated food. These experimental results support AFB_1 as an etiological factor for HCCs in AFB_1-contaminated areas (45).

C. Biomarkers of Susceptibility

Biomarkers of susceptibility are mainly concerned with factors in kinetics and dynamics of uptake and metabolism of exogenous chemicals. The enzymes involved in activation and detoxification of these xenobiotics are divided into two categories: phase I enzymes, mainly the superfamily of cytochrome P450 mixed function oxidase enzymes, and phase II enzymes, which act on an oxidized substrate to conjugate them with various moieties, such as glucuronic acid, glutathione, and sulfate (62–64). Genetic differences in the expression of these metabolic enzymes could be a major source of interindividual variation in susceptibility to disease (65–67). Therefore, the determination of genotype and phenotype of these metabolic enzymes in different populations is being studied to determine if an association exists between exogenous exposure and formation of specific disease in specific metabolic genotype subsets. Many studies over the past several years have found that genes involved in xenobiotic metabolisms, including the cytochrome P450 enzymes CYP1A1, CYP1A2, CYP2A6, CYP2D6, CYP2E1, N-acetyltransferase 1 and 2 (NAT1 and NAT2), and glutathione S-transferase µ and theta (GSTM1 and GSTT1), are polymorphic in human populations, and in some cases specific alleles are associated with increased risk of a variety of different cancers. Further, studies on the genotypes for human cytochrome P450 enzymes in diverse populations have also found ethnic differences in transcription and translation of these enzymes (68–70).

The acetylator biomarkers are another group based on metabolic susceptibility genes that could be important in aromatic and heterocyclic amine exposures (71, 72). Acetyltransferases catalyze both N- and O-acetylation reaction. N-acetylation is a detoxification step for arylamines like 2-naphthylamine and 4-aminobiphenyl (4-ABP), but O-acetylation of N-hydroxy arylamine can form esters that are highly reactive, inducing DNA damage (73–75). Through the use of various biomarkers, such as caffeine metabolites, as indices of acetylator phenotype, several epidemiological studies have found an association between the slow acetylation phenotype and risk for developing bladder cancer, particularly for those people who are occupationally exposed to aromatic amines (76, 77). Additionally, several studies suggest a relationship between the rapid acetylator phenotype and colon cancer (78, 79).

N-acetyltransferases are coded by two distinct genes located in humans on chromosome 8 and designated as NAT1 and NAT2 (80). Polymorphism in NAT2 results from point mutations in the coding regions of this intronless gene, which phenotypically leads to slow and rapid metabolism (81). Slow acetylators are homozygous for the slow acetylator gene, which was found in 5% of Canadian Eskimos, in 10% to 20% of Japanese, in 50% to 60% of whites, and in 90% of Northern Africans (82), whereas rapid acetylators are either heterozygous or homozygous for the rapid acetylator gene. N-acetyltransferase 1, which codes for the acetyltransferase activity originally thought to be monomorphic, has also been shown to be a polymorphic gene (83, 84). A gene-gene-environmental exposure three-way interaction was recently found in N-acetylation polymorphisms in smoking-associated bladder cancer (85).

Among the most studied cytochrome P450s with respect to polymorphism is CYP1A2, which has also been found to be associated with increased risk to human colorectal and bladder cancer (86). CYP1A2 catalyzes the N-oxidation of several aromatic and heterocyclic amines to DNA reactive species. Although no polymorphic sequences in the structural CYP1A2 gene have been found, the metabolic phenotype of the enzyme has been evaluated by using caffeine metabolism, phenacetin O-deethylation, and theophylline 1-demethylation (87). The use of urinary caffeine metabolites as a biomarker for the activity of the enzyme has been validated in several epidemiological studies. The rationale for developing this method was that the initial step in the biotransformation of caffeine (caffeine 3-demethylation) is catalyzed by CYP1A2. The ratio of either [1,7-dimethylxanthine (17X) + 1,7-dimethylurate (17U)]/[caffeine (137X)], examined in urine 4 to 5 hours after caffeine ingestion (88), or [5-acetylamino-6-amino-3-methyluracil (AAMU) + 1-methylxanthene (1X) + 1-methylurate (1U)]/17U, measured 24 hours after caffeine ingestion (89), has been used as a marker of CYP1A2 phenotype in human studies. Except in a Japanese population, the CYP1A2 phenotype distribution appears to be trimodal (slow, intermediate, and rapid metabolizers).

The simultaneous determination of NAT2 and CYP1A2 phenotypes has also been accomplished by measuring different caffeine metabolites as markers of hepatic CYP1A2 and NAT2. The NAT2 phenotype can be determined using the ratio of 5-acetylamino-6-formylamino-3-methyluracil (AFMU)/1X in urine (87, 88). Polymorphisms in both NAT2 and CYP1A2 are associated with susceptibility for various types of human cancers (86). For example, in a case-control study involving patients with a history of colorectal cancer or polyps, a trend toward an increased proportion of rapid acetylators for NAT2 was observed. In addition, a significantly greater percentage of rapid metabolizers for CYP1A2 was found in the cancer and polyp cases than in the controls (90). When comparing the prevalence of individuals who were both rapid acetylators and rapid metabolizers for CYP1A2, 33% of the patients with colorectal cancer or polyps possessed the rapid/rapid phenotype, compared with only 13% of the controls (72).

Human cytosolic glutathione S-transferases (GSTs) belong to a supergene family of enzymes consisting of at least four distinct classes, named alpha (α), mu (μ), pi (π), and theta (θ) (91). The GSTs are regulated by at least seven gene loci, and genetic polymorphisms have been found in the μ and θ class isoenzymes. The μ class isoenzymes include products of the GSTM1 locus that were found to be deleted in approximately half of populations of diverse ethnic origins (92, 93). Because this class of enzymes catalyzes the conjugation of reduced glutathione to reactive electrophilic substrates, including metabolic activating products of benzo(a)pyrene (94), individuals genetically lacking GSTM1 activity could be at enhanced risk for various carcinogen-related cancers. Several human studies using phenotype/genotype determinations have found that lack of GSTM1 activity or its gene is, at least in part, related to the genetic susceptibility to tobacco-related lung cancers, particularly those resulting in adenocarcinomas (95–97). Weak associations with increased cancer risk have also been found for other carcinogen-exposed cohorts. A recent study found that the GSTM1 null genotype is associated with susceptibility to gastric adenocarcinoma and distal colorectal adenocarcinoma in a Japanese population (98). Moreover, the GSTM1 null genotype was found to be associated with a high inducibility of CYP1A1 gene transcription (99). A similar genetic polymorphism has been reported for the GST θ GSTT1) gene locus (100, 101). The GSTT1 null genotype has been linked to induced genetic damage by 1,3-butadiene and alkylhalide exposure (102). A recent case-control study found that the frequency of the GSTT1 null type was significantly increased in colorectal cancer cases and was also associated with an increased susceptibility to total ulcerative colitis (103).

In addition to NAT, CYP1A2, and GST, polymorphisms in CYP1A1 are associated with higher risk for human lung (104–107), laryngeal (108), and colon (109) cancers in specific population. Specific CYP2D6 genotypes and phenotypes were noted to be associated with higher cancer risks in lung and larynx (110–112). Different polymorphisms in CYP2E1 were also linked with gastrointestinal, lung, and nasopharyngeal cancers (16, 113–116). The molecular bases for most of these polymorphisms are mutations in the structural genes that result in unstable enzymes or enzymes with altered catalytic activities. However, the polymorphism may also be caused by other regulatory factors that control gene expression.

Another type of biomarker of susceptibility is DNA repair capacity (DRC). The recognition that genetically determined individual DRC may influence the rate of removal of DNA damage and of fixation of mutations prompts the development of methods to measure this important host-susceptibility factor. Several assays, including the unscheduled DNA synthesis (UDS) and measurement of DNA adduct persistence, have been applied previously to detect carcinogen-induced DNA repair or the capacity of global DNA repair (117). A host cell reactivation assay for examining the proficiency of DNA repair in human lymphocytes has been developed and evaluated in epidemiological studies (118). In this assay, a chemically damaged plasmid DNA containing a reporter gene, chloramphenicol acetyltransferase (CAT), is transfected into human lymphocyte samples obtained from an individual. After an incubation period to allow for repair and expression of the reporter gene, DNA excision-repair capacity is scored by determining the amount of reactivated CAT enzyme activity. Human cells with decreased ability to repair the damaged DNA will have lower expression of the reporter gene and, hence, lower enzymatic activity. This technique has been used

Molecular Epidemiology and Biomarkers

to compare the DNA-repair capacities of basal cell carcinoma (BCC) in skin cancer patients and controls (119). An age-related decline in DNA-repair capacity was detected, and reduced repair capacity was a particularly important risk factor for young individuals with BCC and for those with a family history of skin cancer. Younger individuals with BCC repaired DNA damage poorly when compared with controls. With increasing age, however, differences between cases and controls gradually disappeared (120). The normal decline in DNA repair observed with increasing age may account for the increased risk of skin cancer that begins in middle age, which suggests that the occurrence of skin cancer in the young may represent a biochemical manifestation of decreased repair capacity. The same technique has been applied in another case-control study (121) of 51 newly diagnosed lung cancer patients and 56 age-, sex-, and ethnicity-matched controls. The mean level of DRC in cases (3.3%) was significantly lower than that in controls (5.1%) ($P < 0.01$). Only nine cases (18%) had DNA repair levels higher than the controls. Using the median level of DRC in controls as the cutoff value for calculating the odds ratio (OR), the cases were almost six times more likely than the controls to have reduced DRC after adjustment for age, sex, ethnicity, and smoking status.

In addition, a study (122) of 21 lung cancer patients and 41 healthy controls evaluated induced DNA adducts profiles in peripheral lymphocytes exposed to benzo[a]pyrene-7,8-diol-9,10-epoxide (BPDE), the ultimate carcinogenic metabolite of benzo[a]pyrene. The peripheral lymphocytes of cancer patients tended to accumulate higher levels of BPDE-DNA adducts than those of controls. Using the relative adduct labeling value of controls (10 adducts/10^7 nucleotides) as the cutoff point, 18 of 21 cases and 23 of 41 controls distributed above this level (OR 4.7). In a logistic regression analysis, the level of induced adduct was an independent risk factor (OR 6.4) after adjustment for potential confounding factors, that is, age, sex, ethnicity, and smoking. The significant association between accumulated BPDE-induced DNA adducts and risk for lung cancer suggests that this assay may also be applicable to other carcinogen-related cancer studies.

Mutagen sensitivity as a susceptibility biomarker has also been reported in triple primary cancers (123) and in HCCs (124). A reported study of 28 patients with HCC and 110 healthy controls found that the mean numbers of bleomycin-induced breaks per cell for cases and controls were 0.92 and 0.55, respectively ($P < 0.001$). For BPDE sensitivity, the values were 0.90 for cases and 0.46 for controls ($P < 0.001$). Nearly 68% of the cases, but only 27% of the controls, exhibited bleomycin sensitivity. Eighty percent of the case group, but only 22% of the control group, exhibited BPDE sensitivity. Multivariate analyses found that both bleomycin sensitivity and BPDE sensitivity were associated with significantly elevated risks for HCC, with ORs (95% CI) of 5.63 (2.30, 13.81) and 14.13 (3.52, 56.68) respectively. For individuals who were sensitive to both assays, the risk was 35.88 (123).

III. APPLICATION OF BIOMARKERS FOR HUMAN RISK ASSESSMENT

A. Complex Mixture Exposures

Many studies have used DNA and protein adducts to assess potential sources of environmental carcinogen exposure. One important study examined a variety of molecular biomarkers to assess human exposure to complex mixtures of environmental pollution in Poland (125). Measurement of genotoxic damage in peripheral blood samples from residents of high-exposure regions indicated that environmental pollution was associated with significant increases in carcinogen-DNA adducts (polynuclear aromatic hydrocarbon [PAH]-DNA and other aromatic adducts), SCEs, chromosomal aberrations, and frequency of increased *ras* oncogene expression. Perera and colleagues (125) found that the presence of aromatic adducts on DNA was significantly correlated with chromosomal mutation, providing a possible link between environmental exposure and genetic alterations relevant to disease.

Tobacco smoke, the primary cause of lung cancer, contains several types of known carcinogens. The most abundant are PAHs, arylamines, and the tobacco-specific nitrosamines, including the lung-specific carcinogen NNK. These carcinogens are metabolically activated to reactive species that form specific DNA adducts. Smokers usually have significantly elevated levels of aromatic or hydrophobic adducts, as compared with nonsmokers, and some studies found that DNA-adduct levels are linearly related to total smoking exposure (126). One investigation measured the level of bulky, hydrophobic DNA adducts in lung parenchyma of smokers and exsmokers by the ^{32}P-postlabeling method. Smokers had fivefold higher levels of DNA adducts than exsmokers. A positive linear correlation between bulky adduct levels and CYP1A1 (AHH) activity was found in smokers. A statistically significant correlation was determined when comparing pulmonary microsomal AHH activity and the level of BPDE-DNA adducts ($r = 0.91$; $P < 0.01$) (34). Additionally, BPDE-DNA adducts have been detected in oral mucosa cells of smokers and nonsmokers, and levels of DNA damage were elevated in each of 16 smokers compared with 16 age-, race-, and sex-matched nonsmokers. There was about a threefold range between smokers and nonsmokers (127).

One study (128) detected PAH-DNA adducts in specific subsets of peripheral white blood cells (WBCs) and found that DNA combined from lymphocyte and monocyte fractions of smokers exhibited elevated levels of DNA adducts when compared with nonsmokers. The elevated levels of PAH-DNA adducts in DNA obtained from the WBCs of smokers compared with nonsmokers suggest that only certain subsets of WBCs are a valid, readily accessible source for monitoring genotoxicity from cigarette smoke.

There was a decline of PAH-DNA adducts and 4-aminobiphenyl-hemoglobin (4-ABP-Hb) adducts in peripheral blood after smoking cessation, based on serial samples from 40 heavy smokers ($\geq$ 1 pack/day for $\geq$ 1 year). The substantial reduction (50% to 75%) of PAH-DNA and 4-ABP-Hb adduct levels after quitting indicates these carcinogen adducts are reflective of smoking exposure (129). This is essential information in the validation of biomarkers. The estimated half-life of the PAH-DNA adducts in leukocytes was 9 to 13 weeks; the estimated half-life of the 4-ABP-Hb adducts was 7 to 9 weeks. Women had higher levels of 4-ABP-Hb adducts at baseline and after smoking cessation. These results show that PAH-DNA and 4-ABP-Hb adducts can be useful as intermediate biomarkers verifying smoking cessation and possibly for identifying persons who are at increased risk of cancer from exposure to cigarette smoke because of high levels of carcinogen binding.

In another study (130) immunohistochemical quantitation of 4-ABP-DNA adducts and p53 nuclear overexpression in T1 bladder cancer of smokers and nonsmokers was described. Mean relative staining intensity for 4-ABP-DNA adducts was significantly higher in current smokers compared with nonsmokers. There was a linear relationship between the mean level of relative staining and the number of cigarettes smoked, with lower levels in the 1 to 19 cigarette/day group, compared with the 20 to 40 and the more than 40 cigarette/day group. Nuclear overexpression of p53 was observed in 27 (59%) of the 45 stage T1 tumors analyzed. Nuclear staining of p53 was correlated with smoking status, cigarettes/day, and 4-ABP-DNA adducts. The 4-ABP-DNA adducts in 11 human lung and 8 urinary bladder mucosa specimens were analyzed by alkaline hydrolysis and negative chemical ionization GC/MS. Adduct levels were found to be 0.32 to 49.5 adducts/10^8 nucleotides in the lung and 0.32 to 3.94 adducts/10^8 nucleotides in the bladder samples (131).

Carcinogen-DNA adducts in human breast tissue samples have been reported by Perera et al (132), who analyzed a total of 31 breast tissue samples, which included tumor and tumor-adjacent tissues from 15 women with breast cancer and normal tissue samples from four women undergoing breast reduction. Among the breast cancer cases, the mean aromatic/hydrophobic-DNA adduct level assayed was roughly twofold higher, compared with the noncancer patients. Five of 15 tissues from the cases displayed a pattern of adducts associated with tobacco smoke exposure, and all these positive samples were from current smokers. Tissue samples from the eight nonsmoking cases did not exhibit this pattern. This study indicated that biomarkers may be useful in investigating specific environmental exposures that could contribute to breast cancer.

Alkylating agents such as N-nitroso compounds are potential human carcinogens. Humans are known to be exposed to N-nitrosoamines from diet, work place, cigarette smoke, and through endogenous formation. These compounds alkylate DNA, which leads to formation of various types of DNA adducts.

Among them are 7-alkyl-2-deoxyguanosine (dG) adducts, such as 7-methyl-dGp and 7-ethyl-dGp. Several investigations have focused on the levels of 7-methyl-dG adducts in human lung tissue (133). Higher levels have been found in smokers compared with nonsmokers (134). Separately, 7-methyl-dG levels in lung tissues have been associated with cytochrome P4502D6 and 2E1 genetic polymorphisms (135). One study analyzed N^7-alkylguanine adducts levels in DNA in a group of 46 patients with larynx tumors by the ^{32}P-postlabeling method. The average level of N^7-alkylguanines was $26.2/10^7$ nucleotides in tumor cells, $22.7/10^7$ in nontumor cells, and $13.1/10^7$ in blood leukocytes. Men and smokers had significantly higher levels of adducts than women and nonsmokers (136). In another study, 7-alkyl-2-dG adducts were measured in eight separate lung segments of ten autopsy specimens; 7-methyl-dGp levels were detected in all eight samples. 7-Ethyl-dGp was detected in all but five of the samples (ranging from < 0.1 to 7.1 adducts/10^7 dG). 7-Methyl-dG levels were approximately 1.5-fold higher than 7-ethyl-dG levels and were positively correlated with each other in most individuals. There was no consistent pattern of adduct distribution in the different lung lobar segments (137).

A wide variety of aromatic amines and PAHs have been found to bind at high levels to hemoglobin (138). One of the carcinogen-Hb adducts that has been well characterized is formed by the potent urinary bladder carcinogen, 4-ABP. Several studies have reported 4-ABP-Hb adducts in human blood specimens (139). The results of these studies indicate that the presence of 4-ABP-Hb adducts is closely associated with three major risk factors for bladder cancer: cigarette smoking, the type of tobacco smoked, and acetylator phenotype. One report (140) described the relation between exposure to environmental tobacco smoke (ETS) and levels of 4-ABP-Hb adducts in nonsmoking pregnant women, compared with adduct levels in those women who smoked during pregnancy. A questionnaire on smoking and exposure to ETS was administered to pregnant women. Samples of maternal blood and cord blood were collected during delivery and analyzed for 4-ABP-Hb adducts by GC/MS. The mean adduct level in smokers was approximately ninefold higher than that in nonsmokers. Among nonsmokers, the levels of 4-ABP-Hb adducts increased with increasing ETS level. This relationship between ETS exposure and 4-ABP-Hb adduct levels supports the concept that ETS is a probable hazard during pregnancy.

As previously described, metabolic polymorphism, both in NAT and in CYP1A2, is also expected to affect the formation of 4-ABP-DNA- and Hb-adducts. One study (141) determined levels of DNA adducts in bladder cells and 4-ABP-Hb adducts in 79 individuals, together with the acetylator phenotype and genotype. Among the slow acetylators, levels of 4-ABP-Hb adducts were significantly higher compared with those present in rapid acetylators. This study indicated that clearance of low-dose carcinogen is decreased in the slow acetylator phenotype. Because the highest levels of adducts were found in individuals with

Molecular Epidemiology and Biomarkers

rapid N-oxidation (CYP1A2) and slow N-acetylation (NAT2) phenotype, determination of phenotypes and genotypes may provide a better prediction and assessment of human cancer risk.

Yu et al (142) found that mean 3- and 4-ABP-Hb adduct levels in 151 subjects were statistically significantly higher in cigarette smokers compared with nonsmokers and that the level increased with increasing number of cigarettes smoked/day. Again, slow acetylators consistently exhibited higher mean levels of ABP-Hb adducts compared with rapid acetylators. The mean level of 4-ABP-Hb adducts was higher in subjects possessing the GSTM1-null versus GSTM1-non-null genotype (46.5 vs 36.0 pg/g Hb; $P = 0.037$). In another study (143), the polymorphic distribution of the CYP1A2 and NAT2 phenotypes was examined in relation to ABP-Hb adduct formation in 97 healthy men. Rapid oxidizers and subjects with the combined slow acetylator-rapid oxidizer phenotype showed the highest ABP-Hb adduct levels at a low smoking dose. However, in a subset of 45 available samples, no association was seen between ABP-Hb adduct levels and GSTM1 genotype.

B. Occupational Carcinogen Exposures

Occupational exposure to many chemical carcinogens has been well documented. Since the initial reports of Rehn in aniline dye workers, many laboratories have applied technologies for measuring these carcinogens and their metabolites to confirm exposure in the workplace (9). The occurrence of DNA adducts in exfoliated urothelial cells of a worker exposed to the aromatic amine, 4,4-methylene-bis(2-chloroaniline) (MOCA), an agent that induces lung and liver tumors in rodents and urinary bladder tumors in dogs, has been reported (144). ^{32}P-postlabeling analysis revealed the presence of a single, major DNA adduct that cochromatographed with the major N-hydroxy-MOCA-DNA adduct, N-(deoxyadenosin-8-yl)-4-amino-3-chlorobenzyl alcohol, formed in vitro. In a recent study, PAH-DNA adducts in WBC and 1-hydroxypyrene in urine were examined in a group of 105 workers from an aluminum plant with different PAH exposures. The exposure to PAH was measured by personal monitoring, and ranged from 0.4 to 150 µg/m^3. High exposure to PAH in the work atmosphere was associated with increased concentration of 1-hydroxypyrene in the urine. Polynuclear aromatic hydrocarbon-DNA adducts were detected in 93% of the worker samples. Workers with a high PAH exposure had significantly higher adduct levels than those with a low PAH exposure. A good correlation was found between PAH exposure and the average PAH-adduct values in blood. A statistically significant correlation was also found between the average adduct values and the concentration of 1-hydroxypyrene in the urine of smokers (145).

In another report (146), PAH-DNA adduct levels were examined in workers exposed to ambient air pollution. Significantly higher adduct levels were found in bus drivers working in central Copenhagen, compared with those driving in suburban areas. The urban drivers had higher adduct levels than rural controls. No significant influence on adduct level by potential confounders, including smoking and diet, was observed. A separate study (147) measured BPDE-DNA adducts in 39 coke oven workers (exposed to PAH) and 39 nonexposed controls (each group consisting of smokers and nonsmokers). Adducts were detected in 51% of workers and in 18% of controls. The mean level in workers (15.7×10^8 nucleotides) was 15 times higher than in nonexposed controls. Although large interindividual variations were noted, smoking workers had 3.5 times more adducts than nonsmokers.

C. Cancer Risks For Environmental Carcinogen Exposures

The DNA and protein adducts can serve as biomarkers for both exposure and for cancer risk. This approach has been applied in several published human epidemiological studies. A nested case-control study initiated in 1986 in Shanghai was started to examine the relationship between biomarkers for aflatoxin and HBV and the development of liver cancer (148, 149). In this study, over 18,000 urine samples were collected from healthy men between the ages of 45 and 64. In the subsequent 7 years, 50 of these individuals had liver cancer develop. The urine samples for cases were age-matched and residence-matched with controls and analyzed for both aflatoxin biomarkers and hepatitis B surface antigen (HBsAg) status. A highly significant increase in the relative risk (RR) (3.5) was observed for those liver cancer cases where urinary aflatoxin biomarkers (AFB_1-N^7-guanine + other AFB_1 metabolites) were detected. The RR for people who tested positive for the HBsAg was about 8, but individuals with both urinary aflatoxin biomarkers and positive HBsAg status had an RR for developing liver cancer of 57. These results showed for the first time a relationship between the presence of carcinogen-specific biomarkers and cancer risk. Moreover, these findings provided the first demonstration of a multiplicative interaction between these two major risk factors for liver cancer. Also, when individual aflatoxin metabolites were stratified for liver cancer outcome, the presence of the AFB_1-N^7-guanine in urine always resulted in a two- to threefold elevation in risk of having liver cancer develop (149).

Another study investigated PAH-DNA adducts in peripheral leukocytes from 119 non-small-cell lung cancer patients and 98 controls. Among them, 31 cases had adduct measurements in leukocytes, lung tumor, and nontumor specimens collected at surgery and 34 had paired leukocyte and tumor specimens. After adjustment for age, gender, ethnicity, season, and smoking, DNA adducts in leukocytes were significantly higher in cases than controls; the OR was 7.7 (95% CI: 1.7–34; $P < 0.01$). The DNA adducts in leukocytes were increased significantly in smokers and exsmokers compared with nonsmokers among cases and controls (separately and combined) after adjusting for age, gender, ethnicity, and season (150).

Two studies used aflatoxin-albumin adducts as biomarkers for monitoring liver cancer risks. One nested case-control study (151) was carried out in Taiwan. A cohort of 8,068 men was followed up for 3 years, and 27 cases of HCC were identified and matched with 120 healthy controls. Serum samples were analyzed for AFB_1-albumin adducts by ELISA. The proportion of subjects with detectable serum AFB_1-albumin adducts was higher for HCC cases (74%) than matched controls (66%), giving an OR of 1.5. There was a statistically significant association between detectable level of AFB_1-albumin adduct and HCC risk among men younger than age 52, showing a multivariant-adjusted OR of 5.3, although no association was observed between AFB_1-albumin adduct level and HBsAg carrier status. Another prospective nested case-control study (152) was instituted in Qidong County, Jiangsu Province, China, where liver cancer accounts for 10% of all adult deaths, and both HBV and aflatoxin exposures are common. Serum samples from 804 healthy HBsAg-positive individuals (728 men, 76 women, aged 30 to 65, were obtained and stored frozen in 1991. Between the years 1993 to 1995, 38 of these individuals developed liver cancer. The serum samples for 34 of these patients were matched by age, gender, residence, and time of sampling to 170 controls. Serum AFB_1-albumin adduct levels were determined by RIA. The RR for HCC among AFB_1-albumin positive individuals was 2.4.

McGlynn et al (153) have reported that mutant alleles at two AFB_1 detoxification gene loci (epoxide hydrolase [EPHX] and GSTM1) were significantly overrepresented in individuals with detectable serum AFB_1-albumin adducts in a cross-sectional study. Mutant alleles of EPHX were also significantly overrepresented in persons with HCC in a case-control study. The relationship of EPHX to HCC varied with HBsAg status and suggested that a synergistic effect may exist. Mutations of *p53* at codon 249 were only observed among HCC patients with one or both mutated EPHX and GSTM1 genotypes. These results indicate that individuals with mutant genotypes at EPHX and GSTM1 may be at greater risk of developing AFB_1 adducts, *p53* mutations, and HCC when exposed to AFB_1. Hepatitis B surface antigen carriers with these high-risk genotypes may be an even greater risk than carriers with low-risk genotypes. These findings further support the existence of genetic susceptibility in humans to AFB_1 exposure and indicate that there is a synergistic increase in risk of HCC with the combination of HBV infection and susceptible genotype.

IV. SUMMARY

Epidemiologic research has significantly profited from the development and application of biomarkers of exposure, dose, response, and susceptibility. Such biomarkers have facilitated the investigation of associations between exposure to toxicants and subsequent development of disease in exposed populations by providing qualitative or quantitative indices in individuals of the status of different stages of the toxicological process that leads from exposure to disease. As a result, the use of biomarkers has uncovered new associations, sped up the process of providing evidence of associations for long latent period diseases, and offered opportunities for individual human risk assessment. As the development, validation, and application of biomarkers continues and expands, even greater strides may be possible. This expansion will require vigorous efforts at identification of potential new biomarkers, enhancement of analytic measurement techniques, animal studies for validation, and careful study design and execution in selected human populations.

REFERENCES

1. B Hulka. Epidemiological studies using biological markers: Issues for epidemiologists. Cancer Epidemiol Biomarkers Prev 1:13–19, 1991.
2. JD Groopman, TW Kensler. The light at the end of the tunnel for chemical-specific biomarkers: Daylight or headlight? Carcinogenesis, in press, 1999.
3. PA Schulte. A conceptual and historical framework for molecular epidemiology. In: PA Schulte, FP Perera, eds. Molecular Epidemiology: Principles and Practices. San Diego: Academic Press, Inc., 1993, pp 3–44.
4. D Hattis, K Silver. Use of biomarker in risk assessment. In: PA Schulte, FP Perera, eds. Molecular Epidemiology: Principles and Practices. San Diego: Academic Press, Inc., 1993, pp 251–273.
5. FP Perera. Molecular epidemiology: Insights into cancer susceptibility, risk assessment, and prevention. JNCI 88:496–509, 1996.
6. National Research Council Committee on Biological Markers. Biologic markers in environmental health research. Environ Health Perspect 74:3–9, 1987.
7. JM Links, TW Kensler, JD Groopman. Biomarkers and mechanistic approaches in environmental epidemiology. Annu Rev Public Health 16:83–103, 1995.
8. GN Wogan. Markers of exposure to carcinogens. Environ Health Perspect 81:9–17, 1989.
9. JD Groopman, TW Kensler. Molecular biomarkers for human chemical carcinogen exposure. Chem Res Toxicol 6:764–770, 1993.
10. PA Schulte, FP Perera. Validation. In: PA Schulte, FP Perera, eds. Molecular Epidemiology: Principles and Practices. San Diego: Academic Press, Inc., 1993, pp 79–107.
11. SS Hecht, SG Carmella, PG Foiles, SE Murphy. Biomarkers for human uptake and metabolic activation of tobacco-specific nitrosamines. Cancer Res 54:1912s–1917s, 1994.
12. TW Kensler, JD Groopman, BD Roebuck. Use of aflatoxin adducts as intermediate endpoints to assess the efficacy of chemopreventive interventions in animals and man. Mutat Res 402:165–172, 1998.
13. ME Mattson, ES Pollack, JW Cullen. What are the odds that smoking will kill you? Am J Pub Health 77:425–431, 1987.

14. FF Kadlubar, MA Butler, KR Kaderlik, HC Chou, NP Lang. Polymorphisms for aromatic amine metabolism in humans: Relevance for human carcinogenesis. Environ Health Perspect 98:69–74, 1992.
15. S Ikawa, F Uematsu, K Watanabe T Kimpara, M Osada, A Hossain, I Sagami, H Kikuchi, M Watanabe. Assessment of cancer susceptibility in human by use of genetic polymorphisms in carcinogen metabolism. Pharmacogenetics 5:s154–s160, 1995.
16. H Sugimura, GS Hamada, I Suzuki, T Iwase, E Kiyokawa, I Kino, S Tsugane. CYP1A1 and CYP2E1 polymorphism and lung cancer, case-control study in Rio de Janeiro, Brazil. Pharmacogenetics 5:145s–148s, 1995.
17. T Sugimura, R Inoue, H Ohgaki, T Ushijima, F Canzian, M Nagao. Genetic polymorphisms and susceptibility to cancer development. Pharmacogenetics 5:161s–165s, 1995.
18. JC Barrett, H Vainio, D Peakall, BD Goldstein. 12th meeting of the scientific group on methodologies for the safety evaluation of chemicals: Susceptibility to environmental hazards. Environ Health Perspect 105 (suppl 4):699–737, 1997.
19. JD Groopman, PR Donahue, J Zhu, J Chen, GN Wogan. Aflatoxin metabolism in humans: Detection of metabolites and nucleic acid adducts in urine by affinity chromatography. Proc Natl Acad Sci U S A 82:6492–6497, 1985.
20. JQ Zhu, LS Zhang, X Hu, Y Xiao, JS Chen, YC Xu, J Fremy, FS Chu. Correlation of dietary aflatoxin B1 levels with excretion of aflatoxin M_1 in human urine. Cancer Res 47:1848–1852, 1987.
21. SG Carmella, SA Akerkar, SS Hecht. Metabolites of the tobacco-specific nitrosamine 4-(methylnitrosamino)-1-(3-pyridyl)-1-butanone in smokers' urine. Cancer Res 53:721–724, 1993.
22. SG Carmella, SA Akerkar, JP Richie, SS Hecht. Intraindividual and interindividual differences in metabolites of the tobacco-specific lung carcinogen 4-(methylnitrosamino)-1-(3-pyridyl)-1-butanone (NNK) in smokers' urine. Cancer Epidemiol Biomarkers Prev 4:35–62, 1995.
23. I Romieu, T Carreon, L Lopez, E Palazuelos, C Rios, Y Manuel, M Hernandez-Avila. Environmental urban lead exposure and blood lead levels in children of Mexico City. Environ Health Perspect 103:1036–1040, 1995.
24. WA Anwar. Biomarkers of human exposure to pesticides. Environ Health Perspect 105 (suppl 4): 801–806, 1997.
25. MS Wolff, PG Toniolo, EW Lee, M Rivera, N Dubin. Blood levels of organochlorine residues and risk of breast cancer. J NCI 85:648–652, 1993.
26. N Krieger, MS Wolff, RA Hiatt, M Rivera, J Vogelman, N Orentreich. Breast cancer and serum organochlorines: A prospective study among white, black, and Asian women. J NCI 86:589–599, 1994.
27. CC Harris. Chemical and physical carcinogenesis: Advances and perspectives. Cancer Res 51: 5023s–5044s, 1991.
28. CC Harris. Molecular epidemiology: Overview of biochemical and molecular basis. In: JD Groopman, PL Skipper, eds. Molecular Dosimetry and Human Cancer. Boca Raton: CRC Press, 1991, pp 15–26.
29. A Dipple. DNA adducts of chemical carcinogens. Carcinogenesis 16:437–441, 1995.

30. HC Pitot, III, YP Dragan. Chemical carcinogenesis. In: CD Klaassen, ed. Casarett and Doull's Toxicology: The Basic Science of Poisons, 5th ed, New York: McGraw-Hill, 1996, pp 201–267.
31. KR Kaderlik, DX Lin, NP Lang, FF Kadlubar. Advantages and limitations of laboratory methods for measurement of carcinogen-DNA adducts for epidemiological studies. Toxicol Lett 64/65:469–475, 1992.
32. PT Strickland, MN Routledge, A Dipple. Methodologies for measuring carcinogen adducts in humans. Cancer Epidemiol Biomarkers Prev 2:607–619, 1993.
33. MC Poirier, H Shamkhani, E Reed, RE Tarone, S Gupta-Burt. DNA adducts induced by platinum drug chemotherapeutic agents in human tissues. Prog Clin Biol Res 374:197–212, 1992.
34. MC Poirier, E Reed, CL Litterst, D Katz, S Gupta-Burt. Persistence of platinum-amine-DNA adducts in gonads and kidneys of rats and multiple tissues from cancer patients. Cancer Res 52:149–153, 1992.
35. BN Ames, LS Gold. Endogenous mutagens and the causes of aging and cancer. Mutat Res 250:3–16, 1991.
36. LJ Marnett, PC Burcham. Endogenous DNA adducts: Potential and paradox. Chem Res Toxicol 6:771–785, 1993.
37. EM Park, MK Shihenaga, P Degan, TS Korn, JW Kitzler, CM Wehr, P Kolachana, BN Ames. Assay of excised oxidative DNA lesions: Isolation of 8-oxoguanine and its nucleoside derivatives from biological fluids with a monoclonal antibody column. Proc Natl Acad Sci U S A 89:3375–3379, 1992.
38. International Programme on Chemical Safety. Biomarkers and risk assessment: Concepts and principles. In: World Health Organization, ed. Environmental Health Criteria 155:3–40, 1993.
39. JN Lucas, T Tenjin, T Straume, D Pinkel, D Moore, M Litt, JW Gray. Rapid human chromosome aberration analysis using fluorescence in situ hybridization. Int J Radiat Biol 56:35–44, 1989.
40. DS Rupa, L Hasegawa, DA Eastmond. Detection of chromosomal breakage in the 1cen-1q12 region of interphase of human lymphocytes using multicolor fluorescence in situ hybridization with tandem DNA probes. Cancer Res 55:640–645, 1995.
41. JD Tucker, A Auletta, MC Cimino, KL Dearfield, D Jacobson-Kram, RT Tice, AV Carrano. Sister-chromatid exchange: Second report of the Gene-Tox program. Mutat Res 297:101–180, 1993.
42. G Ballarin, F Sarto, C Giacomelli, GB Bartolucci, E Clonofero. Micronucleated cells in nasal mucosa of formaldehyde-exposed workers. Mutat Res 280:1–7, 1992.
43. H Norppa, S Luomahaara, H Heikanen, S Roth, M Sorsa, L Renzi, C Lindholm. Micronucleus assay in lymphocytes as a tool to biomonitor human exposure to aneuploidogens and clastogens. Environ Health Perspect 101 (suppl 3):139–143, 1993.
44. CC Harris. 1995 Deichman lecture—p53 tumor suppressor gene: At the crossroads of molecular carcinogenesis, molecular epidemiology and cancer risk assessment. Toxicol Lett 82/83:1–7, 1995.
45. SP Hussain, CC Harris. Molecular epidemiology of human cancer: Contribution of mutation spectra studies of tumor suppressor genes. Cancer Res 58:4023–4037, 1998.

46. MW Anderson, SH Reynolds, M You, RM Maronpot. Role of protooncogen activation in carcinogenesis. Environ Health Perspect 98:13–24, 1992.
47. T Sugimura, M Terada, J Yokota, S Hirohashi, K Wakabayashi. Multiple genetic alterations in human carcinogenesis. Environ Health Perspect 98:5–12, 1992.
48. MS Greenblatt, WP Bennett, M Hollstein, CC Harris. Mutations in the p53 tumor suppressor gene: Clues to cancer etiology and molecular pathogenesis. Cancer Res 54:4855–4878, 1994.
49. CC Harris. p53 tumor suppressor gene: From the basic research laboratory to the clinic, an abridged historical perspective. Carcinogenesis 17:1187–1198, 1996.
50. M Hollstein, D Sidransky, B Vogelstein, CC Harris. p53 mutations in human cancers. Sciences 253:49–53, 1991.
51. B Bressac, M Kew, J Wands, M Ozturk. Selective G to T mutations of p53 gene in hepatocellular carcinoma from Southern Africa. Nature 350:429–431, 1991.
52. IC Hsu, RA Metcalf, T Sun, JA Wesh, NJ Wang, CC Harris. Mutational hotspot in the p53 gene in human hepatocellular carcinomas. Nature 350:427–428, 1991.
53. D Li, Y Gao, L He, NJ Wang, J Gu. Aberrations of p53 gene in human hepatocellular carcinoma from China. Carcinogenesis 14:169–173, 1993.
54. M Ozturk and collaborators. p53 Mutation in hepatocellular carcinoma after aflatoxin exposure. Lancet 338:1356–1359, 1991.
55. Y Fujimoto, LL Hampton, PJ Wirth, NJ Wang, JP Xie, SS Thorgeirsson. Alterations of tumor suppressor genes and allelic losses in human hepatocellular carcinoma in China. Cancer Res 54:281–285, 1994.
56. P Coursaget, N Depril, M Chabaud, R Nandi, V Mayelo, P LeCann, B Yvonnet. High prevalence of mutations at codon 249 of the p53 gene in hepatocellular carcinomas from Senegal. Br J Cancer 67:1395–1397, 1993.
57. F Aguilar, CC Harris, T Sun, M Hollstein, P Cerutti. Geographic variation of p53 mutational profile in nonmalignant human liver. Science 264:1317–1319, 1994.
58. P Cerutti, P Hussain, C Pourzand, F Aguilar. Mutagenesis of the H-ras proto-oncogene and the p53 tumor suppressor gene. Cancer Res 54:1934s–1938s, 1994.
59. PL Foster, E Eisenstadt, JH Miller. Base substitution mutation induced by metabolically activated aflatoxin B_1. Proc Natl Acad Sci U S A 80:2695–2698, 1983.
60. A Puisieux, S Lim, JD Groopman, M Ozturk. Selective targeting of p53 gene mutational hotspots in human cancers by etiologically defined carcinogens. Cancer Res 51:6185–6189, 1991.
61. F Aguilar, SP Hussain, P Cerutti. AflatoxinB_1 induces the transversion of G→T in codon 249 of the p53 tumor suppressor gene in human hepatocytes. Proc Natl Acad Sci U S A 90:8586–8590, 1993.
62. FP Guengerich. Roles of cytochrome P-450 enzymes in chemical carcinogenesis and cancer chemotherapy. Cancer Res 48:2946–2954, 1988.
63. FP Guengerich. Metabolic activation of carcinogens. Pharmacol Ther 54:17–61, 1992.
64. A Parkinson. Biotransformation of xenobiotics. In: CD Klaassen, ed. Casarett and Doull's Toxicology: The Basic Science of Poisons, 5th ed. New York: McGraw-Hill, 1996, pp 113–186.
65. N Caporaso, MT Landi, P Vineis. Relevance of metabolic polymorphisms to human carcinogenesis: Evaluation of epidemiologic evidence. Pharmacogenetics 1:4–19, 1991.

66. AK Daly, S Cholerton, M Armstrong, JR Idle. Genotyping for polymorphisms in xenobiotic metabolism as a predictor of disease susceptibility. Environ Health Perspect 102(suppl 9):55–61, 1994.
67. K Kawajiri, J Watanabe, H Eguchi, S-I Hayashi. Genetic polymorphisms of drug-metabolizing enzymes and lung cancer susceptibility. Pharmacogenetics 5:s70–s73, 1995.
68. F Broly, A Gaedigk, M Heim, M Eichelbaum, K Morike, UA Meyer. Debrisoquine/sparteine hydroxylation genotype and phenotype: Analysis of common mutations and alleles of CYP2D6 in a European population. DNA Cell Biol 10:545–558, 1991.
69. G Cosma, F Crofts, E Taioli, P Toniolo, S Garte. Relationship between genotype and function of human CYP1A1 gene. J Toxicol Environ Health 40:309–316, 1993.
70. KT Kelsey, JK Wiencke, MR Spitz. A race-specific genetic polymorphism in the CYP1A1 gene is not associated with lung cancer in African Americans. Carcinogenesis 15:1121–1124, 1994.
71. DM Grant, F Lottspeich, UA Meyer. Evidence for two closely related isozymes of arylamine N-acetyltransferase in human liver. FEBS Lett 244:203–207, 1989.
72. RJ Turesky, NP Lang, MA Butler, CH Teitel, FF Kadlubar. Metabolic activation of carcinogenic heterocyclic amines by human liver and colon. Carcinogenesis 12:1839–1845, 1991.
73. T Deguchi, M Mashimo, T Suzuki. Correlation between acetylator phenotypes and genotypes of polymorphic arylamine N-acetyltransferase in human liver. J Biol Chem 265:12757–12760, 1990.
74. M Blum, A Demierre, DM Grant, M Heim, UA Meyer. Molecular mechanism of slow acetylation of drugs and carcinogens in humans. Proc Natl Acad Sci U S A 88:5237–5241, 1991.
75. D Hickman, E Sim. N-acetyltransferase polymorphism. Comparison of phenotype and genotype in humans. Biochem Pharmacol 42:1007–1014, 1991.
76. RA Cartwright, RW Glasha, HJ Rogers, RA Ahmad, D Barham-Hall, E Higgins, MA Kahn. Role of N-acetyltransferase phenotypes in bladder carcinogenesis: A pharmacogenetic epidemiological approach to bladder cancer. Lancet 2:842–845, 1982.
77. AE Karakaya, I Cok, S Sardas, O Gogus, OS Sardas. N-acetylransferase phenotype of patients with bladder cancer. Human Toxicol 5:333–335, 1986.
78. NP Lang, DZJ Chu, CF Hunter, DC Kendall, TJ Flamming, FF Kadlubar. Role of aromatic amine acetyltransferase in human colorectal cancer. Arch Surg 121:1259–1261, 1986.
79. WG Kirlin, F Ogolla, AF Andrews, A Trinidad, RJ Ferguson, T Yorokun, M Mpezo, D Hein. Acetylator genotype-dependent expression of arylamine N-acetyltransferase in human colon cytosol from noncancer and colorectal cancer patients. Cancer Res 51:549–555, 1991.
80. M Blum, DM Grant, W McBride, M Heim, UA Meyer. Human arylamine N-acetyltransferase genes: Isolation, chromosomal localization, and functional expression. DNA Cell Biol 9:193–203, 1990.
81. KP Vatsis, KJ Martell, WW Weber. Diverse point mutations in the human gene for polymorphic N-acetyltransferase. Proc Natl Acad Sci U S A 88:6333–6337, 1991.
82. LL Marchand, L Sivaraman, AA Franke, LJ Custer, LR Wilkens, AF Lau, RV Cooney. Predictors of N-acetyltransferase activity: Should caffeine phenotyping and NAT2

genotyping be used interchangeably in epidemiological studies? Cancer Epidemiol Biomarkers Preven 5:449–455, 1996.

83. DA Bell, EA Stephens, T Castranio, DM Umbach, M Weston, M Deakin, J Elder, C Hendrickse, H Duncan, RC Strange. Polyadenylation polymorphism in the acetyltransferase 1 gene (NAT1) increase risk of colorectal cancer. Cancer Res 55:3537–3542, 1995.

84. DA Bell, AF Badawi, NP Lang, KF llett, FF Kadlubar, A Hirvonen. Polymorphism in the N-acetyltransferase 1 (NAT1) polyadenylation signal: Association of NAT1 10 allele with higher N-acetylation activity in bladder and colon tissue. Cancer Res 55:5226–5229, 1995.

85. JA Taylor, DM Umbach, E Stephens, T Castranio, D Paulson, C Robertson, JL Mohler, DA Bell. The role of N-acetylation polymorphisms in smoking-associated bladder cancer: Evidence of a gene-gene-exposure three-way interaction. Cancer Res 58:3603–3610, 1998.

86. FF Kadlubar. Biochemical individuality and its implication for drug and carcinogen metabolism: Recent insights from acetyltransferase and cytochrome P4501A2 henotyping and genotyping in humans. Drug Metabol Rev 26:37–46, 1994.

87. MA Butler, M Iwasaki, FP Guengerich, FF Kadlubar. Human cytochrome P450PA (P-4501A2), the phenacetin O-deethylase, is primarily responsible for the hepatic 3-demethylation of caffeine and N-oxidation of carcinogenic amines. Proc Natl Acad Sci U S A 86:7696–7700, 1989.

88. MA Butler, NP Lang, JF Young, NE Caporaso, P Vineis, RB Hayes, CH Teitel, JP Massengill, MF Lawsen, FF Kadlubar. Determination of CYP1A2 and acetyl-transferase phenotype in human populations by analysis of caffeine urinary metabolites. Pharmacogenetics 2:116–127, 1992.

89. W Kalow, BK Tang. The use of caffeine for enzyme assays: A critical appraisal. Clin Pharmacol Ther 53:503–514, 1993.

90. MA Butler, NP Lang, JP Massengill, M Lawson, FF Kadlubar. CYP1A2 and AT phenotypes in human colorectal cancer or polyps. Proc Am Assoc Cancer Res 33:1749, 1992.

91. B Mannervik, UH Danielson. Glutathione transferases-structure and catalytic activity. CRC Crit Rev Biochem 23:281–334, 1988.

92. J Seidegard, WR Vorachek, RW Pero, WR Pearson. Hereditary differences in the expression of the human glutathione S-transferase active on *trans*-stilbene oxide are due to a gene deletion. Pro Natl Acad Sci U S A 85:7293–7297, 1988.

93. S Zhong, AF Howie, B Ketterer, J Taylor, JD Hayes, GJ Beckett, CG Wathen, CR Wolf, NK Spurr. Glutathione S-transferase mu locus: Use of genotyping and phenotyping assays to assess association with lung cancer susceptibility. Carcinogenesis 12:1533–1537, 1991.

94. SR Heckbert, NS Weiss, SK Hornung, DL Eaton, AG Motulsky. Glutathione S-transferase and epoxide hydrolase activity in human leukocytes in relation to risk of lung cancer and other smoking-related cancers. J NCI 84:414–422, 1992.

95. M Kihara, K Noda, M Kihara. Distribution of GSTM1 null genotype in relation to gender, age and smoking status in Japanese lung cancer patients. Pharmacogenetics 5:s74–s79, 1995.

96. JE McWilliams, BJS Sanderson, EL Harris, KE Richert-Boe, WD Henner. Glutathione S-transferase M1 (GSTM1) deficiency and lung cancer risk. Cancer Epidemiol Biomarkers Prev 4:589–594, 1995.
97. J To-Figueras, M Gene, J Gomez-Catalan, C Galan, J Firvida, M Fuentes, M Rodamilans, E Huguet, J Estape, J Corbella. Glutathione-S-transferase M1 and codon 72 p53 polymorphisms in a northwestern Mediterranean population and their relation to lung cancer susceptibility. Cancer Epidemiol Biomarkers Prev 5:337–342, 1996.
98. T Katoh, N Nagata, Y Kuroda, H Itoh, A Kawahara, N Kuroki, R Ookuma, DA Bell. Glutathione S-transferase M1 (GSTM1) and T1 (GSTT1) genetic polymorphism and susceptibility to gastric and colorectal adenocarcinoma. Carcinogenesis 17:1855–1859, 1996.
99. C Vaury, R Laine, P Noguiez, P de Coppet, C Jaulin, F Praz, D Pompon, M AmorGueret. Human glutathione S-transferase M1 null genotype is associated with a high inducibility of cytochrome P450 1A1 gene transcription. Cancer Res 55:5520–5523, 1995.
100. FP Guengerich, R Thier, M Persmark, JB Taylor, SE Pemble, B Ketterer. Conjugation of carcinogens by q class glutathione S-transferases: Mechanisms and relevance to variations in human risk. Pharmacogenetics 5:s103–s107, 1995.
101. EJD Lee, JYY Wong, PN Yeoh, NH Gong. Glutathione S transferase-q (GSTT1) genetic polymorphism among Chinese, Malays and Indians in Singapore. Pharmacogenetics 5:332–334, 1995.
102. JK Wiencke, S Pemble, B Ketterer, KT Kelsey. Gene deletion of glutathione S-transferase q: Correlation with induced genetic damage and potential role in endogenous mutagenesis. Cancer Epidemiol Biomarkers Prev 4:253–259, 1995.
103. M Deakin, J Elder, C Hendrickse, D Peckham, D Balwin, C Pantin, N Wild, P Leopard, DA Bell, P Jones, H Duncan, K Brannigan, J Alldersea, AA Fryer, RC Strange. Glutathione S-transferase GSTT1 genotypes and susceptibility to cancer: Studies of interactions with GSTM1 in lung, oral, gastric and colorectal cancers. Carcinogenesis 17:881–884, 1996.
104. K Kawajiri, K Nakachi, K Imai, A Yoshii, N Shinoda, J Watanabe. Identification of genetically high risk individuals to lung cancer by DNA polymorphisms of the cytochrome P4501A1 gene. FEBS Lett 263:131–133, 1990.
105. S Hayashi, J Watanbe, K Nakachi, K Kawajiri. Genetic linkage of lung cancer-associated *MspI* polymorphisms with amino acid replacement in the heme binding region of the human cytochrome P4501A1 gene. J Biochem 110:407–411, 1991.
106. GS Hamada, H Sugimura, I Suzuki, K Nagura, E Kiyokawa, T Iwase, M Tanaka, T Takahashi, S Watanabe, I Kino, S Tsugane. The heme-binding region polymorphism of cytochrome P4501A1 (CypIA1), rather than the RsaI polymorphism of IIE1 (CypIIE1), is associated with lung cancer in Rio de Janeiro. Cancer Epidemiol Biomarkers Prev 4:63–67, 1995.
107. X Xu, KT Kelsey, JK Wiencke, JC Wain, DC Christiani. Cytochrome P450CYP1A1 *NspI* polymorphism and lung cancer susceptibility. Cancer Epidemiol Biomarkers Prev 5:687–692, 1996.
108. A Hildesheim, C-J Chen, NE Caporaso, Y-J Cheng, RN Hoover, M-M Hsu, PH Levine, IH Chen, J-Y Chen, C-S Yang, AK Daly, JR Idle. Cytochrome P4502E1 genetic

polymorphisms and risk of nasopharyngeal carcinoma: Results from a case-control study conducted in Taiwan. Cancer Epidemiol Biomarkers Prev 4:607–610, 1995.
109. L Sivaraman, MP Leatham, LR Wilkens, AF Lau, LL Marchand. CYP1A1 genetic polymorphisms and *in situ* colorectal cancer. Cancer Res 54:3692–3695, 1994.
110. NE Caporaso, MA Tucker, RN Hoover, RB Hayes, LW Pickle, HJ Issaq, GM Muschik, L Green-Gallo, D Buivys, S Aisner, JH Resau, BF Trump, D Tollerud, A Weston, CC Harris. Lung cancer and the debrisoquine metabolic phenotype. J NCI 82:1264–1272, 1990.
111. SJ London, AK Daly, DC Thomas, NE Caporaso, JR Idle. Methodological issues in the interpretation of studies of the CYP2D6 genotype in relation to lung cancer risk. Pharmacogenetics 4:107–108, 1994.
112. S Benhamou, C Bouchardy, C Paoletti, P Dayer. Effects of CYP-2D6 activity and tobacco on larynx cancer risk. Cancer Epidemiol Biomarkers Prev 5:683–686, 1996.
113. S Hayashi, J Watanabe, K Kawajiri. Genetic polymorphisms in the 5´-flanking region change transcriptional regulation of the human cytochrome P450IIE1 gene. J Biochem 110:559–565, 1991.
114. F Uematsu, S Ikawa, H Kikuchi, I Sagami, R Kanamaru, T Abe, K Satoh, M Motomiya, M Watanabe. Restriction fragment length polymorphism of human CYP2E1 (cytochrome P450IIE1) gene and susceptibility to lung cancer: Possible relevance to low smoking exposure. Pharmacogenetics 4:58–63, 1994.
115. S Kato, M Onda, N Matsukura, A Tokunaga, T Tajiri, DY Kim, H Tsuruta, N Matsuda, K Yamashita, PG Shields. Cytochrome P4502E1 (CYP2E1) genetic polymorphism in a case-control study of gastric cancer and liver disease. Pharmacogenetics 5:s141–s144, 1995.
116. AF Badawi, SJ Stern, NP Lang, FF Kadlubar. Cytochrome P-450 and acetyltransferase expression as biomarkers of carcinogen-DNA adduct levels and human cancer susceptibility. In: C Walker, JD Groopman, TJ Slaga, A Klein-Szanto, eds. Progress in Clinical and Biological Research, Vol 395: Genetics and Cancer Susceptibility: Implications for Risk Assessment, New York: Wiley-Liss, Inc., 1996, pp 109–140.
117. GB Sancar, W Siede, AA van Zeeland. Repair and processing of DNA damage: A summary of recent progress. Mutat Res 362:127–146, 1996.
118. WF Athas, MA Hedayati, GM Matanoski, ER Farmer, L Grossman. Development and field-test validation of an assay for DNA repair in circulating human lymphocytes. Cancer Res 51:5786–5793, 1991.
119. Q Wei, GM Matanoski, ER Farmer, MA Hedayati, L Grossman. DNA repair and aging in basal cell carcinoma: A molecular epidemiology study. Proc Natl Acad Sci U S A 90:1614–1618, 1993.
120. Q Wei, GM Matanoski, ER Farmer, MA Hedayati, L Grossman. DNA repair related to multiple skin cancers and drug use. Cancer Res 54:437–440, 1994.
121. Q Wei, L Chen, WK Hong, MR Spitz. Reduced DNA repair capacity in lung cancer patients. Cancer Res 56:4103–4107, 1996.
122. D Li, M Wang, L Chen, MR Spitz, WN Hittelman, Q Wei. In vitro induction of benzo(a)pyrene diol epoxide-DNA adducts in peripheral lymphocytes as susceptibility market for human lung cancer. Cancer Res 56:3638–3641, 1996.

123. DG Miller, R Tiwari, S Pathak, VL Hopwood, F Gilbert, TC Hsu. DNA repair and mutagen sensitivity in patients with triple primary cancers. Cancer Epidemiol Biomarkers Prev 7:321–327, 1998.
124. X Wu, J Gu, Y Patt, M Hassan, MR Spitz, RP Beasley, L-Y Hwang. Mutagen sensitivity as a susceptibility market for human hepatocellular carcinoma. Cancer Epidemiol Biomarker Prev 7:567–570, 1998.
125. FP Perera, K Hemminki, E Gryzbowska, G Motykiewicz, J Michalska, RM Santella, TL Young, C Dickey, P Brandt-Rauf, I DeVivo. Molecular and genetic damage in humans from environmental pollution in Poland. Nature 360:256–258, 1992.
126. H Bartsch. DNA adducts in human carcinogenesis: Etiological relevance and structure-activity relationship. Mutat Res 340:67–79, 1996.
127. YJ Zhang, TM Hsu, RM Santella. Immunoperoxidase detection of polycyclic aromatic hydrocarbon-DNA adducts in oral mucosa cells of smokers and nonsmokers. Cancer Epidemiol Biomarkers Prev 4:133–138, 1995.
128. RM Santella, RA Grinberg-Funes, TL Young, C Dickey, VN Singh, LW Wang, FP Perera. Cigarette smoking related polycyclic aromatic hydrocarbon-DNA adducts in peripheral mononuclear cells. Carcinogenesis 13:2041–2045, 1992.
129. LA Mooney, RM Santella, L Covey, AM Jeffrey, W Bigbee, MC Randall, TB Cooper, R Ottman, W-Y Tsai, L Wazneh, AH Glassman, T-L Young, FP Perera. Decline of DNA damage and other biomarkers in peripheral blood following smoking cessation. Cancer Epidemiol Biomarker Prev 4:627–634, 1995.
130. G Gurigliano, Y-J Zhang, L-Y Wang, G Flamini, A Alcini, C Ratto, M Giustacchini, E Alcini, A Cittadini, RM Santella. Immunohistochemical quantitation of 4-aminobiphenyl-DNA adducts and p53 nuclear overexpression in T1 bladder cancer of smokers and nonsmokers. Carcinogenesis 17:911–916, 1996.
131. D Lin, JO Lay Jr, MS Bryant, C Malaveille, M Friesen, H Bartsch, NP Lang, FF Kadlubar. Analysis of 4-aminobiphenyl-DNA adducts in human urinary bladder and lung by alkaline hydrolysis and negative ion gas chromatography-mass spectrometry. Environ Health Perspect 102(suppl 6):11–16, 1994.
132. FP Perera, A Estabrook, A Hewer, K Channing, A Rundle, LA Mooney, R Whyatt, DH Phillips. Carcinogen-DNA adducts in human breast tissue. Cancer Epidemiol Biomarkers Prev 4:233–238, 1995.
133. PG Shields, AC Povey, VL Wilson, A Weston, CC Harris. Combined high performance liquid chromatography/^{32}P-postlabeling assay of N^7-methyldeoxy-guanosine. Cancer Res 50:6580–6584, 1990.
134. A Izzotti, S de Flora, GL Petrilli, J Gallagher, M Rojas, K Alexandrov, H Bartsch, J Lewtas. Cancer biomarkers in human atherosclerotic lesions: Detection of DNA adducts. Cancer Epidemiol Biomarkers Prev 4:105–110, 1995.
135. S Kato, ED Bowman, AM Harrington, B Blomeke, PG Shields. Human lung carcinogen-DNA adduct levels mediated by genetic polymorphisms in vivo. JNCI 87: 902–907, 1995.
136. K Szyfter, K Hemminki, W Szyfter, Z Szmeja, J Banaszewski, M Pabiszczak. Tobacco smoke-associated N7-alkylguanine in DNA of larynx tissue and leucocytes. Carcinogenesis 17:501–506, 1996.

Molecular Epidemiology and Biomarkers

137. B Blomeke, MJ Greenblatt, VD Doan, ED Bowman, SE Murphy, CC Chen, S Kato, PG Shields. Distribution of 7-alkyl-2-deoxyguanosine adduct levels in human lung. Carcinogenesis 17:741–748, 1996.
138. SR Tannenbaum. Hemoglobin-carcinogen adducts as molecular biomarkers in epidemiology. Princess Takamatsu Symposia, vol 21. Tokyo: Japan Scientific Society Press, 1990, pp 351–360.
139. PL Skipper, SR Tannenbaum. Molecular dosimetry of aromatic amines in human populations. Environ Health Perspect 102 (suppl 6): 17–21, 1994.
140. SK Hammond, J Coghlin, PH Gann, M Pau, K Taghizadeh, PL Skipper, SR Tannenbaum. Relationship between environmental tobacco smoke exposure and carcinogen-hemoglobin adduct levels in nonsmokers. JNCI 85:474–478, 1993.
141. P Vineis, H Bartsch, N Caporaso, AM Harrington, FF Kadlubar, MT Laudi, C Malaveille, PG Shields, P Skipper, G Talaska, SR Tannenbaum. Genetically based N-acetyltransferase metabolic polymorphism and low-level environmental exposure to carcinogens. Nature 369:154–156, 1994.
142. MC Yu, RK Ross, KK Chan, BE Henderson, PL Skipper, SR Tannenbaum, GA Coetzee. Glutathione S-transferase M1 genotype affects aminobiphenyl-hemoglobin adduct levels in white, black, and Asian smokers and nonsmokers. Cancer Epidemiol Biomarkers Prev 4:861–864, 1995.
143. MT Landi, C Zocchetti, I Bernucci, FF Kadlubar, S Tannenbaum, P Skipper, H Bartsch, C Malaveille, P Shields, NE Caporaso, P Vineis. Cytochrome P4501A2: Enzyme induction and genetic control in determining 4-aminobiphenyl-hemoglobin adduct levels. Cancer Epidemiol Biomarkers Prev 5:693–698, 1996.
144. KR Kaderlik, G Talaska, DG DeBord, AM Osorio, FF Kadlubar. 4,4-Methylene-bis(2-chloroaniline)-DNA adduct analysis in human exfoliated urithelial cells by ^{32}P-postlabeling. Cancer Epidemiol Biomarkers Prev 2:63–69, 1993.
145. FJ van Schooten, FJ Jongeneelen, MJX Hillebrand, FE van Leeuwen, AJA de Looff, APG Dijkmans, JGM van Rooij, L den Engelse, E Kriek. Polycyclic aromatic hydrocarbon-DNA adducts in white blood cell DNA and 1-hydroxypyrene in the urine from aluminum workers: Relation with job category and synergistic effect of smoking. Cancer Epidemiol Biomarkers Prev 4:69–77, 1995.
146. PS Nielsen, N de Pater, H Okkels, H Autrup. Environmental air pollution and DNA adducts in Copenhagen bus drivers—effect of GSTM1 and NAT2 genotypes on adduct levels. Carcinogenesis 17:1021–1027, 1996.
147. M Rojas, K Alexandrov, G Auburtin, A Wastiaux-Denamur, L Mayer, B Mahieu, P Sebastien, H Bartsch. Anti-benzo[a]pyrene diolepoxide-DNA adduct levels in peripheral mononuclear cells from coke oven workers and the enhancing effect of smoking. Carcinogenesis 16:1373–1376, 1995.
148. R Ross, J-M Yuan, M Yu, GN Wogan, G-S Qian, J-T Tu, JD Groopman, Y-T Gao, BE Henderson. Urinary aflatoxin biomarkers and risk of hepatocellular carcinoma. Lancet 339:943–946, 1992.
149. G-S Qian, MC Yu, R Ross, J-M Yuan, Y-T Gao, GN Wogan, JD Groopman. A follow-up study of urinary markers of aflatoxin exposure and liver cancer risk in Shanghai, P.R.C. Cancer Epidemiol Biomarkers Prev 3:3–11, 1994.

150. D Tang, RM Santella, AM Blackwood, T-L Young, J Mayer, A Jaretzki, S Grantham, W-Y Tsai, FP Perera. A molecular epidemiological study of lung cancer. Cancer Epidemiol Biomarkers Prev 4:341–346, 1995.
151. M-W Yu, C-J Chen, L-W Wang, RM Santella. Aflatoxin B_1 adduct level and risk of hepatocellular carcinoma. Proc Am Assoc Cancer Res 36:1644, 1995.
152. S-Y Kuang, X Fang, P-X Lu, Q-N Zhang, Y Wu, J-B Wang, Y-R Zhu, JD Groopman, TW Kensler, G-S Qian. Aflatoxin-albumin adducts and risk for hepatocellular carcinoma in residents of Qidong, People's Republic of China. Proc Am Assoc Cancer Res 37:1714, 1996.
153. KA McGlynn, EA Rosvold, ED Lustbader, Y Hu, ML Clapper, T Zhou, CP Wild, XL Xia, A Baffoe-Bonnie, D Ofori-Adjei, T London. Susceptibility to hepatocellular carcinoma is associated with genetic variation in the enzymatic detoxification of aflatoxin B_1. Proc Natl Acad Sci USA 92:2384–2387, 1995.

13

Quantitative Cancer Risk Assessment of Nongenotoxic Carcinogens

Anna M. Fan and Robert A. Howd
California Environmental Protection Agency, Oakland, California

I. INTRODUCTION

For cancer risk assessment of chemicals, efforts have been made to distinguish between genotoxic mechanisms (involving direct DNA changes, such as alkylation or chromosome breakage) and nongenotoxic mechanisms (acting at some other site in the cell that does not involve genetic material). The rationale for this distinction is the assumption that genotoxic actions may have no effective threshold (a linear low-dose response), whereas nongenotoxic effects require a suprathreshold dose to disturb a homeostatic system sufficient to elicit a toxic response. This also implies that different scientific methods may be appropriate to assess cancer risk at low doses, depending on the chemical's mechanism of action. Methods that assume no threshold imply a potential risk at any dose, whereas the opposite assumption is that a safe dose can be estimated. Thus, the question of potential genotoxicity is important for risk assessment of a chemical shown to be carcinogenic in experimental animals or humans. For those chemicals for which one or more bioassays show an increase in tumors at one or multiple sites, the question of potential genotoxicity-mediated carcinogenesis becomes the overriding factor in the toxicity evaluation. The significance is that if a chemical is assumed to be genotoxic, the de minimis risk level might be estimated at a level much lower than the estimated safe level based on a non-linear dose-response. This difference may, in turn, make an enormous difference in regulation of the chemical in air, water, food, or toxic waste. The approach for cancer risk assessment has been to assume a linear low-dose response for carcinogens unless there is sufficient evidence to prove otherwise. In the past 10 years, increased efforts have focused on the study of carcinogens that may act by way of a nongenotoxic mechanism. This chapter discusses recent developments in the area of cancer risk assessment based on nongenotoxic or epigenetic modes of action, with various interpretations of the adequacy of data on low-dose responses for risk assessment.

II. DEFINITION OF NONGENOTOXIC CARCINOGENS

Nongenotoxic carcinogens are generally considered to be those that do not react, directly or through a reactive metabolite, with nuclear DNA. This can be measured as incapability to form DNA adducts, to induce DNA repair, and to produce a positive response in genotoxicity assays, including mutagenicity tests. Experimentally, genotoxicity is commonly evaluated in a battery of short-term tests using a variety of test systems and endpoints, both in vivo and in vitro (1). When all the short-term tests show negative responses, the chemical is believed to be nongenotoxic. But the overall evidence would depend on the adequacy and quality of the short-term tests used, and there are often equivocal responses. Any positive responses would be evidence that the chemical may be genotoxic in humans. However, even when the short-term tests are negative, it does not preclude the possibility that a chemical can have both genotoxic and nongenotoxic properties, for example, reaction with DNA and stimulation of cell proliferation.

Nongenotoxic carcinogens can affect the cell reproductive process by acting as cytotoxicants, inducing regenerative cell proliferation, and producing secondary critical effects such as inflammation (2). Cytotoxicity releases nucleases that may induce DNA damage. Cytolethality can induce regenerative cell proliferation, which can influence spontaneous tumor rates that result from pre-existing mutations. Inflammation can increase the generation of oxygen radicals that may induce mutations. Cell proliferation can also occur through more direct growth-stimulatory mechanisms. Chemicals that interact with hormone receptors (estrogens, androgens, growth hormone) or increase the synthesis of these hormones can increase cell division in responsive tissues and thus lead to increased growth of pre-existing tumor cells. Induced cell division may also result in conversion of DNA adducts (from endogenous or exogenous sources) to mutations before DNA repair can occur and affect expression of growth control genes (oncogenes and tumor suppressor genes). Increased circulating growth factors can also affect tissue regeneration, as precancerous cells may be more responsive, thus providing a selective growth to neoplastic cells. These modes of action are considered as tumor promotion rather than tumor initiation, and most of these are believed to exhibit non-linear (threshold) dose-responses.

III. PUBLIC HEALTH IMPLICATIONS

The challenge for risk assessment is to decide when the evidence is adequate to consider a particular nongenotoxic action of a chemical as the *cause* of increased tumor growth, rather than merely a concomitant effect, and to exclude all possible linear low-dose causes. Risk assessment based on this nongenotoxic mode of action could then incorporate quantitative models that do not rely on the default, linear low-dose extrapolation.

Utilization of alternative risk assessment models for nongenotoxic carcinogens is likely to result in an estimated concentration or dose of regulatory concern that is higher than estimated with standard cancer models such as the linearized multistage (LMS) model. But if the judgment about mode of action is incorrect, than the risk to the exposed population may be underestimated. Faced with this uncertainty, prudence in making regulatory judgements is required. However, toxic chemicals, both natural and synthetic, abound in our environment. The regulatory challenge is to adequately estimate risk so we can expend our economic resources and efforts at controlling the toxicants associated with the greatest public health impact for the maximum enhancement of public safety. Quantitative risk assessment provides us with the tools to do this, although the methodology is still evolving, especially with regard to adequacy of the documentation of possible carcinogenesis mechanisms. Agreement must be reached on the type of evidence and level of confidence required to assume that a chemical produces tumors by a nongenotoxic mode of action.

The quantitative data needed for this risk assessment may include in vitro studies, animal bioassays, and human studies. Interpretation of the results should result in public health regulatory decisions that protect the most sensitive populations and prevent the greatest or most serious risks. Therefore, the risk assessment should consider a variety of endpoints and mechanistic interpretations that explore the ramifications of the various risk assessment judgments. If, after the complete body of evidence has been explored, the decision of the risk assessor or risk manager is that a non-linear, nongenotoxic mechanism for any observed tumors has not been adequately demonstrated, then the default linear assumption may still be used.

In the following sections, the implications of this public health mandate to the practice of risk assessment for several different types of chemicals is further explored. Methods for dose-response extrapolation are presented, followed by discussion of various chemicals representing different putative nongenotoxic carcinogenic mechanisms.

IV. QUANTITATION PRINCIPLES AND METHODS

Cancer risk assessment has traditionally used the linear multistage model for low-dose risk estimation. However, if a nongenotoxic mode of action is accepted for a chemical, linear low-dose extrapolation methods may not necessarily be appropriate. Risk assessment based on non-cancer endpoints has generally involved determining the lowest observed effect level (LOEL) or no observed effect level (NOEL) and dividing by uncertainty factors (UFs) to estimate a dose likely to protect humans from adverse effects. This extrapolation considers uncertainty and variability, including sensitive population groups. Safe levels can be estimated for acute, subchronic, and chronic exposures by the same principles. Thus, if there is a good study in healthy adult humans that demonstrates a safe chronic dose, dividing that dose by 10 may be adequate to protect sensitive populations (3). Additional factors of 10 for other extrapolations (estimating safe chronic dose from acute or subchronic data, estimating human dose from animal data, accounting for an incomplete database or severity of adverse effects) can account for further uncertainties. The "factor of 10" convention is undoubtedly a very crude estimate of total uncertainty (4–7). This approach also does not calculate the actual risk of an adverse effect; it only provides a rough estimate of maximum safe dose.

To improve this situation, dose-response and cross-species extrapolation models have been developed and applied to risk assessment. Physiologically based pharmacokinetic (PBPK) models (8, 9), benchmark dose methods (10–12), and advanced threshold curve-fitting models for continuous data (13) have all contributed to the development of more quantitatively based risk assessment methods. Although there has been tremendous progress in modeling physiological processes and toxic responses, consensus has by no means been reached on how to estimate safe dose in humans from animal data (14, 15). The debates are even more heated when the endpoint is increased tumors, with a proposed nongenotoxic mode of action.

Risk assessments for nongenotoxic carcinogens may be carried out using any of the above methods, from NOEL with uncertainty factors, through complex threshold data-fitting methods, to linearized multistage. It is common for several different quantitation methods to be explored. The most defensible method, or sometimes the most health-protective conclusion, is chosen.

The methods must follow the conclusions reached in the hazard assessment as to the potential mode(s) of action. The assessment begins by analyzing the data in the range of observation and evaluating how best to extrapolate to low-dose, health-protective risk levels. If standard cancer models are used, the de minimis risk criteria are usually set at the 95% lower confidence limit of dose yielding a 10^{-6} risk level, as for genotoxic carcinogens (see chapter on genotoxic carcinogens). When UF methods are used, the methods are applied as for noncarcinogens except that an additional tenfold safety factor may be applied for severity of effect. When animal data are the basis of analysis, estimation of a human dose may use PBPK modeling if appropriate data are available.

For most toxicity tests, about 10% of the animals responding (the ED_{10}) represents a practical observational threshold, the lower limit for statistical significance. Alternatively, a 10% average change in a measured parameter may be used. The estimated dose corresponding to a 10% response by either definition represents a common "benchmark" from which an effective safe dose may be extrapolated, either with probabilistic methods or by application of uncertainty factors. Depending on the magnitude of the combined uncertainty factors, the effect of the threshold assumption is usually a higher estimated safe exposure level than if traditional low-dose linear extrapolation is used. The difference in dose response extrapolation between nonthreshold and threshold mechanisms is illustrated in Figure 1 below. In this figure, the squares represent percent of tumor-bearing animals in a lifetime cancer study. The solid line represents a fitted curve through zero dose, and the dotted line represents a straight line drawn through the two data points with a significantly increased proportion of tumors.

Using the linear multistage model (the solid line), the upper 95% confidence limit on the dose that produces a 10^{-6} cancer risk for these data is 0.01 mg/kg. Using a 100-fold uncertainty factor from the threshold of 500 mg/kg, defined by the dotted line, provides an estimated safe dose of 5 mg/kg. Specific corrections for differences among species (such as metabolism, distribution, and

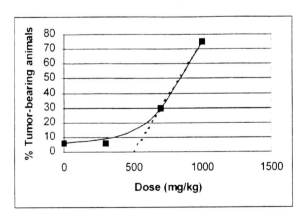

FIG. 1. Theoretical tumor data, illustrating extrapolation of cancer risk using a curve-fitting model (**solid line**) drawn through zero dose, and a straight line drawn through the points with significantly increased tumor incidence (**the dotted line**), defining a threshold at a dose of 500 mg/kg.

tissue sensitivity) and experimental uncertainties usually retain this large difference between nonthreshold and threshold estimates of a safe dose.

Quantitative estimates of a safe dose are usually affected more by the decision to apply a non-linear low-dose assumption than by the method chosen among the linear or non-linear models. The examples provided below show how these principles have been applied in some recent risk assessment decisions that involve primarily the U.S. Environmental Protection Agency (EPA) and the State of California.

V. EXAMPLES OF EVALUATIONS REGARDING NONGENOTOXIC MECHANISMS

A. Kidney Cancer in Male Rats and α-2-Microglobulin Nephropathy

Alpha(α)-2-microglobulin (α-2mG) associated nephropathy has been evaluated as a mechanism of renal tubular cell carcinogenesis in male rats (16–18). The hypothesis supporting this mechanism is based on the finding that chemicals that induce α-2mG accumulation and renal carcinogenesis in male rats have not been shown to induce kidney tumors in animals that lack the ability to synthesize α-2mG in the liver. However, both data that are consistent and inconsistent with this hypothesis exist (16).

In the livers of male rats, the low-molecular-weight protein α-2mG is synthesized under androgen control. It is not synthesized in the livers of female rats, mice of either sex, humans, and several other species (16). Hydrocarbons that bind to α-2mG do not increase the level of hepatic synthesis of this protein. Its physiological function is not known. The accumulation of α-2mG in kidneys of male rats in response to certain toxic chemicals is hypothesized to cause lysosomal dysfunction that results in cell killing, although the actual cause of cell death is not known. Regenerative proliferation of epithelial cells in the P2 segment of renal tubules occurs in response to cell loss. It is suggested that regenerative hyperplasia causes tumorigenic responses in the male rat kidney by increasing the likelihood that spontaneously initiated cells will divide and eventually produce a detectable tumor. Accumulation of α-2mG in the male rat kidney may occur by ligand binding to this protein, rendering it more resistant to proteolytic degradation, or by direct inhibition of the proteolytic enzymes that degrade this protein. Other hypotheses are that α-2mG facilitates transport of a protease inhibitor or its precursor to the kidney, or protease inhibitors reach the kidney without binding to α-2mG.

The U.S. EPA (19) has defined three criteria that should be met to consider whether the α-2mG-process could be a factor in observed male rat renal tumors. These criteria are:

1. Increased number and size of hyaline droplets in renal proximal tubule cells of treated male rats. The abnormal accumulation of hyaline droplets in the P2 segment of the renal tubule is necessary to attribute the renal tubule tumors to the α-2mG sequence of events.
2. The accumulating protein in the hyaline droplets is α-2mG. Hyaline droplet accumulation is a nonspecific response to protein overload in the renal tubule and may not be due to α-2mG.
3. Additional aspects of the pathological sequence of lesions associated with α-2mG nephropathy are present. Typical lesions include: single-cell necrosis, exfoliation of epithelial cells into the proximal tubular lumen, formation of granular casts, linear mineralization of papillary tubules, and tubule hyperplasia. If the response is mild, all of these lesions may not be observed; however, some elements consistent with the pathological sequence must be demonstrated to be present. (The U.S. EPA stated that if data from adequately conducted studies of male rats do not meet the criteria in any one category, then the available tumor data should be analyzed in accordance with standard risk assessment principles.)

The International Association for Research on Cancer (IARC) (18) has also recently formulated seven necessary criteria for assuming kidney tumors are mediated through an a-2mG-associated response in male rats. These are shown below:

1. Lack of genotoxic activity (agent and/or metabolite) based on an overall evaluation of in vitro and in vivo data
2. Male rat specificity for nephropathy and renal tumorigenicity
3. Induction of the characteristic sequence of histopathological changes in shorter-term studies, of which protein droplet accumulation is obligatory
4. Identification of the protein accumulating in tubular cells as α-2mG
5. Reversible binding of the chemical or metabolite to α-2mG
6. Induction of sustained increased cell proliferation in the renal cortex
7. Similarities in dose-response relationship of the tumor outcome with the histopathological endpoints (protein droplets, α-2mG accumulation, cell proliferation).

Methyl tertiary butyl ether (MTBE), widely used in gasoline as an oxygenate, is a chemical of current interest for which the possible role of α-2mG in induction of kidney tumors has been intensely debated. Evaluation by our program for the development of a public health goal (PHG) for MTBE in drinking water (20) has considered both the U.S. EPA and IARC criteria. The evaluation concluded that MTBE did not meet IARC criteria numbers 2, 4, 5, and 7.

The issues involving evaluation of kidney tumors in rats given MTBE have also been addressed by other scientific bodies. These are reviewed and summarized by the Office of Environmental Health Hazard Assessment (OEHHA) (20) and presented below. The Interagency Assessment of Oxygenated Fuels (National Science and Technology Council [NSTC]) (21, 22) concluded that the available data on MTBE do not support the view that the male kidney tumor response resulted solely from accumulation of α-2mG. The Health Effects Institute report (23) indicated that the mechanisms are not understood. Although it appears that the hyaline droplet nephropathy in male F344 rats may be rightfully implicated as a potential factor in the pathogenesis of the renal tumor response, other factors may also be involved. A National Research Council (NRC) panel (24), based on reviews of abstracts from the Chemical Industry Institute of Technology (CIIT) available in 1996, concluded that the research conducted at CIIT appears to have fulfilled the EPA criteria for causation with respect to α-2mG. The panel also concluded that the data for MTBE-induced kidney tumors in male rats should not be used for human risk assessment until the more recent data on mechanism of action are reviewed and evaluated. The NSTC (22) reviewed the NRC panel's conclusions and evaluated all available data at that time and concluded that the second and third U.S. EPA criteria for α-2mG had not been fulfilled for MTBE. The report listed several key issues that remain to be resolved and concluded that it is reasonable to believe that other modes of action are operating. The NSTC concluded that the prudent public health approach is to use the kidney tumor response for both human hazard identification and quantitative estimation of cancer risk for MTBE.

The OEHHA's scientific review of MTBE supports the NSTC judgment. Depending on several technical considerations involved in using the U.S. EPA decision criteria and the IARC criteria, the evidence for an α-2mG role in the rat kidney tumor response does not reasonably fulfill the criteria. The evaluation of kidney tumor response after MTBE inhalation is difficult because the evidence shows mild α-2mG accumulation and symptomatic nephropathy, with some of the nephropathy intermingled with a background of non-α-2mG nephropathy in both males and females. Based on available experimental data, it is reasonable to believe that other modes of action are operating. Because evidence of an influence of α-2mG on the rat kidney tumor response is not established, the prudent public health approach is to include the use of this tumor response for both hazard identification and quantitative estimation of cancer risk. It was acknowledged that other views on this subject differ, and the decision will be re-evaluated as further evidence is accumulated.

The cancer potency estimates for MTBE involved a polynomial fit to the experimental data to establish the lower 95% confidence bound on the dose associated with a 10% increased risk of cancer (LED10). The final potency estimate was derived from the geometric mean of the cancer slope factors of the combined rat kidney adenomas and carcinomas, the male rat Leydig cell tumors, and the combined leukemia and lymphomas in female rats, giving a value of 1.8×10^{-3} $(mg/kg\text{-}day)^{-1}$. The corresponding public health goal is 13 ppb. In comparison, the U.S. EPA (25) has noted that a nonlinear mode of action has not been established for MTBE, and it is prudent to assume a linear dose-response. It also stated that the weight of evidence indicates that MTBE is an animal carcinogen, and the chemical poses a carcinogenic potential to humans. The agency recommended that MTBE concentrations in the range of 20 to 40 ppb or below would assure both consumer acceptance of the water (because of taste) and a large margin of safety from any toxic effects, based on the margin of exposure (MOE) approach.

B. Mouse Liver Tumors

The view has often been expressed that chemicals that produce only mouse liver tumors, particularly in male B6C3F1 mice, may be acting as tumor promotors rather than complete carcinogens. Male B6C3F1 mice have a high natural incidence of liver tumors (and elevated peroxisome levels) compared with females and other mouse strains. Evidence to support the view that most carcinogens that only cause mouse liver tumors are not primary carcinogens came from the work of Ashby and Tennant (26) and Gold et al (27). Contradictory evidence was presented by the National Toxicology Program (NTP) (28), showing that for 20 chemicals that produced only B6C3F1 mouse liver tumors, there was no correlation of tumor occurrence with sex of the animals.

Ashby and Tennant (26) and Tennant et al (29) also reported a difference in distribution of tumor sites for agents that showed a positive response in the Ames *Salmonella* assay, compared with those that were negative. Using data on 222 carcinogens tested by NCI/NTP in both rats and mice, the investigators divided the chemicals in two classifications: (1) genotoxic—chemicals that are mutagenic in *Salmonella* and have certain structural attributes associated with mutagenicity, and (2) nongenotoxic—negative in *Salmonella* and absence of structural attributes associated with mutagenicity. The genotoxic chemicals produced all 27 tumor types being examined, and the nongenotoxic chemicals produced 5 of the tumor types. The liver was the most common tumor site for both genotoxic and nongenotoxic chemicals, but the nongenotoxic chemicals were approximately twice as likely to cause liver tumors as the genotoxic ones. One notable chemical in this evaluation is benzene, because it would be classified as nongenotoxic under this classification. However, although it does not induce point mutation in the *Salmonella* assay, it is strongly positive in most in vivo and in vitro short-term assays (30).

Chlorinated hydrocarbon pesticides represent classic examples of the mouse liver tumor response. This is relevant to OEHHA evaluations of chlordane, heptachlor, and lindane, for instance, for the development of public health goals for these chemicals in drinking water (31–33). However, for these chemicals, findings of liver tumors in male mice are further supported by tumors in female mice and in rats. The mouse liver tumors were then considered as part of the weight of evidence for carcinogenicity, and the risk assessment dose-response extrapolations were based on the linear multistage model.

C. Peroxisome Proliferation

Peroxisomes are subcellular organelles that contain oxidase enzymes that produce hydrogen peroxide (H_2O_2) and catalase, the enzyme that converts H_2O_2 to water and oxygen. This conversion presumably prevents toxicity that might occur from H_2O_2 accumulation. There is an apparent correlation between rat liver peroxisome proliferation (associated with induction of hepatic cytochrome P4A) and hepatocarcinogenesis. This observation has been used as a basis for the proposal that the hypolipidemic peroxisome proliferators may represent a novel class of nongenotoxic chemical carcinogens that act by a threshold mechanism (34). Some hypolipidemic drugs, such as clofibrate, and industrial plasticizers, such as the phthalates, produce increased liver size, a significant increase in size and amount of hepatic peroxisomes, induction of specific cytochrome enzymes, a decrease in serum lipid levels, and increased incidences of hepatocellular neoplasms in rats and mice. The nonneoplastic changes revert back to control levels after cessation of the chemical exposure. However, such correlation between the two responses does not prove a cause-and-effect relationship. In addition, exceptions have been observed: When di(2-ethylhexyl)phthalate (DEHP) and di(2-ethylhexyl)adipate (DEHA) were administered at similar doses to rats in a 2-year study, similar levels of peroxisome induction occurred, but only DEHP induced a significant liver tumor response (35, 36).

Although most peroxisome proliferators lack genotoxic activity, some, such as nafenopin and ciprofibrate, do produce genotoxic effects (37). It has also been proposed that an imbalance between production and degradation of H_2O_2 caused by enhanced peroxisome oxidation could lead to an increase of H_2O_2-mediated oxidative damage. Increased levels of hydroxyl radical generated from H_2O_2 are hypothesized to react with DNA, which results in carcinogenesis (38). This would presumably be a threshold mechanism. Another potential mechanism of peroxisome proliferator-mediated carcinogenicity is increased replicative DNA synthesis and cell division (39, 40). However, early high levels of DNA synthesis and peroxisome proliferation do not appear to be a sufficient stimulus to explain the hepatocarcinogenesis (41).

Hypolipidemic effects of the peroxisomal proliferators clofibrate and ciprofibrate are prominent in both rats and humans. The mechanism of this response is not fully understood. The liver peroxisome proliferator-activated receptor (PPAR) provides a mechanistic basis for understanding how peroxisome proliferators modulate gene expression that leads to induction of peroxisome enzymes (42). The PPAR is a ligand-activated intracellular transcription factor. The receptor is activated by ligand binding and forms a heterodimer with the retinoid X receptor. This ternary complex binds to specific DNA response elements, causing transcriptional activation of genes coding for peroxisome enzymes, such as cytochrome P4A (in rats). Humans have PPAR subtypes, among which is one that shows high homology with rodent PPAR-alpha and can be activated by peroxisome proliferators (43). Although this may justify the hypothesis that the mechanistic pathway can be relevant to humans, neither peroxisomal proliferation nor hepatic tumors have been observed in humans from the hypolipidemic agents or phthalates.

This issue was most relevant to OEHHA for development of the DEHP public health goal for drinking water (44). For this chemical, the rat bioassays appeared to show a linear dose-response for liver tumors. Our evaluation concluded that relevant biochemical pathways for tumor production do occur in humans, and that the evidence was inadequate to conclude that DEHP acts to produce tumors by a threshold (nongenotoxic) mechanism. Dose-response models were applied using both linear and non-linear assumptions. The estimated health protective concentrations are 12 ppb by the linearized multistage model and 230 ppb by the UF method, based on the 95% lower bound on the dose that yields a 10% tumor rate, and a combined UF of 1000, respectively. The chemical remains under active evaluation as more data are being generated.

D. Dioxin and the Aryl Hydrocarbon Receptor

2,3,7,8-Tetrachlorodibenzo-p-dioxin (TCDD) is a highly potent animal carcinogen that has been implicated in the etiology of soft tissue cancers in humans (45). It does not have a convincing database on genotoxicity and is thought to act as a tumor promotor. The major biological activities of TCDD are mediated by binding to and activating the Ah (aryl hydrocarbon) receptor in different tissues. The activated receptor forms a heterodimer with another transcription factor (Ah receptor nuclear translocator, [arnt]). This ternary complex binds to regulatory sequences on DNA and alters the expression of several proteins, some of which may be involved in the carcinogenic process. If TCDD carcinogenesis is based on Ah receptor occupancy, then the carcinogenic dose-response would exhibit a threshold. To test this threshold hypothesis, a PBPK model has been produced to study dosimetry and the proposed carcinogenic mechanism (46). The model is consistent with a large number of observed responses under several dosing regimens, but its predictions do not support the hypothesis that mediation of dioxin's effects by the Ah receptor imposes threshold dose-response behavior.

The U.S. EPA's evaluation of TCDD toxicity concluded that its tumorigenic effects appeared to be linear at low doses. This does not constitute evidence for lack of an Ah-receptor-mediated tumorigenicity threshold but merely reflects the fact that tumors were found at all doses tested. At present, despite the agreement on a receptor-mediated mechanism of action and no evidence for genotoxicity, all regulatory risk assessments have used linear low-dose extrapolation methods. However, estimated safe levels based on the potent noncarcinogenic effects on immune and endocrine systems (after including uncertainty factors) are nearly as low as those estimated based on a linear extrapolation. Environmental TCDD levels are perceived to be a problem, irrespective of extrapolation method, so in this case the quantification method used for risk assessment may be of less direct importance. The critical factor for risk assessment of dioxins may be the method(s) of estimating combined risk of the many dioxin and furan analogues, including planar polychlorinated biphenyls.

E. Regenerative Hyperplasia

Chloroform has been used as an illustrative approach for assessing the potential cancer risk for nongenotoxic/cytotoxic carcinogens (2). The carcinogencity of chloroform is proposed to be secondary to the cytotoxicity and compensatory cell replication in target tissues. When given by gavage to male and female B6C3F1 mice in corn oil at 138–477 mg/kg/day, chloroform induced liver tumors. Chloroform also produced kidney tumors in Osborne-Mendel rats at 90 and 180 mg/kg/day (47). When administered in drinking water, chloroform did not increase liver tumors in the mice and male Osborne-Mendel rats, and the incidence of male rat kidney tumors was much lower, even though the daily chloroform intake was similar to that in the gavage study (48). Using the mouse liver cancer incidence data from the gavage study, the investigators indicated that the linearized multistage risk model suggests that a 1 in 100,000 lifetime increased human cancer risk would be achieved at 0.004 ppm chloroform in drinking water.

A detailed dose-response study was conducted in female B6C3F1 mice in which chloroform was given daily by gavage, under conditions similar to those in the cancer study, for periods up to 3 weeks to study the cytopathological correlates (49). At the doses of 238 and 477 mg/kg/day, substantial necrosis, inflammation, and regenerative cell proliferation were observed. The NOAEL for histopathological changes in the liver was 10 mg/kg/day. Using a total uncertainty factor of 1000 (10 each for intraspecies extrapolation, individual variation, and subchronic to chronic duration), a human dose of 0.01 mg/kg/day was derived by the investigators as a level not anticipated to increase cancer risk, assuming nongenotoxic threshold criteria. This corresponds to an estimated safe level in drinking water of 0.07 ppm.

Larson et al (49) also administered chloroform to the mice in drinking water at up to 1800 ppm under conditions similar to the cancer bioassay. No hepatic histopathological changes or cell proliferation were observed. The cumulative daily amount of chloroform ingested in the highest dose group was similar to that given in the gavage study. This lack of toxicity was considered by the investigators as consistent with the lack of female mouse liver tumors in the drinking water cancer bioassay and used as evidence that chloroform-induced mouse liver tumors are secondary to cytolethality and cell proliferation. The investigators postulated that the difference observed between the gavage and drinking water studies can be explained solely by toxicokinetics. When given as a single bolus dose by gavage, the treatment resulted in a high rate of production of toxic metabolites in the liver that overwhelmed cellular detoxification mechanisms and killed the hepatocytes. When given in drinking water, ingestion was in small sips throughout the day. The chemical was probably not cytotoxic or carcinogenic under these conditions, as delivery to the target tissue was at rates low enough to allow detoxification, even though the cumulative dose was 329 mg/kg/day. The authors concluded that a model that assumes that tumor formation is secondary to proliferative effects describes the pattern of tumors more accurately than the linearized multistage model. They derived an NOAEL of 1800 ppm for induced cell proliferation and mouse liver tumors. Using a UF of 1000, a level of 0.18 ppm was considered a level below which no increased risk of tumors would be expected.

Under the 1996 *U.S. EPA Proposed Guidelines for Carcinogen Risk Assessment*, both linear and nonlinear approaches are acceptable (50). To use the nonlinear method, the agent's mode of action in causing tumors must be reasonably understood and support an assumption of nonlinearity. Also, in 1996, U.S. EPA (51, 52) cosponsored an International Life Sciences Institute (ILSI) project in which an expert panel reviewed data on the carcinogenicity of chloroform and considered how endpoints related to its mode of carcinogenic action can be applied in the hazard and dose-response assessment. After analysis of the data, the panel concluded the following (51, 52):

1. The weight of evidence for understanding of the mode of action indicated that chloroform was not acting through direct reaction with DNA.
2. The evidence suggested that exposure to chloroform resulted in recurrent or sustained toxicity as a consequence of oxidative generation of highly tissue reactive and toxic metabolites (i.e., phosgene and hydrochloric acid), which in turn would lead to regenerative cell proliferation.
3. Oxidative metabolism is the predominant metabolic pathway for chloroform, and this mode of action was the key influence of the chemical on the carcinogenic process.
4. The weight of evidence for the mode of action was stronger for the mouse kidney and liver responses and more limited, but still supportive, for rat kidney tumor responses.
5. Chloroform is a likely carcinogen to humans above a certain dose range but an unlikely carcinogen below this range.
6. The mechanism is expected to involve a dose-response relationship that is nonlinear and probably exhibits an exposure threshold.

7. The nonlinear default or margin-of-exposure approach is recommended for quantifying the cancer risk associated with exposure to chloroform.

In its reassessment of chloroform in 1998 (52) after the 1994 proposed rule and 1997 NODA, U.S. EPA considered the ILSI report and the new science available and concluded the following:

1. Chloroform is a likely human carcinogen by all routes of exposure.
2. Chloroform's carcinogenic potential is indicated by animal tumor evidence (liver tumors in mice and renal tumors in both mice and rats from inhalation and oral exposures), as well as metabolism, toxicity, mutagenicity, and cellular proliferation data that contribute to the understanding of the mode of action.
3. Although a precise mechanism of action is not known, U.S. EPA agrees with the ILSI panel that a DNA-reactive mutagenic mechanism is not likely to be the predominant influence of chloroform on the carcinogenic process.
4. There is a reasonable scientific basis to support a predominant mode of carcinogenic action that involves cytotoxicity produced by the oxidative regeneration of highly reactive metabolites, phosgene and hydrochloric acid, followed by regenerative proliferation.
5. The EPA agrees with an ILSI report that the chloroform dose-response should be considered nonlinear.

In its dose-response assessment, U.S. EPA used three approaches to estimate the maximum contaminant level goal (MCLG) for chloroform: the LED10 for tumor response, and the risk reference dose (RfD) for hepatotoxicity. The linear multistage approach was also considered because of the remaining uncertainties associated with the understanding of chloroform's mode of carcinogenic action. These included lack of data on cytotoxicity and cell proliferation responses in Osborne-Mendel rats, lack of mutagenicity data on chloroform metabolites, and the lack of comparative metabolic data between humans and rodents. Overall, the U.S. EPA viewed the linear multistage approach as overly conservative in estimating low-dose risk and concluded that the nonlinear default or margin of exposure approach is preferred for quantifying the cancer risk, because the evidence is stronger for a nonlinear mode of action.

Based on the reassessment, the U.S. EPA considered revising its MCLG for chloroform from zero to 0.30 mg/L, using hepatotoxicity rather than carcinogenicity as the endpoint, using the following:

$$MCLG = (RfD \times BW \times RSC) / CR$$

where:

RfD = LOAEL for hepatotoxicity from a chronic oral dog study (53) and an uncertainty factor of 1000 (10 each for inter- and intraspecies and for using LOAEL rather than NOAEL)
BW = body weight of 70 kg
RSC = drinking water relative source contribution of 80%
CR = drinking water consumption rate of 2 liters/day

Thus,

MCLG = (0.01 mg/kg d × 70 kg × 0.8)/2 L/day = 0.3 mg/L (rounded)

While considering the carcinogenicity data, the kidney tumor response data in Osborne-Mendel rats from the study of Jorgenson et al (47) was used by the U.S. EPA as the basis for the point of departure (i.e., LED10 and ED10), because a relevant route of exposure (oral) and multiple doses (a total of five, including zero) were used in the study. The animal data were adjusted to equivalent human doses using body weight raised to the ¾ power as the interspecies scaling factor. The ED10 of 37 mg/kg/day and LED10 of 23 mg/kg/day were divided by an MOE of 1000. This yielded dose estimates of 0.37 and 0.23 mg/kg/day for carcinogenicity, respectively. These translate into MCLGs of 1.0 and 0.6 mg/L, respectively. The U.S. EPA considered these MCLGs protective of susceptible groups, including children, because of the following: chloroform is not biopersistent, and humans are exposed to relatively low levels of chloroform in the drinking water (generally under 100 μg/L), which are below exposures needed to induce a cytotoxic response. The mode of action for chloroform's cytotoxic and carcinogenic effects involves a generalized mechanism of cytotoxicity that is consistent across species. The activity of the enzyme (i.e., CYP2E1) involved in generating metabolites key to chloroform's mode of action is not greater in children than adults, and is probably less.

The level of 0.3 mg/L was considered the MCLG because hepatotoxicity is a more sensitive effect than tumor induction by chloroform. Even if low-dose linearity were assumed, as it was in the 1994 proposed rule, the value of 0.3 mg/L would be equivalent to a 5×10^{-5} cancer risk level (54).

However, in its "final rule" (Federal Register 63:69390-475, 1998) U.S. EPA declined to follow this reasoning, considering it to be premature, and continued use of an MCLG of zero. It should be noted that Melnick et al (54) have cast considerable doubt on the cytotoxicity-regenerative hyperplasia theory, concluding that regenerative hyperplasia is not required for liver tumor induction in female B6C3F1 mice exposed to trihalomethanes (THMs). All these conclusions, and data interpretations will be considered by OEHHA in its evaluation of THMs.

F. Thyroid Tumors

Thyroid tumors are the second tumor type for which the U.S. EPA has developed generic guidance for risk assessment (55). The agency efforts have been directed specifically at tumors of the thyroid follicular cells, but some of the information gained from analysis of these tumors will likely be applicable to other endocrine tumor sites in the future. Clinical manifestation of thyroid cancer in humans in the United States is uncommon and largely nonfatal. The only known human thyroid carcinogen is X-radiation, which causes an increase in papillary tumors. Human thyroid follicular cell cancer is most often diagnosed histologically as a papillary tumor. The rodent thyroid neoplasms rarely metastasize, whereas the human cancers frequently metastasize. The molecular events that lead to thyroid follicular cell tumors are not fully known. In rodents, these tumors arise from mutation, perturbation of thyroid and pituitary hormone status with increased stimulation of thyroid cell growth by thyroid stimulating hormone (TSH), or a combination of both. In spite of the potential qualitative similarities, humans may not be as quantitatively sensitive to thyroid cancer development from thyroid-pituitary disruption as rodents. Evidence has shown that although rodents readily respond to reduced iodide intake by developing cancer, humans develop profound hyperplasia with adenomatous changes with only suggestive evidence of malignancy.

In the assessment of thyroid follicular cell tumors, and in considering thyroid carcinogenesis mode of action, the U.S. EPA has adopted the following three science policy positions:

- Chemicals that produce rodent thyroid tumors may cause tumors in humans
- In the absence of chemical-specific data, humans and rodents are presumed to be equally sensitive to thyroid cancer due to thyroid-pituitary disruption.
- Adverse rodent noncancer thyroid effects (e.g., thyroid gland enlargements), caused by chemically induced thyroid-pituitary disruption, are presumed to be relevant to humans.

The U.S. EPA has further adopted the following five science policy positions for conducting dose-response assessments of chemical substances that have produced thyroid follicular cell (and related pituitary) tumors in experimental animals:

    1. A linear dose-response procedure should be assumed when needed experimental data to understand the cause of thyroid tumors are absent and the mode of action is unknown.

    2. A linear dose-response procedure should be assumed when the mode of action underlying thyroid tumors is judged to involve mutagenicity alone.

    3. A nonlinear dose-response relationship (MOE) should be used when thyroid-pituitary disruption is judged to be the sole mode of action of the ob-

served thyroid and related pituitary tumors. Thyroid-pituitary perturbation is not likely to have carcinogenic potential in short-term or highly infrequent exposure conditions. The MOE procedure generally should be based on thyroid-pituitary disruptive effects themselves, in lieu of tumor effects, when data permit.

4. Both linear and MOE considerations should be assumed when both mutagenic and thyroid-pituitary disruption modes of action are judged to be potentially at work. The weight of evidence for emphasizing one over the other should also be presented.

5. Dose-response relationships for neoplasms other than the thyroid (or pituitary) should be evaluated using mode of action information bearing on their induction and principles laid out in current cancer risk assessment guidelines.

The above policy positions (reproduced directly from the report of Hill et al, [55]) are based on current knowledge of thyroid carcinogenesis and will be updated, should significant new information become available.

Overall the data required for risk assessment are information on mode of action, mutagenicity, increase in follicular cell growth (cell size and number) and thyroid gland weight, thyroid-pituitary hormones, site of action, correlation between doses that produce thyroid effects, and cancer and reversibility effects when dosing ceases (55). The U.S. EPA Risk Assessment Forum evaluation, which formed the basis of this report, focused on the use of a threshold for risk assessment for thyroid follicular tumors. Tumor promotion studies with thiourea, potassium thiocyanate, 4,4′-methylenebis(N,N-dimethylbenzenamine), and phenobarbital have provided support to the hypothesis that antithyroid compounds exert effects secondarily on the thyroid through the chronic stimulation of persistently elevated levels of TSH (56). A practical threshold is therefore expected for thyroid cancer based on studies that provided information that there will be no antithyroid activity until certain intracellular concentrations of the chemical are reached.

Hurley et al (57) reported on 240 pesticides screened for carcinogenicity by the U.S. EPA. Among these, 24 produced thyroid follicular tumors in rodents, and at least some thyroid tumor mode of action information is available for each. These pesticides are acetochlor, amitrole, bromacil, clofentezine, DCPA, ethiozin, ethofenprox, ethylenethiourea, etridiazole, fenbuconazole, fipronil, mancozeb, N-OBHD, pendimethalin, pentachloronitrobenzene, piperonyl butoxide, prodiamine, pronamide, pyrethrins, pyrimethanil, terbutryn, thiazopyr, triadimefon, and trifluralin. No further information was given on the conduct of risk assessments on these pesticides.

VI. SUMMARY AND CONCLUSIONS

Carcinogens can act through a variety of modes of action, and risk assessment of carcinogens should consider all the available mechanistic information. Differentiating between genotoxic and nongenotoxic carcinogens is only a first step. Identification of a carcinogen as nongenotoxic should be followed by utilization and defense of an alternative low-dose extrapolation methodology. The NOEL/UF method of estimating a safe level remains the usual default. However, actual risk modeling would be preferable to this method, because we have come to expect numerical risk estimates for carcinogens. The U.S. EPA model-free linear low-dose extrapolation approach, described in the 1996 proposed cancer risk assessment guidelines, is usually applied as a straight line drawn from the lower confidence limit for a dose causing a defined benchmark response (e.g., 10% response level), through zero dose. This is as applicable to genotoxic as to nongenotoxic carcinogens. The real challenge for risk estimation is whether nongenotoxic carcinogenesis mechanisms can be well enough established to define a nonlinear extrapolation through zero. Chemical effects that involve a dose-response threshold, which would shift the zero-response level above zero dose, are more acknowledged in theory than in fact.

At present, it may be easier to use the NOEL/UF method for risk assessment of nongenotoxic carcinogens. One benefit of falling back to this nonquantitative methodology is that we do not have to reach agreement on either the exact mechanism or the threshold for the toxic response. The controversy engendered by the recent U.S. EPA proposed risk assessment of chloroform provides a good example of the problems. The evidence for a nongenotoxic mode of action based on mouse liver toxicity (sustained cytolethality and regenerative cell proliferation response in the liver of female B6C3F1 mice) forms a substantial, although not conclusive, package—and an appropriate vehicle for furthering the discussion of the issues in risk assessment of nongenotoxic carcinogens.

Other nongenotoxic mechanisms of tumor induction that have been extensively studied include kidney tumors in male rats associated with α-2mG, liver tumors associated with peroxisome proliferation and regenerative cell proliferation, and thyroid tumors associated with induction of hepatic cytochrome P450. Increasing attention to mechanism of action in risk assessment and regulatory decision-making may result in improved assessment of true risk for chemicals that produce these and other toxic effects. More accurate estimation of risks would enable more attention to be paid to the largest risks, for better allocation of resources for protection of health and the environment.

Although much progress has been made in the field of carcinogenesis, the lack of consensus on the modes of action of specific carcinogenic chemicals and the reliability of available data impairs agreement on the underlying toxicological processes and the principles to be applied to risk assessment. Reaching agreement will be time-consuming and technically demanding. The need for better data may motivate research in this area that will provide a better understanding and assessment of both tumor induction and human risks from exposure to chemical carcinogens in our environment. Coordination between

REFERENCES

1. U.S. EPA. Guidelines for mutagenicity risk assessment. Fed Regis 51:34006–34012, 1986.
2. BE Butterworth, RB Conolly, KT Morgan. A strategy for establishing mode of action of chemical carcinogens as a guide for approaches to risk assessment. Cancer Lett 93:129–146, 1995.
3. JL Cicmanec, ML Dourson, RC Hertzberg. Noncancer risk assessment: Present and emerging issues. In: AM Fan, LW Chang, eds. Toxicology and Risk Assessment: Principles, Methods, and Applications. New York: Marcel Dekker, 1996, pp 293–300.
4. DW Gaylor. The use of safety factors for controlling risk. J Toxicol Environ Health 11:329–336, 1983.
5. DG Barnes, ML Dourson. Reference Dose (RfD): Description and use in health risk assessments. Regul Toxicol Pharmacol 8:471–486, 1988.
6. D Hattis, P Banati, R Goble. Distributions of individual susceptibility among humans for toxic effects—for what fraction of which kinds of chemicals and effects does the traditional 10-fold factor provide how much protection? Presented at the International Workshop, Uncertainty in the Risk Assessment of Environmental and Occupational Hazards, Bologna, Italy, September 25–26, 1998.
7. JC Swartout, PS Price, M Dourson, H Carlson-Lynch, RE Keenan. A probabilistic framework for the reference dose (probabilistic RfD). Risk Anal 18:271–282, 1998.
8. RB Connolly, ME Andersen. Biologically based pharmacodynamic models: Tools for toxicological research and risk assessment. Annu Rev Pharmacol Toxicol 31:503–523, 1991.
9. RS Yang, RS Thomas, DL Gustafson, J Campain, SA Benjamin, HJ Verhaar, MM Mumtaz. Approaches to developing alternative and predictive toxicology based on PBPK/PD and QSAR modeling. Environ Health Perspect 106 (suppl 6): 1385–1393, 1998.
10. BC Allen, RJ Kavlock, CA Kimmel, EM Faustman. Dose-response assessment for developmental toxicity 2: Comparison of generic benchmark dose estimates with no observed adverse effect levels. Fund Appl Toxicol 23:487–495, 1994.
11. CA Kimmel, RJ Kavlock, BC Allen, EM Faustman. The application of benchmark dose methodology to data from prenatal development toxicity studies. Toxicol Lett 82–83: 549–554, 1995.
12. JA Murrell, CJ Portier, RW Morris. Characterizing dose-response I: Critical assessment of the benchmark dose concept. Risk Anal 18:13–26, 1998.
13. KS Crump. Calculation of benchmark doses from continuous data. Risk Anal 15:79–89, 1995.
14. PS Price, RE Keenan, JC Swartout, CA Gillis, M Dourson. An approach for modeling non-cancer dose responses with an emphasis on uncertainty. Risk Anal 17:427–437, 1997.

15. SM Bartell, EM Faustman. Comments on "An approach for modeling noncancer dose responses with an emphasis on uncertainty" and "A probabilistic framework for the Reference Dose (Probabilistic RfD)." Risk Anal 18:663–664, 1998.
16. RL Melnick, MC Kohn, CJ Portier. Implications for risk assessment of suggested nongenotoxic mechanisms of chemical carcinogenesis. Environ Health Perspect 104 (suppl 1):123–134, 1996.
17. JA Swenberg, B Short, S Borghoff, J Strasser, M Charbonneau. The comparative pathology of α2μ-globulin nephropathy. Toxicol Appl Pharmacol 97:35–46, 1989.
18. IARC 1998. Species differences in thyroid, kidney and urinary bladder carcinogenesis. Consensus Report. Final draft, 8/1/98.
19. U.S. EPA. Alpha 2-μ-globulin: Association with chemically induced renal toxicity and neoplasia in the male rat. Risk Assessment Forum, United States Environmental Protection Agency, Washington, D.C., 1991. EPA/625/3-91/019F.
20. OEHHA. Public Health Goals for Chemicals in Drinking Water: Methyl tertiary butyl ether (MTBE). Office of Environmental Health Hazard Assessment, California Environmental Protection Agency, Oakland/Sacramento, California, 1999. (http://www.oehha.ca.gov).
21. NSTC. Interagency Assessment of Potential Health Risks Associated with Oxygenated Gasoline. White House Office of Science and Technology Policy (OSTP) through the President's National Science and Technology Council (NSTC), Committee on Environment and Natural Resources (CENR) and Interagency Oxygenated Fuels Assessment Steering Committee. Washington DC: National Science and Technology Council, February 1996.
22. NSTC. Interagency Assessment of Oxygenated Fuels. National Science and Technology Council (NSTC), Committee on Environment and Natural Resources (CENR) and Interagency Oxygenated Fuels Assessment Steering Committee. White House Office of Science and Technology Policy (OSTP) through the CENR of the Executive Office of the President. Washington, DC: National Science and Technology Council, June 1997. (http://www.whitehouse.gov/WH/EOP/OSTP/html/OSTP_Home.html).
23. HEI. The potential health effects of oxygenates added to gasoline: A review of the current literature. A special report of the Health Effects Institute's Oxygenated Fuels Evaluation Committee. Cambridge, Massachusetts: Health Effects Institute, April, 1996. (http://www.healtheffects.org/oxysum.htm).
24. NRC. Toxicological and performance aspects of oxygenated motor vehicle fuels. Committee on Toxicological and Performance Aspects of Oxygenated Motor Vehicle Fuels, Board on Environmental Studies and Toxicology, Committee on Life Sciences, National Research Council, National Academy of Sciences, Washington DC: National Academy Press, 1996.
25. U.S. EPA. Drinking water advisory: Consumer acceptability advice and health effects analysis on methyl tertiary-butyl ether (MTBE). Fact sheet and advisory. Health and Ecological Criteria Division, Office of Science and Technology, Office of Water, U.S. Environmental Protection Agency, Washington D.C. EPA-822-F-07-009. (http://www.epa.gov/oust/mtbe/index.htm).

26. J Ashby, RW Tennant. Chemical structure, Salmonella mutagenicity and extent of carcinogenicity as indicators of genotoxic carcinogenesis among 222 chemicals tested in rodents by the U.S. NCI/NTP. Mutat Res 204:17–115, 1988.
27. LS Gold, L Bernstein, R Magaw, TH Slone. Interspecies extrapolation in carcinogenesis: Prediction between rats and mice. Environ Health Perspect 81:211–219, 1989.
28. RR Maronpot, JK Haseman, GA Boorman, SE Eustis, GN Rao, JE Huff. Liver lesions in B6C3F1 mice: the National Toxicology Program, experience and position. Arch Toxicol 10(suppl):10–26, 1987.
29. RW Tennant, S Stasiewicz, JW Spalding. Comparison of multiple parameters of rodent carcinogenicity and in vitro genetic toxicity. Environ Mutagen 8:205–227, 1986.
30. J Ashby. Fundamental structural alerts to potential carcinogenicity or non-carcinogenicity. Environ Mutagen 7:919–921, 1985.
31. OEHHA. Public Health Goals for Chemicals in Drinking Water: Chlordane. Office of Environmental Health Hazard Assessment, California Environmental Protection Agency, Oakland/Sacramento, California, 1997. (http://www.oehha.ca.gov).
32. OEHHA. Public Health Goals for Chemicals in Drinking Water: Heptachlor and heptachlor epoxide. Office of Environmental Health Hazard Assessment, California Environmental Protection Agency, Oakland/Sacramento, California, 1997. (http://www.oehha.ca.gov).
33. OEHHA. Public Health Goals for Chemicals in Drinking Water: Lindane. Office of Environmental Health Hazard Assessment, California Environmental Protection Agency, Oakland/Sacramento, California, 1999. (http://www.oehha.ca.gov).
34. JK Reddy, DL Azarnoff, C Hignite. Hypolipidaemic hepatic peroxisome proliferators from a novel class of chemical carcinogens. Nature 283:397–398, 1980.
35. NTP. Carcinogenesis bioassay of di(2-ethylhexyl)phthalate (CAS No. 117-81-7) in F344 rats and B6C3F1 mice (feed study). National Toxicology Program technical report 217. Bethesda, MD: National Institute of Health, 1982.
36. NTP. Carcinogenesis bioassay of di(2-ethylhexyl)adipate (CAS No. 103-23-1). National Toxicology Program technical report 212. Bethesda, MD: National Institute of Health, 1981.
37. H Reisenbichler, PM Eckl. Genotoxic effects of selected peroxisome proliferators. Mutat Res 286:135–144, 1993.
38. JK Reddy, ND Lalwani. Carcinogenesis by hepatic peroxisome proliferators: Evaluation of the risk of hypolipidemic drugs and industrial plasticizers to humans. CRC Crit Rev Toxicol 12:1–58, 1983.
39. DS Marsman, RC Cattley, JG Conway, JA Popp. Relationship of hepatic peroxisome proliferation and replicative DNA synthesis to the hepatocarcinogenicity of the peroxisome proliferators di(2-ethylhexyl)phthalate and [4-chloro-6-(2,3-xylidino)-2-pyrimidinylthiol-acetic acid (Wy-14643) in rats. Cancer Res 48:6739–6744, 1988.
40. BE Butterworth. Consideration of both genotoxic and nongenotoxic mechanisms in predicting carcinogenic potential. Mutat Res 239:117–132, 1990.
41. C Peraino, RJM Fry, E Staffeldt, WE Kisieleski. Effects of varying the exposure to phenobarbital on its enhancement of 2-acetylaminofluorene-induced hepatic tumorigenesis in the rat. Cancer Res 33:2701–2705, 1973.

42. I Issemann, S Green. Activation of a member of the steroid hormone receptor superfamily by peroxisome proliferators. Nature 347:645–650, 1990.
43. T Sher, HF Yi, OW McBride, FJ Gonzalez. CDNA cloning, chromosome mapping, and functional characterization of human peroxisome proliferator activated receptor. Biochemistry 32:5598–5604, 1993.
44. OEHHA. Public Health Goals for Chemicals in Drinking Water: Di(ethylhexyl) phthalate (DEHP). Office of Environmental Health Hazard Assessment, California Environmental Protection Agency, Oakland/Sacramento, California, 1999. (http//:www/oehha.ca.gov).
45. M Eriksson, L Hardell, HO Adami. Exposure to dioxins as a risk factor for soft tissue sarcoma: A population-based case control study. J Natl Cancer Inst 82:486–490, 1990.
46. MC Kohn, GW Lucier, CJ Portier. The importance of biological realism in dioxin risk assessment models. Risk Anal 14:993–1000, 1994.
47. NCI. Carcinogen Bioassay of Chloroform. National Cancer Institute, Bethesda, MD. National Technical Information Service No. PB264018/AS, 1976.
48. TA Jorgenson, EF Meierhenry, CJ Rushbrook, RJ Bull, M Robinson. Carcinogenicity of chloroform in drinking water to male Osborne-Mendel rats and female B6C3F1 mice. Fund Appl Toxicol 5:760–769, 1985.
49. JL Larson, DC Wolf, BE Butterworth. Induced cytotoxicity and cell proliferation in the hepatocarcinogenicity of chloroform in female B6C3F1 mice: Comparison of administration by gavage in corn oil vs ad libitum in drinking water. Fund Appl Toxicol 22:90–102, 1994.
50. U.S. EPA. Proposed Guidelines for Carcinogen Risk Assessment. Washington, DC: United States Environmental Protection Agency, Federal Register 61 (B79):17960-18011, April 23, 1996.
51. ILSI. An evaluation of EPA's proposed guidelines for carcinogen risk assessment using chloroform and dichloroacetate as case studies: Report of an expert panel. International Life Sciences Institute, Health and Environmental Sciences Institute, November, 1997.
52. U.S. EPA. Health risk assessment/characterization of the drinking water disinfection byproduct chloroform. Office of Science and Technology, Office of Water, Mar 13, 1998.
53. R Heywood. Safety evaluation of toothpaste containing chloroform. III. Long-term study in beagle dogs. J Environ Pathol Toxicol 2:835–851, 1979.
54. R Melnick, M Kohn, JK Dunnick, JR Leninger. Regenerative hyperplasia is not required for liver tumor induction in female B6C3F1 mice exposed to trihalomethanes. Toxicol Appl Pharmacol 148:137–147, 1998.
55. RN Hill, TM Crisp, PM Hurley, SL Rosenthal, DV Singh. Risk assessment of follicular cell tumors. Environ Health Perspect 106:447–457, 1998.
56. GC Hard. Recent developments in the investigation of thyroid regulation and thyroid carcinogenesis. Environ Health Perspect 106:427–436, 1998.
57. PM Hurley, RN Hill, RJ Whiting. Mode of carcinogenic action of pesticides inducing thyroid follicular cell tumors in rodents. Environ Health Perspect 106:437–445, 1998.

14

Risk Assessment of Genotoxic Carcinogens

Lauren Zeise
California Environmental Protection Agency, Oakland, California

I. INTRODUCTION

Prediction of cancer risk is typically undertaken to guide decision-making with the overall aim of reducing risk. Assessments performed for these purposes necessarily take a practical tack. Despite considerable research and much insight, our understanding about mechanism of action for nearly all carcinogens is limited. It has been the practice within the United States to assume that any exposure to a carcinogen produces some risk, and in predicting risk at low doses the dose-response relationship has been assumed to be linear. Many now believe that use of this approach for certain "nongenotoxic" carcinogens results in a substantial overstatement of risk at low doses. Thus attempts are made to identify those carcinogens that are "low-dose" nonlinear or have thresholds above environmentally relevant exposure levels. This chapter emphasizes empirical approaches to the prediction of risk of carcinogens with genotoxic activity; a separate chapter in this monograph is devoted to biologically based mathematical models of carcinogenesis. Because of limited scientific knowledge, by necessity pragmatic approaches are taken in predicting risk beyond the doses and species for which observations have been made. It is within this context that the suggested modifications given in this chapter to the standard approaches to risk prediction are made.

The basic framework for interpreting biological data for quantitative risk prediction of carcinogens is as described by Moolgavkar et al (1), and the reader is referred for detailed discussion to that work. Cancer is a multistage process that can arise when a normal stem cell acquires two or more heritable genetic alterations. Relevant target genes for genetic modification include tumor suppressor genes and proto-oncogenes. Genetic alterations that play an important role in human cancers include chromosomal translocation, gene amplification, and point mutation.

Within this basic framework, the kinetics of cell division and death play a major role in determining risk: fixation of promutagenic lesions occurs when DNA synthesis precedes DNA repair; spontaneous induction of DNA lesions results from errors in DNA replication; genomic rearrangement or loss can occur during cell division. Also, increased cell division can increase critical target populations, unless compensated by increased differentiation and apoptosis. Inhibition of apoptosis can influence the process by decreasing the rate of elimination of altered, "initiated" cells.

There is no universally accepted definition of the terms "genotoxic" and "nongenotoxic" (2). For simplicity of the current discussion, chemicals that increase the rate of genetic alterations by some means other than by increasing the rate of cell proliferation, differentiation, or death will be considered genotoxic. Thus, chemicals that alter DNA through direct interactions, producing point mutation and chromosome damage, are here treated as genotoxic, along with those that, for example, alter DNA repair. This is a somewhat broader definition of genotoxicity than is used in Moolgavkar et al (1).

II. BASIC FORMULATION

A. Low Dose Linearity

The hypothesis driving risk predictions for genotoxic carcinogens since the mid-1970s is that at low doses, the dose-response relationship is linear. Crump et al (3) observed that, "in environments containing significant amounts of carcinogenic processes," the carcinogenic effect will increase proportionately to the amounts of carcinogen added. They noted that, if individual cancers arise from a single transformed cell and the agent in question acts additively with any ongoing process, then under almost any model the response will be low-dose linear. Figure 1 illustrates this concept. It applies primarily to carcinogenic processes in which the agent acts at the cellular level to produce heritable genetic or epigenetic change, but may not apply to indirect influences (e.g., pH changes or dietary modifications that lead to modifications of gut flora) (3).

Genotoxic Carcinogens

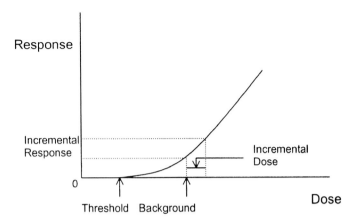

FIG. 1. Small, proportional increase in response from a small increment in dose above background.

Although at low doses, the dose-response relationship is assumed to be linear, at high doses nonlinearities are common. For any particular set of observations, the nonlinearity can often be attributed to one or two factors, such as nonlinear pharmacokinetics, cell proliferation, or competing, differential mortality among groups exposed at different dose rates. The procedures used to predict environmental risk involve adjustments and analyses to account for such nonlinearity, to obtain an upper-bound estimate on the low-dose slope, also called "cancer potency" or "unit risk." Risk can be predicted by multiplying the cancer potency by daily dose, averaged over the lifetime (e.g., in units of milligrams per kilogram bodyweight per day).

As illustrated in Figure 1, in the absence of background exposures, the risk from small doses typically encountered might be negligible, or even nonexistent. For common cancers, the effective background dose may be such that experimentally observed slopes may not significantly overstate the low-dose slope, whereas for rare cancers, greater differences are expected on these theoretical grounds. There is some limited experimental support for this concept in the very large cancer bioassays of diethylnitrosamine in rats (4, 5) and 2-acetylaminofluorene (2-AAF) in mice (6, 7). Figure 2 shows that the dose-response relationship for 2-AAF-induced bladder tumors—rare in the mouse strain studied—is highly nonlinear, with the slope of the dose-response relationship in the high-dose region substantially greater than any estimate that could be derived at lower doses. The figure also illustrates the opposite case for liver tumors, which are relatively common in the mouse strain used.

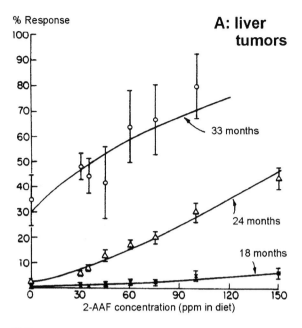

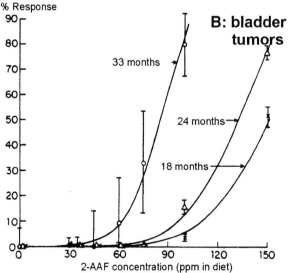

FIG. 2. Dose-response relationships for liver (**A**) and bladder (**B**) tumors in female BALB/c mice fed 2-aminoacetylfluorene (2-AAF) in the diet (6, 7). Observations were made at 18 (×), 24 (Δ), and 33 (○) months. The dose-response relationship for rare bladder tumors is highly nonlinear, whereas the opposite is true for the more common liver tumors. (Adapted from Ref. 8.)

Similarly, as shown in Figure 3, the dose-response relationship for liver tumors induced by diethylnitrosamine is low-dose linear. The liver tumors are relatively uncommon in the rat strain used, with an observed background rate of roughly 0.5%. However, the rarer esophageal tumors exhibited a nonlinear dose response relationship in the low-dose region. Nonetheless, it remains possible that in a still larger experiment in which background incidence was marginally observable, proportionate increases in response could be detected (8).

The shape of the dose-response relationship was found to be correlated with background tumor rates for multiple dose-level cancer bioassays (9) compiled in the Cancer Potency Database (10). The presence of the linear term in the multistage polynomial (see section B, below) was correlated with background tumor rate, and the presence of higher degree terms in dose decreased as the background rate increased. However, a linear term was found for two thirds of the cases with background rates estimated to be zero; the prevalence increased with increasing background parameter estimates. The findings held for chemicals positive or negative in *Salmonella* tests.

The possibilities that DNA damage can result in the delay of cell cycle and that low levels of genotoxic chemicals may induce (saturable) repair lend hypothetical support to the notion of hormesis—that small doses of genotoxic agents might have a protective effect against cancers caused by "background" processes. Hormesis that results from these and other mechanisms has been simulated using two-stage clonal expansion and other models (11–13). However, hormesis has not been observed in the large and sensitive cancer bioassays performed to date on genotoxic chemicals (e.g., refs. 4–7; Fig. 2 and 3), nor has it been demonstrated in large, analytical epidemiological studies of genotoxic chemicals or ionizing radiation (e.g., benzene [14]; radon [1]; aflatoxin [15]). In fact, the results of these studies generally support the use of low-dose linear relationships for predicting overall risk. Crawford and Wilson (16) note that hormesis could not occur "where the background had already brought the total dose to a region of positive slope of the dose response curve." Ultimately, whether hormesis occurs with genotoxic chemicals to a degree that impacts on human cancers remains a matter of scientific inquiry and debate (17–19).

B. Multistage Analysis

The standard model used by regulatory agencies in the United States to predict cancer potency at low doses from animal cancer bioassay data is the multistage polynomial. Under this model, the probability of developing cancer from exposures equivalent to a daily dose d is:

$$p(d) = 1 - \exp(-(q_0 + q_1 d + q_2 d^2 + \ldots)), \text{ with } q_i \geq 0 \text{ for all i.}$$

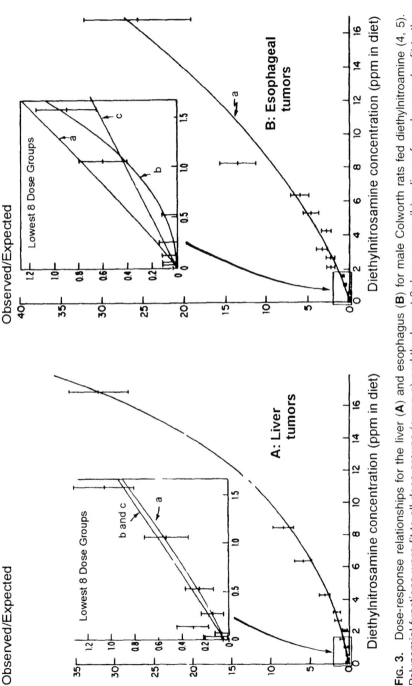

FIG. 3. Dose-response relationships for the liver (A) and esophagus (B) for male Colworth rats fed diethylnitroamine (4, 5). Polynomial functions were fit to all dose groups (curve a) and the lowest 8 dose groups (b); a linear formula was also fit to the lowest 8 groups (c). The dose-response relationship appears linear at low doses for the liver, as evidenced by the similar fit in the low-dose region for curves a, b, and c. This result can be contrasted with the curve fits for the rarer esophageal tumors. (Adapted from Ref. 8.)

Cancer potency is frequently taken as the upper 95% confidence bound on the linear term (q_1), derived using maximum likelihood techniques (20). The animal bioassay data that provides the basis for cancer potency derivation typically involve exposure levels orders of magnitude above environmental levels. Thus, the accuracy of prediction at environmental levels is unknown. For this reason a variant of this approach, the "benchmark dose" procedure, is beginning to receive greater use in risk assessments associated with environmental regulation. It involves the estimation of a benchmark dose associated with a certain risk level, say the effect dose (ED) causing 10% lifetime cancer risk in those exposed, or "ED10." Below the benchmark dose, risk is assumed to decrease proportionately with decreasing dose for genotoxic and other agents assumed to have a linear dose-response relationship at low doses.

The multistage polynomial formulation, although a derivative of a biologically based approach put forward in the mid-1950s, garnered, and even today maintains, extensive use because it can be applied to a wide range of data sets to address upward-curving nonlinearity, as well as possible low-dose linear behavior. Although the model parameters (q_i in the formula above) cannot be interpreted in terms of biological events (21), the upper bound on the linear term (q_1) represents a reasonable, plausible upper limit on cancer risk at low doses (20, 21), based on the full set of experimental data from which it was derived (20).

The multistage polynomial can be derived, with simplifying assumptions, from the multistage model of Armitage and Doll (22). Following the earliest attempts to describe the origination of cancer from a single cell in a multistep process (23–25), these authors, in 1954, published their seminal work describing the incidence of various human cancers as proportional to some power of age. The rationale for the formulation was that cancer arose from a single cell undergoing a series of successive mutations, each of which could occur with different probability, and some of which could be dose dependent. Although practical and straightforward to apply, much biology clearly was not addressed in the basic 1954 formulation, in particular, the kinetics of cell division and differentiation. In 1957, Armitage and Doll proposed a two-stage process, with cells in the first stage growing exponentially and those in the second escaping control and becoming malignant (26). Two- and three-stage models have undergone considerable development and refinement by Moolgavkar, Knudson, and others since that early work (1, 27–30). Experimental evidence suggests that human cancers can involve many mutations, and the two-stage model can be generalized to an arbitrary number of stages, each with potentially different division, death, and transition rates (8). Yet such models are clearly impractical, given the large numbers of parameters and limited data available to define them. Thus, Moolgavkar and others have developed two-stage clonal expansion models as a matter of convenience and parsimony, acknowledging that the number of rate-limiting mutations is not known and that the models are essentially a mathematical formulation for the initiation-promotion-progression paradigm (1).

Fundamental biological data are needed to make reliable predictions from multistage clonal expansion models, and data solely from standard bioassays are inadequate for this purpose (1, 31). For this reason, the multistage polynomial, with its ability to accommodate various dose-response data from the bioassay, is frequently used. Routine application of the multistage and similar models has led to databases of cancer potency values for use in risk predictions under a variety of scenarios. Examples are the cancer unit risk or guidelines values of the U.S. Environmental Protection Agency in its Integrated Risk Information System (IRIS) database (32), air quality and drinking water guidelines values of the World Health Organization (33, 34), and no significant risk levels of California's Proposition 65 (35). For genotoxic chemicals, some regulatory and advisory institutions decline using the approach out of concern for error in low-dose prediction (Table 1). Still, nonthreshold approaches, such as the multistage procedure, have been applied in the United States, Canada, certain European countries, and international programs (Table 1).

Any particular chemical may influence the development of cancer by multiple mechanisms. In cases in which the chemical is weakly genotoxic and nonlinear exposure-effect processes (e.g., toxicity-induced cell proliferation) are likely to play a role, characterization of the dose-response relationship is a challenge. An example of such a case is chloroform-induced kidney cancer. Sustained cell proliferation in response to persistent cell injury is implicated as a causal factor in kidney carcinogenesis for this chemical (37). Although results from standard studies of mutagenicity are not strong, several findings suggest chloroform may be weakly genotoxic. These include in vivo increases in micronucleated kidney cells (38), dose-dependent chromosomal aberrations in vivo in rats treated orally or intraperitoneally with chloroform (39), positive mouse micronuclei assay, dose-related induction of intrachromosomal recombination in yeast (40) and reduction in recombination in the presence of a free radical scavenger, DNA binding in vivo (41), and reductive metabolism of chloroform in vivo (42–45) leading to dichloromethyl radicals at some level. Thus, the dose-response relationship may be determined by both genotoxic and nongenotoxic mechanisms. Because cell proliferation may play a strong role at high doses, the potency predicted through use of the multistage procedure may be a significant overestimate, whereas the alternative safety factor approach may understate low-dose hazard. Two- and three-stage clonal expansion models that examine the extent to which genotoxicity might contribute to risk at high and low doses for chemicals such as chloroform are worthy of exploration.

Genotoxic Carcinogens

TABLE 1. Approaches Used by Various Institutions to Characterize Dose-Response Relationships for Carcinogens*

Procedure	Country or Institution†
Nongenotoxic—NOEL/UF	European Union
Genotoxic—no quantitative assessment	United Kingdom
	WHO/FAO (JEFCA and JMPR)‡
Nonthreshold dose-response model	UNSCEAR (radiation)
	NCRP (radiation)
	ICRP (radiation)
	USA§
Genotoxic—nonthreshold modeling	WHO Drinking Water
Nongenotoxic—NOEL/UF	Netherlands‖
Carcinogens with mode of action:	U.S. EPA 1996 (*Proposed Guidelines*)
Low-dose linear—linear extrapolation below the point of departure (*e.g.*, ED_{10})	
Low-dose nonlinear—margin of exposure comparison between point of departure and environmental level	
Nonthreshold effect—"characterize the dose response relationship" (by mathematical modeling, relative potencies ranking, or division of effect levels by (large) uncertainty factor)	WHO IPCS
Threshold effect—NOEL/UF approach	
Known human or probable human and genotoxic—considered nonthreshold toxicant	Canada
Possible human or nongenotoxic probable human carcinogen—considered threshold toxicant	
Rank carcinogens on the basis of their TDx, and other factors¶	Norway

*Adapted from ref. 36.
†WHO: World Health Organization; FAO: Food and Agriculture Organization of the United Nations; JEFCA: Joint FAO/WHO Expert Committee on Food Additives; JMPR: Joint FAO/WHO Meeting on Pesticide Residues; UNSCEAR: United Nations Scientific Committee on the Effects of Atomic Radiation; NCRP: National Council on Radiological Protection; ICRP: International Council on Radiological Protection; IPCS: International Programme on Chemical Safety
‡In a few instances, WHO (JEFCA/JMPR) has used a UF/NOEL for agents found to act via genotoxic mechanisms
§When data are sufficient to support different dose-response models, exceptions occur.
‖For agents with an unknown mode of action, nonthreshold modeling is applied.
¶TDx is the lowest daily dose inducing increased tumor incidence.

C. Analyses of Epidemiological Data

With few exceptions, the multistage polynomial is applied in making risk predictions from animal cancer bioassay data. Epidemiological data are typically analyzed, with empirical models of absolute or relative risk as a function of exposure. There is, however, a growing number of applications of mechanistic models, such as the two-stage clonal expansion model to epidemiological data (e.g., radon in hard rock miners [12, 46], atomic bomb survivors [47]). The estimates can be age- and time-dependent and are derived using a variety of empirical approaches. The reader is referred to the recent review of Krewski et al (21).

An example of empirical approaches to predicting risk from epidemiological data is provided for the mixed exposure case of the mycotoxin aflatoxin and hepatitis B virus (HBV). Aflatoxins are recognized carcinogens for the human liver but often occur at high levels in areas of high hepatitis B infection prevalence, another causative factor in human liver cancer. Aflatoxin B_1 (AFB_1) is clearly genotoxic, with DNA adduct formation increasing with dose in a linear fashion in multiple species (48) and the urinary excretion of the metabolite AFB_1-N7-guanine also proportionate to human intake (46). An epidemiological study (prospective cohort) measuring both aflatoxin levels and HBV infection status (137) provides the basis for estimating cancer potency. Wu-Williams et al (15) fit various empirical models to the data corresponding to assumptions regarding (a) whether liver cancer, from aflatoxin and HBV combined, adds to or multiplies with the background of liver cancer (absent these influences), and (b) whether aflatoxin and HBV themselves add or interact (both add and multiply). A final possibility considered was a multiplicative relative risk model wherein the increases in relative risk were due to HBV multiplying the relative risk from AFB_1. The mathematical expressions corresponding to these assumptions and the results of the modeling exercise are shown in Table 2. Note that in this case, the relative and absolute risk forms of the model are mathematically equivalent (reparameterizations of one another) and cannot be distinguished on the basis of goodness of fit to the data. However, when the models were used to make predictions for the United States, substantially different risks were predicted because the background of HBV infection in the Chinese cohort is much greater than in the United States (15). Thus, under certain assumptions regarding the prevalence of HBV in the United States, models could be distinguished in terms of the extent to which the predictions were realistic.

An obvious advantage of the use of epidemiological data for risk prediction is that no interspecies extrapolation is required. Nonetheless, there can be great uncertainty associated with such risk predictions when applied to populations different from the epidemiological study. A frequent difficulty encountered in using occupational studies to predict risk for the general population is understanding the extent to which results may predict risk from exposures of the young and el-

TABLE 2. Model Fits to Data on Daily Aflatoxin Dose (d), HBV Infection (h), and Annual Liver Cancer Incidence (y) from Guangxi, China*

	Risk models		
Model forms and parameters	Additive	Interactive	Multiplicative in relative risk
Excess risk form (ER)	$y = a + b_1 h + b_2 d$	$y = a + b_1 h + b_2 d + b_3 h d$	—
Relative risk form (RR)	$y = a(1 + b_1'h + b_2'd)$	$y = a(1 + b_1'h + b_2'd + b_3'hd)$	$y = a(1 + b_1'h)(1 + b_2'd)$
a, background incidence (per year, per 100,000)	2.87 (7.77 - 22.5)	7.17 (0 - 22.5)	1.35 (7.55 - 22.5)
b_1, HBV effect (annual incidence due to HBV)	ER: 0.0055(0.0045,0.0067) RR: 191 a (NC)	ER: 0.29 (0.069 - 0.55) RR: 4690 a (369a-15681a)	25.10 a (15.47a-43.64a)
b_2, aflatoxin effect, (annual incidence per mg/kg-day)	ER: 0.42 (0.72-1.76) RR: 14520 a (NC)	ER: 0.20 (0.069, 0.55) RR: 4690 a (369a-15681a)	1242 a (666 a, 2150 a)
Goodness of fit: χ^2	17.27 (5 df)	6.11 (4 df)	6.68 (5 df)

*HBV status—negative; $h = 0$; positive $h = 1$.
Maximum likelihood estimates with upper and lower 95% confidence bounds given in parentheses.
NC: confidence limits infinite.
Source of data: ref. 137; analysis: ref. 15.

derly. Also, the occupational group studied can differ from the population of interest in geographical location, gender distribution, cultural habits, and other factors. Finally, the exposures for the occupational groups are often not well defined and exposure estimates correspondingly uncertain. Predictions can differ by orders of magnitude, depending on assumptions made about these factors (see, for example, the effort to examine the potency of benzidine by the International Agency for Research on Cancer [49] and the comparisons of prediction of lifetime risk of solid tumors derived from atomic bomb survivor data [47]).

III. FACTORS TO CONSIDER IN RISK PREDICTION

There are a number of sources of error and bias in low-dose risk prediction, some of which can be partially addressed quantitatively through adjustment factors or through modeling. These include the age at exposure and observation, species specificity, interspecies and dose differences in pharmacokinetics, competing risks, and influences of background exposures. Factors that can contribute to nonlinearity at high doses are depicted in Figure 4.

A. Age

One important determinant of an individual's risk can be the age at which he or she is exposed. Yet in risk prediction, age at exposure is not often routinely addressed. The same dose given to people of different ages is often assumed to confer the same lifetime risk of cancer. A complication in evaluating the importance of age for any particular substance is that the studies used to evaluate carcinogenic potential miss critical exposure periods. Exposure is initiated in the standard rodent bioassay during a period of sexual maturation, and the study is usually

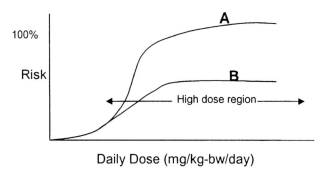

FIG. 4. Nonlinearities in the dose-response relationship can sometimes be attributable to saturable detoxification or cell proliferation (curve A), saturable activation, or competing causes of death (curve B).

truncated well before the end of the natural life of the animal. Most evaluations based on human data derive from occupational studies, corresponding to exposures from young adulthood through middle age. Of particular recent concern in the United States is the degree to which in utero and childhood exposures are adequately addressed by the risk assessment process, and various initiatives at the state and federal level are directed at improving policy and methodology in this regard (e.g., 1998 U.S. Food Quality Protection Act; 1999 California Senate Bill 25).

Age can impact on cancer risk in different ways. First, there can be inherent differences in susceptibility at different ages resulting, for example, from the changes that tissues undergo through the course of development, including differences in cell proliferation rates, hormone responsiveness, immunological activity, and development and maturation of enzyme systems that activate or detoxify chemicals. Independent of susceptibility is the issue of timing or latency; an individual exposed early in life simply has a longer period for the damage to be expressed. Another consideration is the sequencing of the exposure in question with other endogenous and exogenous agents that affect the carcinogenic process. Finally, some differences in risk with age can be explained by differences in exposure associated with, for example, food consumption patterns, behavioral factors (e.g., hand-to-mouth, sports), and physiological differences (e.g., breathing rates, fat distribution).

An example of inherent early-in-life susceptibility is the marked age dependence observed in the cancer potency of vinyl chloride (50, 51). Where 5 days of exposure to newborn rats led to high incidence of liver angiosarcoma and hepatoma, minimal effects were observed when 11-week-old rats were exposed under the same protocol (52). These differences are hypothesized (50) to be from the comparatively greater formation of vinyl chloride-induced DNA adducts in the livers of younger animals (53) at a time when hepatic tissue is undergoing rapid proliferation, resulting in considerably greater conversion of DNA lesions to mutations. Quantitative risk predictions related to increased mutation that results from increased DNA insult during a period of heightened proliferation in early life can be explored through two-and three-stage clonal expansion models. The significantly greater breast cancer susceptibility of teenage and prepubescent girls compared with adults exposed to ionizing radiation has been noted for atomic bomb survivors (54), and patients treated for cancer (55), ankylosing spondylitis (56; radium-224), and ringworm (57). Degree of differentiation of the mammary gland as a major explanatory factor for the age effect has been explored and observed experimentally with chemical carcinogens (e.g., 7,12-dimethylbenz[a]anthracene [58]).

Growth and development of the thyroid gland appear to explain age at exposure as an important host factor that influences thyroid cancer risk from radiation observed epidemiologically. Risks appear significantly greater in terms of frequency (59, 60) and degree of malignancy (59, 61) when exposure occurs in young children. The unexpectedly large increase in radiation-induced thyroid cancers in children after the Chernobyl accident (31) suggests a need for a better understanding of quantitative relationships. On the other hand, inherent age-dependent tissue susceptibility was not found to explain the age effect observed for colon and lung cancers in atomic bomb survivors by one group of researchers (47). Although these researchers found substantially greater risk associated with exposure during childhood in contrast to adulthood, the effect could be explained in terms of two major factors: the greater number of remaining years of life and increasing spontaneous (as opposed to radiation-induced) initiation rate with age.

The above examples illustrate various explanatory factors for differences in cancer risk observed with age at exposure. Qualitatively, a specific agent does not affect sensitive cancer sites equally, and differences are observed from one agent to another. A large number of such examples exist in the experimental, clinical, and epidemiological literature. For some agents, the type of cancer observed depends on the age at exposure. The clear cell adenocarcinomas of the cervix seen in offspring whose mothers took diethylstilbestrol during pregnancy were not observed in the mothers themselves, who instead (as expected of estrogens) are at risk for increased endometrial and breast cancer (62–64). Diethylslilbestrol serves as the quintessential example of the influence of age on the cellular site of tumor development.

Quantitatively, a number of agents show relatively large differences in risk from exposure at different ages at certain sites (4, 65–75), but minor or negligible differences are also noted for several chemicals. Systematic reviews to quantitatively describe the large literature relevant to this topic have not been published. Nonetheless, it is clear that the standard dose metrics used to predict risk—average daily dose (e.g., daily milligram per kilogram bodyweight) or cumulative dose (e.g., ppm-years)—need to be re-evaluated in light of the considerable evidence. This is especially important when dose rate differs substantially with age.

The multistage model provides the basic framework for mathematically modeling age-dependent carcinogenesis. From the Armitage and Doll multistage model (22; see above), one can develop a weighting function to address exposure at different ages. Figure 5 shows the relative effectiveness of doses applied at different ages for a six-stage model, with the first stage dose-dependent. Correspondingly, Figure 6 shows for a dose given during the first 2 years of life, the average lifetime dose and an equivalent chronic dose that would produce same effect under the Doll Armitage model (for derivation, see 21, 36, 76). Under this model, use of average daily dose can result in underestimation of risk by a factor of k, the number of stages in the Doll Armitage model theoretically required for a cancer to occur. The model can be used to take into account the simple fact that early exposures are necessarily correlated with greater remaining years of life. Because the formulation is reasoned on mutation rates, it has been applied to

Genotoxic Carcinogens

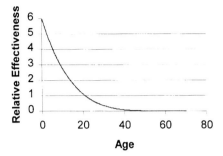

FIG. 5. Relative effectiveness of dosing at different ages under a six-stage Doll-Armitage model, when the first stage is dose-dependent.

genotoxic carcinogens to adjust for age at exposure and account for intercurrent mortality. For example, Doll Armitage-type formulations have been applied to the bioassay of ethylene dibromide to account for changing doses rates and high mortality (77), to account for age at exposure and differential mortality in evaluating interexperimental variability in studies of benzo(a)pyrene (78, 79), and to normalize results from different bioassays on urethane (80). The considerably greater consumption of apple juice by infants coupled with the potentially greater dose effectiveness of genotoxic carcinogens indicated by the Doll Armitage

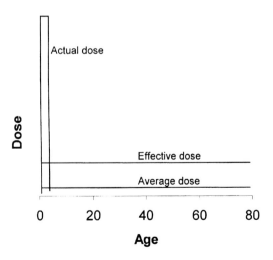

FIG. 6. Comparison of actual, average, and time-weighted dose for a chemical given during the first year of life that affects the first stage in a six-stage Doll-Armitage carcinogenesis process.

model led to concern about use in apple production of the plant growth regulator Alar, which breaks down to unsymmetrical dimethylhydrazine (82, 86). Although the formulation does not address intrinsic age susceptibility (81) (e.g., immunological surveillance or hormone status), it has been discussed as an improvement to the estimation of risk because it takes into account the longer period available for mutations that occur early in life to be expressed as cancer (82). Similar logic provides the basis for corrections for the length of bioassays or early mortality by the third power of age in the Cancer Potency Database of Gold et al. (83) and state of California (84) and EPA risk assessments (20), and in calculations of "effective number of animals" in NTP bioassays by the "poly-3" method (85, 138).

Murdoch and Krewski (76) explored the issue of age at exposure by using a two-stage clonal expansion model. In cases where the first stage was dose-dependent and initiated cells proliferated rapidly, risk could be considerably underestimated by using average dose as the dose metric. The work of Kai et al (47) and the National Research Council BEIR V Committee (87) indicate the substantial degree by which results can differ with the model chosen for predicting age-dependent risk from single exposures: Predictions of lung cancer risk at ages 70 and 100 from radiation exposure in survivors of the atomic bomb detonations differed significantly, depending on the model used.

B. Pharmacokinetics

Nonlinearity in cancer dose-response relationships can result from saturation of a metabolic pathway. When an activation pathway is saturated, the dose response relationship can plateau at a level less than 100% lifetime risk, as in the frequently noted (1, 8, 89) case of vinyl chloride (88). When a detoxification or deactivation pathway saturates, the classic hockey stick dose-response curve can result. Data permitting, these effects can be formally taken into account through physiologically based pharmacokinetic (PBPK) modeling. The aim of the PBPK model in cancer risk prediction is to obtain a better estimate of effective dose than the administered dose, using model parameters measured in or estimated from experiments subsidiary to the bioassay. A cancer dose-response relationship for the bioassay or occupational dataset can then be expressed in terms of effective dose. If sufficient information and understanding permits the development of a PBPK model for the exposure circumstance and individuals for which risk is to be predicted, exposure can be related to effective dose. A third relationship can then be developed—risk as a function of exposure. This relationship is theoretically more reliable than that obtained through analyses based solely on administered dose, and for this reason, PBPK models have been used in cross-species, route, and dose extrapolations. However, because of incomplete understanding of critical factors (e.g., identity of the active metabolite or activation pathway) and appropriate model structure, or uncertain parameter estimates, the reliability of predictions has been questioned for a variety of PBPK applications (36, 90–92).

The PBPK models represent the body as compartments, related to one another in series or in parallel. Figure 6 depicts the PBPK model that has been used for various organic solvents (e.g., perchloroethylene [93–95], styrene [96]). The model in this case is composed of compartments for fat tissue, richly perfused tissue, poorly perfused tissue, a (volumeless) gas/blood exchange unit, and a liver metabolizing group. Tissues in which the chemical concentration changes at roughly the same rate are often grouped together (1, 94), with some tissues tracked separately because of degree of metabolism (e.g., liver) or bioaccumulation. Compartments are typically characterized by a partition coefficient describing the affinity of a chemical for two different media (such as blood/fat or blood/air), tissue volume, blood flow, and, if applicable, metabolic parameters (Fig. 7).

Mathematically, a series of differential equations describe the change in tissue concentration over time, due to transport to tissue by blood flow, uptake by and elimination from tissue according to partition coefficients, and metabolism within the tissue. Use of PBPK modeling in cancer risk prediction requires the identification of the activation pathway associated with carcinogenesis and of a kinetic model (for example, Michaelis-Menten). Data needed to populate the model include compartment partition coefficients, blood flows, tissue volumes, and metabolic parameters.

Physiologically based pharmacokinetics models have the advantage over classical pharmacokinetic models of being interpretable in biological terms. However, experience has shown that extrapolation with these models (across dose levels, routes, or species) involves considerable uncertainty, with the appropriate model structure and activation pathway frequently unknown despite considerable study and critical parameters uncharacterized. An illustration of these points is provided by early models of perchloroethylene (93–95). These models assumed that activation occurred through cytochrome P450 oxidation, confined to the liver, with the active metabolite unstable and not transported far; yet leukemia and kidney cancer are observed experimentally in the rat and there is suggestive evidence for non-liver target sites in humans (97). Although it had been suspected earlier (95), glutathione conjugation is now recognized as an important metabolic pathway, with the formation of S-(trichlorovinyl)glutathione, which is cleaved to S-(trichlorovinyl)-L-cysteine, and again cleaved in the kidney to dichlorothioketene, which can react with cellular macromolecules (98). S-(trichlorovinyl)-L-cysteine may also react with water to form dichloroacetic acid, an animal liver carcinogen. With respect to model structure, for 10 different PBPK models of perchloroethylene, discrepancies were noted between data and predicted levels associated with the P450 pathway (91). It was proposed that more sophisticated models accounting for heterogeneity of the fat compartment or intertissue diffusion between fat

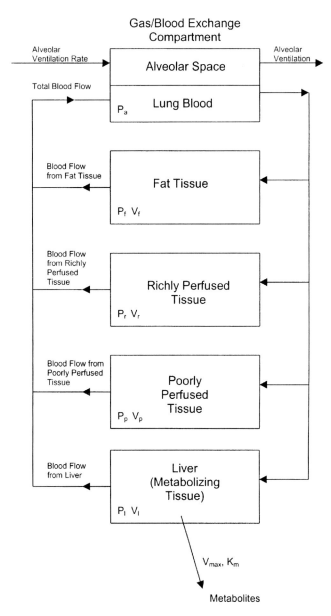

FIG. 7. Schematic representation of a physiologically based pharmacokinetic model. For each compartment i, change in chemical concentration is described mathematically by mass balance differential equations defined by blood flow rates, partition coefficient (P_i), volume (V_i), and if metabolism occurs in the compartment, metabolic coefficients (e.g., V_{max} and K_m).

Genotoxic Carcinogens

and muscle groups would result in better predictions (91). Various researchers have noted the sensitivity of the PBPK models for perchloroethylene to selection of metabolic parameters (90, 93, 99) and partition coefficients (90, 93).

Partition coefficients for PBPK models are frequently determined in vitro by using tissue homogenates and are often assumed to be the same in different species. However, they may not reflect partitioning into the organ in vivo, and significant species differences have been observed for some tissues (100, 101). Metabolic parameters, Km and Vmax, often cannot be estimated independently, introducing further error. Formal statistical procedures are usually not used to derive parameters from in vivo studies, and in vitro estimates can be inaccurate. Finally, the models typically are not independently validated. Some of these limitations have been addressed experimentally and through uncertainty analyses within a formal statistical (92–94, 102) or qualitative (91, 95) framework. The general problem of model validation can be addressed in part through the introduction of biomarker components in occupational studies. The application of a Bayesian framework to PBPK analyses for the integration of different types of data with varying levels of uncertainty is a promising technique under development (92, 94). The large range of estimates for perchloroethylene metabolized at relatively low levels of exposure (at 1 ppm, estimates of 2% to 86% metabolism [95]) was partially resolved through application of Bayesian techniques to a controlled study of healthy male volunteers exposed to relatively high levels for 4 hours and subsequently followed up. A characterization of high-dose kinetics in this group was provided by data collected near the time of exposure, whereas low-dose kinetics were characterized over subsequent days. The analyses described the varying degree of metabolism among the individual volunteers and the uncertainty in those estimates.

Although the accuracy of PBPK models for low dose and across species predictions can often be questioned, these models play a valuable role in hypothesis testing and exploratory analyses.

C. Human Heterogeneity

Individuals differ in susceptibility to cancer because of genetic, environmental, occupational, and lifestyle factors. A variety of genetic disorders have been identified as conferring substantially greater risk for certain types of cancer (1, 103). Examples include inherited cancer syndromes (such as familial retinoblastoma [103]), genetically determined immunodeficiency diseases (such as ataxia telangiectasia [104]), and recessive syndromes of DNA-deficient repair (such as xeroderma pigmentosum [105]). Taken together, though, these syndromes explain only a small fraction of human cancer.

Large interindividual differences in activities of a number of P450, glutathione-S-transferease, and other enzymes involved in carcinogen activation and detoxification have been described. The differences may be due to race, gender, genetic predisposition, and exposure to inducers (106). Differences in physiological status, DNA repair efficiency, and other factors (including DNA methyltransferase activity [107]) contribute to human heterogeneity in response to any particular carcinogen exposure. As an exercise in modeling interindividual variability in risk that results from metabolism and physiological factors, human variability in response to 4-aminobiphenyl (ABP) was simulated (108). A PBPK model was used to quantify the formation of proximate carcinogen (N-hydroxy-4-ABP) and DNA-binding in the bladder. Metabolic parameters (for example, for N-oxidation and N-acetylation), and physiological factors (including, urine pH and frequency of urination) were varied, via Monte Carlo simulation, according to published values described for humans. Figure 8 shows the variation among individuals in a fraction of 4-aminobiphenyl bound to DNA, in simulations representing 500 individuals. The most and least susceptible individuals in this simulation differed by several orders of magnitude, with urine pH, ABP N-oxidation, N-acetylation, and urination frequency appearing as the factors most affecting individual sensitivity. The theoretical extremes for a large population were two orders of magnitude

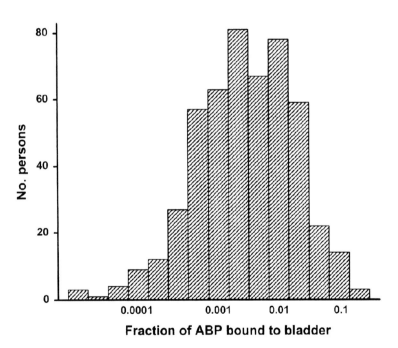

FIG. 8. Distribution of proportion of administered 4-aminobiphenyl bound to bladder DNA simulated for 500 human subjects. (From Ref. 108.)

more disparate, suggesting that very large differences in susceptibility are possible in the general population. As an example of another approach to describe human heterogeneity, a hierarchical Bayesian modeling approach was applied in a PBPK modeling framework to data in healthy human volunteers exposed to perchloroethylene (109, 110). Estimates of variability in metabolism along with uncertainty in those estimates were derived. These two examples suggest ways differences in human risk might be systematically addressed in risk predictions. Clearly, characterization of the variability in cancer risk among humans is a critical need in risk prediction exercises.

D. Across Species Risk Prediction

When human data are insufficient for risk prediction, estimates of human risk are made from animal data. There is good theoretical and empirical support for qualitatively predicting human carcinogenic activity from the animal bioassay. All chemicals identified as known human carcinogens also have been found to produce cancer experimentally in animals (111). Also, systematic reviews of the literature on chronic exposure bioassays find strong cross-species correlations (112, 113). The finding of carcinogenicity at one or more sites in the rat is concordant with findings in the mouse, and vice versa (112, 113), although the degree to which true concordance may be under or overestimated from analyses of animal bioassay data is a matter of debate (114–116). Slightly greater species concordance has been observed for *Salmonella*-identified mutagens (113). Still, species differences in the sites observed as affected are frequent, with the principal sites affected in humans not always predicted by the bioassay. This suggests caution when predicting risk for specific target sites from bioassay and pharmacokinetic data, particularly in the absence of knowledge of affected sites in humans.

Quantitatively, although measures of carcinogenic activity in different species are highly correlated, the best means of calibrating the potency in one species for prediction of potency in a second species remains unclear, despite 25 years of discussion and research. The efforts have been limited by the ability to obtain precise estimates of cancer potency for the known human carcinogens. The problem of cross-species prediction is often framed as the selection of the best dose metric. Conceptually, when dose is expressed in the correct metric, exposure to the same dose in different species results in the same risk. The most commonly used metrics are (a) average daily dose, for example, in units milligrams per kilogram bodyweight (mg/kg^1/day); (b) mg per surface area, also called two thirds scaling, or $mg/kg^{2/3}$/day; (c) $mg/kg^{3/4}$/day (three-quarters scaling), which corresponds roughly to scaling with metabolic rate; and (d) cumulative dose (e.g., total mg/kg received throughout life).

A number of empirical investigations on this matter have been published. Comparisons of dose-response relationships for secondary cancers resulting from cancer chemotherapy to those observed in animal bioassays found the best correspondence when dose was expressed as cumulative mg/kg received (117, 118). Measures of TD25 (dose causing 25% excess cancer risk) derived from human data and animal bioassays for 23 chemicals were highly correlated, for dose expressed in terms of daily dose or cumulative dose (119). However, the uncertainty in potency measures derived from human data in this study was substantial. A later work by the same research team compared the dose, which when given to humans from ages 20–65 years was estimated to result in a 25% excess cancer risk, to the estimated average daily dose administered throughout an animal bioassay that would give the same level of excess risk (120). They found mg/kg/day to be the dose metric that did not result in overestimation of human risk. Had dose averaged over human lifetime been used instead, this metric would have been found to understate human risk. Comparisons of findings in mice and rats in terms the dose-response slopes derived from a standard set of experiments conducted by the National Toxicology Program supported quantitative prediction between these species on the basis of mg/kg/day (121); the results were also consistent with surface area scaling. Comparisons (122, 123) of average daily doses associated with an excess 50% cancer risk (TD50) compiled in the Cancer Potency Database (10) suggests, on average, that dose in units mg/kg bodyweight may result in underestimates of risk in the larger species, although the uncertainties were such to preclude definitive conclusions. In a number of cases, relatively large species differences in potency were observed.

Conclusions about cross-species scaling drawn from bioassay-derived cancer potencies have been questioned; some correlation in potency is expected on the basis of the experimental design of the standard bioassay (conducted at maximum tolerated doses, with fixed study length and small group sizes) (124, 125). Nonetheless, statistical analyses indicate that experimental interspecies correlation cannot be entirely explained as an artifactual result of study design (126). Overall, the resolution of the available data precludes the selection of a cross-species method on the basis of empirically observed cancer potency (127). Related studies comparing acute and subacute systemic toxicity across species (128–131) suggest a narrower range of scaling possibilities, but because the toxic endpoints differ so distinctly from carcinogenesis in type and time frame, they too provide limited information on the issue (127).

Cross-species differences depend on both pharmacokinetic and pharmacodynamic species differences. In theory, pharmacokinetically equivalent doses for genotoxic agents are those producing the same average daily area under curve (AUC) in concentration versus time plots, as AUC may be predictive of the rate of generation of DNA adducts per cell (132). The ¾ or ⅔ scaling discussed above is conceived of producing the same AUC in different species for the reactive intermediate for various carcinogens. Pharmacodynamics is conceptually more difficult to consider. Theoretically, at the pharmacokinetically equivalent dose, susceptible cells in different species are, on average, exposed to the same concentration of reactive compound. However compared with many test species the larger and longer lived human will then have far greater numbers of target cells exposed, and the transformation of only one of these is required to initiate cancer. These cells will also be exposed longer and subject to more cell division. Clearly the enormous cancer incidences in large species one would expect from predictions from smaller species exposed to the same concentrations are not observed. The overall spontaneous lifetime cancer risks for different sized species are roughly comparable. Presumably, species differences in repair, cell turnover, proportionate numbers of stem cells, and other factors may one day be understood well enough to account for these observations. Meanwhile, species differences in pharmacodynamic factors associated with any particular carcinogen exposure are not known, and this remains a major uncertainty in any cross-species prediction exercise.

E. Background Exposures

Several years ago, Doll and Peto (133) compared cancer incidence in areas of lowest occurrence to that of highest occurrence in different locations worldwide for the major cancer sites. They noted that roughly 90% of human cancer was caused by exogenous, as opposed to endogenous, factors. It is recognized that people are exposed to a myriad of chemicals through the environment, consumer products, and the diet. Yet any particular risk assessment may only characterize risk from a single chemical by a single exposure pathway. The degree that an exposure adds to and modulates background processes is important in assessing risk at low doses (3, 134–136). It is particularly important in evaluating the qualitative relevance of findings indicating strong nonlinearities in dose-response relationships. In cases where a low-dose nonlinear mode of action is suspected, background exposures may be sufficiently large to move certain subgroups into a level at which risk is roughly proportional to dose.

Figure 9 illustrates hypothetical dose-response curves under conditions of high and low background exposure for a chemical with a threshold occurring at a non-zero dose. For the first curve shown, background exposures are high, and risk increases linearly with increased dose. For the second curve, background exposures are low and with incremental increases in dose "d," the exposure remains below the threshold, although the margin of safety is lower (total dose is closer to the critical dose or threshold).

For genotoxic chemicals, there has been the general assumption of dose-response relationships that are linear at low doses, although nonlinear relationships have been proposed for certain cases. Agents that are weakly genotoxic and may affect the process of carcinogenesis by influencing cell kinetics are a special case about which there is some debate. The question for the assessor is where

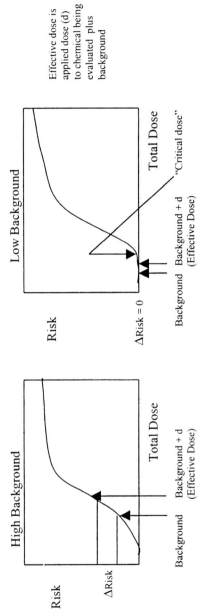

FIG. 9. Incremental increase in risk in the presence of high and low background processes.

individuals fall on the dose-response curve, particularly those who have been highly exposed. Risk will depend on the exposure to the chemical under study as well as other chemicals in the "background" that operate by the same mechanism. Estimates of background exposure are difficult to achieve, but a screening level analysis may be enough to provide a qualitative finding as to whether one should proceed with a nonlinear dose-response assumption.

IV. CONCLUDING REMARKS

This chapter discusses approaches for predicting cancer risk for genotoxic carcinogens. Over the past several decades it has been assumed that at low doses, the dose-response relationship for genotoxins is linear. This remains a reasonable assumption in the absence of detailed data and understanding of the specific mechanism of action of a chemical. Estimates of low-dose slope can be obtained through empirical fits of linear or the multistage polynomials to epidemiological and bioassay data. However, in generating these estimates, sources of high-dose nonlinearity, for example, from intercurrent mortality or pharmacokinetics, should be considered and adjustments made accordingly. This needs to be done judiciously, for over- or underadjustment can result in substantial misestimation.

One dose-response phenomenon seen with radiation that is a potential source of misestimation is the inverse exposure-rate effect—fractionation of a given total dose increases risk. The effect is predicted by the two-stage clonal expansion model (1). This is potentially a factor in the analysis of certain epidemiological and bioassay data sets.

Predicting risk for the general population from either animal or human cancer data sets is challenging. Epidemiological data for risk assessment are typically occupational, and exposures are frequently not well defined, often for relatively short periods, with study subjects of adult white males. The uncertainties in predicting risk for the population can be substantial, even more so when attempting to address exposures to the fetus, children, or the elderly. For animal data, there is the well-recognized problem of across-species prediction, but the window of exposure also does not cover the lifetime. The in utero, childhood, and senior years do not appear to be adequately addressed. By necessity, the bioassay cannot be designed to mirror the heterogeneity of the human population. But due to size, limited geographic scope, and other factors, heterogeneity will not be sufficiently described by data from most occupational studies used in risk prediction. The development of approaches to address the age dependence and interindividual variability of risk is one of the main challenges for risk assessors in the coming years. It is difficult to foresee how some of the stunning advances in the biological sciences, such as gene chip technology, might impact this effort, but they no doubt will prove useful.

REFERENCES

1. S Moolgavkar, D Krewski, M Schwartz. Mechanisms of carcinogenesis and biologically based models for estimation and prediction of risk. In: S Moolgavkar, D Krewski, L Zeise, E Cardis, H Moller, eds. Quantitative Estimation and Prediction of Human Cancer Risk. Lyon: IARC Scientific Publications, No. 131, 1999, pp 179–237.
2. H Vainio, PM Magee, DB McGregor, AJ McMichael. Mechanisms of Carcinogenesis in Risk Identification. Lyon: IARC Scientific Publications, No. 116, 1992.
3. KS Crump, DG Hoel, CH Langley, R Peto. Fundamental carcinogenic processes and their implications for low dose extrapolation. Cancer Res 36:2973–2979, 1976.
4. R Peto, R Gray, P Brantom, P Grasso. Nitrosamine carcinogenesis in 5120 rodents: Chronic administration of sixteen different concentrations of NDEA, NDMA, NPYR and NPIP in the water of 4400 inbred rats with parallel studies on NDEA alone of the effect of age of starting (3, 6, or 20 weeks) and of species (rats, mice, hamsters). In: IK O'Neill, RC Borstell, CT Miller, J Long, H Bartsch, eds. N-Nitroso Compounds: Occurrence, Biological Effects and Relevance to Human Cancer. Lyon: IARC Scientific Publications, No. 57, 1984.
5. R Peto, R Gray, P Brantom, P Grasso. Effects on two tonnes of inbred rats of chronic ingestion of diethyl- or dimethyl-nitrosamine: An unusually detailed dose-response study. Imperial Cancer Research Fund, Cancer Studies Unit. Nuffield Department of Clinical Medicine. Oxford: Ratcliffe Infirmary, April 1982.
6. RL Kodell, JH Farmer, DL Greenman, CH Frith. Estimation of distributions of time to appearance of tumor and time to death from tumor after appearance in mice fed 2-acetylaminofluorene. J Environ Pathol Toxicol 3:89–102, 1979.
7. NA Littlefield, JH Farmer, D Gaylor, WG Sheldon. Effects of dose and time in a long-term low-dose carcinogenesis study. J Environ Pathol Toxicol 3:17–34, 1979.
8. L Zeise, EAC Crouch, R Wilson. Dose response relationships for carcinogens: A review. Environ Health Perspect 73:258–308, 1987.
9. DW Gaylor. Relationship between shape of dose-response curves and background tumor rates. Regul Toxicol Pharmacol 16:2–9, 1992.
10. LS Gold, CB Sawyer, R Magaw, GM Backman, M de Veciana, R Levinson, NK Hooper, WR Havender, L Bernstein, R Peto, MC Pike, BN Ames. A carcinogenic potency database of the standardized results of animal bioassays published through December 1982. Environ Health Perspect 67:161–200, 1986.
11. WK Lutz, A Kopp-Schneider. Threshold dose response for tumor induction by genotoxic carcinogens modeled via cell-cycle delay. Toxicol Sci 49:110–115, 1999.
12. KT Bogen. Mechanistic model predicts a U-shaped relation of radon exposure to lung cancer risk reflected in combined occupational and US residential data. Belle Newslett 7:9–14, 1998.
13. T Downs, R Frankowski. A cancer risk model with adaptive repair. Belle Newslett 7:14–16, 1998.
14. RB Hayes, SN Yin, M Dosemeci, GL Li, S Wacholder, LB Travis, CY Lin, N Rothman, RN Hoover, MS Linet. Benzene and the dose-related incidence of hematologic neoplasms in China. JNCI 89:1065–1071, 1997.
15. AH Wu-Williams, L Zeise, D Thomas. Risk assessment for aflatoxin B1: A modeling approach. Risk Anal 12:559–567, 1992.

16. M Crawford, R Wilson. Low-dose linearity: The rule or the exception? Human Ecol Risk Assess 2:305–330, 1996.
17. CJ Portier, F Ye. U-shaped dose response curves for carcinogens. Belle Newslett 7:23–26, 1998.
18. SH Moolgavkar. Comments on papers on U-shaped dose-response relationships for carcinogens. Belle Newslett 7:21–23, 1998.
19. DG Hoel. Response to the reports of Andersen, Bogen, and Downs. Belle Newslett 7:26–27, 1998.
20. EL Anderson, and the Carcinogen Assessment Group of the US Environmental Protection Agency. Quantitative approaches in use to assess cancer risk. Risk Anal 3:277-295, 1983.
21. D Krewski, E Cardis, L Zeise, V Feron. Empirical approaches to risk estimation and prediction. In: S Moolgavkar, D Krewski, L Zeise, E Cardis, J Moller, eds. Quantitative Estimation and Prediction of Human Cancer Risks. IARC Scientific Publications, No. 131. Lyon: International Agency for Research on Cancer, 1999.
22. P Armitage, R Doll. The age distribution of cancer and a multistage theory of carcinogenesis. Br J Cancer 8:1–12, 1954.
23. P Stocks. A study of the age curve for cancer of the stomach in connection with a theory of the cancer producing mechanism. Br J Cancer 7:407–417, 1953.
24. HJ Muller. Radiation damage to the genetic material. Sci Prog 7:93, 1951.
25. CO Nordling. A new theory on the cancer inducing mechanism. Br J Cancer 7:68–72, 1953.
26. P Armitage, R Doll. A two stage theory of carcinongenesis in relation to the age distribution of cancer. Br J Cancer 11:161–169, 1957.
27. SH Moolgavkar, AG Knudson. Mutation and cancer: A model for human carcinogenesis. JNCI 66:1037–1052, 1981.
28. A Kopp-Schneider, CA Portier. Birth and death/differentiation rates of papillomas in mouse skin. Carcinogenesis 13:973-978, 1992.
29. S Moolgavkar. Carcinogenesis modelling: From molecular biology to epidemiology. Annu Rev Public Health 7:151–169, 1986.
30. SH Moolgavkar, EG Luebeck. Incorporating cell proliferation kinetics into models for cancer risk assessment. Toxicology 102:141–147, 1995.
31. S Moolgavkar, A Woodward, D Krewski, E Cardis, L Zeise. Future perspectives, unresolved issues and research needs. In: S Moolgavkar, D Krewski, L Zeise, E Cardis, H Moller. Quantatitive Estimation and Prediction of Human Cancer Risk. Lyon: IARC Scientific Publications, No. 131, pp. 179–237.
32. US Environmental Protection Agency, Integrated Risk Information System, Office of Research and Development, Washington, DC, 2000.
33. World Health Organization. WHO Guidelines for Drinking Water Quality, Vol. 1, Recommendations. Geneva World Health Organization, 1993.
34. World Health Organization. Air Quality Guidelines for Europe, WHO Regional Office for Europe, Copenhagen, 1987.
35. The California Safe Drinking Water and Toxic Enforcement Act of 1986. Codified in California Health and Safety Code Sections 25249.5 through 25249.12.
36. L Zeise, E Cardis, K Hemminki, M Schwarz. Quantitative estimation and prediction of cancer risk: Review of existing activities. In: S Moolgavkar, D Krewski, L Zeise,

E Cardis, H Moller. Quantiative Estimation and Prediction of Human Cancer Risk. Lyon: IARC Scientific Publications, No. 131, 1999, pp 179–237.
37. US Environmental Protection Agency Science Advisory Board. Chloroform Risk Assessment Review. US EPA Science Advisory Board Chloroform Risk Assessment Subcommittee, Washington DC, 2000.
38. L Robbiano, E Mereto, AM Morando, P Pastore, G Brambilla. Increased frequency of micronucleated kidney cells in rats exposed to halogenated anaesthetics. Mutat Res 413:1–6, 1998
39. K Fujie, T Aoki, M Wada. Acute and subacute cytogenetic effects of the trihalomethanes on rat bone marrow cells in vivo. Mutat Res 242:111–119, 1990.
40. RJ Brennan, RH Schiestl, Chloroform and carbon tetrachloride induce intrachromosomal recombination and oxidative free radicals in Saccharomyces cerevisiae. Mutat Res 397(2):271–278, 1998.
41. A Colacci, S Bartoli, B Bonora, L Guidotti, G Lattanzi, M Mazzullo, A Niero, P Perocco, P Silingardi, S Grilli. Chloroform bioactivation leading to nucleic acids binding. Tumori 77:285–290, 1991.
42. S Gemma, S Faccioli, P Chieco, M Sbraccia, E Testai, L Vitozzi. In vivo ChCl3 bioactivation, toxicokinetics, toxicity, and induced compensatory cell proliferation in B6C3F1 mice. Toxicol Appl Pharmacol 141:394–402, 1996.
43. S Rossi, S Gemma, L Fabrizi, E Testai, L Vittozzi. Time dependence of chloroform-induced metabolic alterations in the liver and kidney of B6C3F1 mice. Arch Toxicol 73:387–393, 1999.
44. E Testai, S Di Marzio, A di Domenico, A Piccardi, L Vittozzi. An in vitro investigation of the reductive metabolism of chloroform. Arch Toxicol 70:83–88, 1999.
45. E Testai, S Di Marzio, L Vittozzi. Multiple activation of chloroform in hepatic microsomes from uninduced B6C3F1 mice. Toxicol Appl Pharmacol 104(3):496–503, 1990.
46. E Cardis, L Zeise, M Schwarz, S Moolgavkar. Review of specific examples of QEP. In: S Moolgavkar, D Krewski, L Zeise, E Cardis, J Moller, eds. Quantitative Estimation and Prediction of Human Cancer Risks. IARC Scientific Publications, No. 131. Lyon: International Agency for Research on Cancer, 1999.
47. M Kai, EG Luebeck, SH Moolgavkar. Analysis of the incidence of solid cancer among atomic bomb survivors using a two-stage model of carcinogenesis. Radiat Res 148:348–358, 1997.
48. WN Choy. A review of the dose response induction of DNA adducts by aflatoxin B1 and its implications to quantitative risk assessment. Mutat Res 296:181–198, 1993.
49. International Agency for Research on Cancer. IARC Monographs on the Evaluation of the Carcinogenic Risk of Chemicals to Humans, Volume 29, Some Industrial Chemicals and Dyestuffs. Lyon: IARC, 1982, pp 394–395.
50. JA Swenberg, N Fedtke, L Fishbein. Age-related differences in DNA adduct formation and carcinogenesis of vinyl chloride in rats. In: PS Guzelian, CJ Henry, SS Olin, eds. Similarities and Differences between Children and Adults: Implications for Risk Assessment. Washington DC: ILSI Press, 1992.
51. VJ Cogliano, GF Hiatt, A Den. Quantitative cancer assessment for vinyl chloride: Indications of early-life sensitivity. Toxicology 111:21–28, 1996.

52. C Maltoni, G Lefemine, A Cilberti, LG Cotti, D Caretti. Experimental research on vinyl chloride carcinogenesis. In: C Maltoni, MA Mehlman, eds. Archives of Research on Industrial Chemicals Volume II. Princeton Scientific: Princeton, NJ, 1984.
53. N Fedke, JA Boucheron, VE Walker, JH Swenberg. Vinyl chloride-induced DNA adducts. II. Formation and persistence of 7-(2´-oxoethyl)guanine and N´-3-etheno-guanine in rat tissue DNA. Carcinogenesis 11:1287, 1990.
54. M Tokunaga, CE Land, S Tokuoka, I Nishimori, M Soda, S Akiba. Incidence of female breast cancer among atomic bomb survivors, Hiroshima and Nagasaki, 1950-1985. Radiat Res 138:209–223, 1994.
55. S Bhatia, AT Meadows, LL Robison. Second cancers after pediatric Hodgkin's disease. J Clin Oncol 16(7):2570–2572, 1998.
56. EA Nekolla, AM Kellerer, M Kuse-Isingschulte, E Eder, H Spiess. Malignancies in patients treated with high doses of radium-224. Radiat Res 152(6 suppl):S3–S7, 1999.
57. B Modan, A Chetrit, E Alfandary, L Katz. Increased risk of breast cancer after low-dose irradiation. Lancet 1(8643):916, 1989.
58. J Russo, G Wilgus, IH Russo. Susceptibility of the mammary gland to carcinogenesis. I. Differentiation of the mammary gland as determinant of tumor incidence and type of lesion. Am J Pathol 96:721–736, 1979.
59. Hall PF. Cancer risks after medical radiation. Med Oncol Tumor Pharmacother 8(3):141–145, 1991.
60. E Ron, B Modan, D Preston, E Alfandary, M Stovall, JD Boice. Thyroid neoplasia following low-dose radiation in childhood. Radiat Res 120(3):516–531, 1989.
61. E Ron, B Modan. Benign and malignant thyroid neoplasms after childhood irradiation for tinea capitis. J Natl Cancer Inst 65(1):7–11, 1980.
62. RM Giusti, K Kwamoto, EE Hatch. Diethylstilbestrol revisted: A review of the long-term health effects. Ann Intern Med 122(10):778-788, 1995.
63. International Agency for Research on Cancer. IARC Monographs on the Evaluation of Carcinogenic Risks to Humans. Supplement 7. Overall Evaluations of Carcinogenicity: An Updating of IARC Monographs Volumes 1 to 42. Lyon: World Health Organization, IARC, 1987, pp 273–278.
64. International Agency for Research on Cancer. IARC Monographs on the Evaluation of Carcinogenic Risks to Humans. Sex Hormones (II). Lyon: World Health Organization, IARC, 1979, pp 173–231.
65. RT Drew, GA Boorman, JK Haseman, EE McConnell, WM Busey, JA Moore. The effect of age and exposure duration on cancer induction by a known carcinogen in rats, mice and hamsters. Tox Appl Pharmacol, 68:120–130, 1983.
66. E Ron, JH Lubin, RE Shore, K Mabuchi, B Modan, LM Pottern, AB Schneider, MA Tacker, JD Boice Jr. Thyroid cancer after exposure to external radiation: a pooled analysis of seven studies. Radiat Res 141:259–277, 1995.
67. J Russo, G Wilgus, IH Russo. Susceptibility of the mammary gland to carcinogenesis. I. Differentiation of the mammary gland as determinant of tumor incidence and type of lesion. Am J Pathol 96:721–736, 1979.
68. GC Hard. Effect of age at treatment on the incidence and type of renal neoplasm induced in the rat by a single dose of dimethylnitrosamine. Cancer Res 39:4965–4970, 1979.

69. DR Meranze, M Gruenstein, MB Shimkin. Effect of age and sex on the development of neoplasms in Wistar rats receiving a single intragastric instillation of 7,12-dimethylbenz(a)anthracene. Int J Cancer 4:480–486, 1969.
70. DA Bosch. Short and long term effects of methyl-and ethylnitrosourea (MNU & ENU) on the developing nervous system of the rat. Acta Neurol Scand 55:85–105, 1977.
71. RFX Noronha, CM Goodall. The effects of estrogen on single dose dimethylnitrosamine carcinogenesis in male inbred Crl/CDF rats. Carcinogenesis 5(8):1003–1007, 1984.
72. VN Anisimov. Effect of age on dose-response relationship in carcinogenesis induced by single administration of N-nitrosomethyl urea in female rats. J Cancer Res Clin Oncol 114:628-635, 1988.
73. T Shirai, A Nakamura, S Fukushima, S Takahashi, K Ogawa, N Ito. Effects of age on multiple organ carcinogenesis induced by 3,2´-dimethyl-4-aminobiphenyl in rats, with particular reference to the prostate. Jpn J Cancer Res 80:312–316, 1989.
74. SD Vesselinovitch. Perinatal hepatocarcinogenesis. Biol Res Pregnancy Perinatol 4: 22–25, 1983.
75. MD Reuber. Influence of age and sex on carcinoma and cirrhosis of the liver in AXC strain rats ingesting 0.025% N-2-fluorenyldiacetamide. Pathol Microbiol 43:31–37, 1975.
76. DJ Murdoch, D Krewski. Carcinogenic risk assessment with time-dependent exposure patterns. Risk Anal 8:521–530, 1988.
77. KS Crump, RB Howe. The multistage model with a time-dependent dose pattern: Applications to carcinogenic risk assessment. Risk Anal 4:163–176, 1984.
78. L Zeise, EAC Crouch. Experimental variation in the carcinogenic potency of benzo(a)pyrene. Energy and Environmental Policy Center, Kennedy School of Government, Harvard University, 1983.
79. L Zeise. Surrogate measures of human cancer risk. Ph.D. dissertation. Harvard University, Cambridge, MA, 1984.
80. L Zeise, AG Salmon, T McDonald, P Painter. Cancer potency estimation. In: AG Salmon, L Zeise, eds. Risks of Carcinogenesis from Urethane Exposure. Boca Raton: CRC Press, 1991.
81. R Peto, FJC Roe, PN Lee, L Levy, J Clack. Cancer and ageing in mice and men. Br J Cancer 32:411–426, 1975.
82. National Research Council (NRC). Pesticides in the Diets of Infants and Children. Committee on Pesticides in the Diets of Infants and Children. Washington, DC: NRC, National Academy Press, 1993.
83. L Zeise, P Painter, PE Berteau, AM Fan, RJ Jackson. Alar in Fruit: Limited regulatory action in the face of uncertain risks. In: BJ Garrick, WC Gekler, eds. Advances in Risk Analysis, vol. 9. The Analysis, Communication, and Perception of Risk. New York: Plenum, 1991.
84. R Peto, MC Pike, L Bernstein, LS Gold LS, BN Ames. The TD50: A proposed general convention for the numerical description of the carcinogenic potency of chemicals in chronic-exposure animal experiments. Environ Health Perspect 58:1–8, 1984.

85. SM Hoover, L Zeise, WS Please, LE Lee, MP Hennig, LB Weiss, C Cranor. Improving the regulation of carcinogens by expediting cancer potency estimation. Risk Anal 15:267–280, 1995.
86. AJ Bailer, CJ Portier. Effects of treatment-induced mortality and tumor-induced mortality on tests for carcinogenicity in small smaples. Biometrics 44:417–431, 1988.
87. National Research Council. Committee on the Biological Effects of Ionizing Radiation. Health Effects of Exposure to Low Levels of Ionizing Radiation. BEIR V. Washington DC: National Academy Press, 1990.
88. C Maltoni, G Lefemine, A Ciliberti, G Cotti, D Carretti. Vinyl chloride carcinogenicity bioassays (BT Project) as an experimental model for risk identification and assessment in environmental and occupational carcinogenesis. Epidemiologie Animale et Epidemiologie Humane: le Cas du Chlorure de Vinyle Monomare. Proceedings of the 20th Meeting of Le Club Cancerogenese Chimique in Paris, 10 November, 1979.
89. T Green, DE Hathaway. The biological fate of vinyl chloride in relation to its oncogenicity. Chem Biol Interact 11:545–562, 1975.
90. JY Jang, PO Droz, HK Chung. Uncertainties in physiologically based pharmacokinetic models caused by several input parameters. Int Arch Occup Environ Health 72:247–254, 1999.
91. D Hattis, P White, P Koch. Uncertainties in pharmacokinetic modeling for perchloroethylene: II. Comparison of model predictions with data for a variety of different parameters. Risk Anal 13:599–610, 1993.
92. FY Bois. Analysis of PBPK models for risk characterization. Ann NY Acad Sci 895:317–337, 1999.
93. FY Bois, L Zeise, TN Tozer. Precision and sensitivity of pharmacokinetic models for cancer risk assessment: Tetrachloroethylene in mice, rats, and humans. Toxicol Appl Pharmacol 102:300–315, 1990.
94. FY Bois, A Gelman, J Jiang, DR Mazle, L Zeise, G Alexeef. Population toxicokinetics of tetrachloroethylene. Arch Toxicol 70:347–355, 1996.
95. D Hattis, P White, L Marmostein, P Koch. Uncertainties in pharmacokinetic modeling for perchloroethylene. I. Comparison of model structure, parameters, and predictions for low-dose metabolism rates for models derived by different authors. Risk Anal 10:449–457, 1990.
96. JC Ramsey, ME Anderson. A physiologically based description of the inhalation pharmacokinetics of styrene in rats and humans. Toxicol Appl Pharmacol 73:169–175, 1984.
97. International Agency for Research on Cancer. IARC Monographs on the Evaluation of Carcinogenic Risks to Humans. Volume 63. Dry Cleaning, Some Chlorinated Solvents and Other Industrial Chemicals. Lyon. IARC, 1995.
98. W Volkel, M Friedewald, W Lederer, A Pahler, J Parker, W Dekant. Biotransformation of perchloroethene: Dose dependent excretion of trichloroacetic acid, dichloroacetic acid, and N-acetyl-S-(trichlorovinyl)-L-cysteine in rats and humans after inhalation. Toxicol Appl Pharmacol 153:20–27, 1998.
99. RH Reitz, ML Gargas, AL Mendrala, AM Schumann. In vivo and in vitro studies of perchloroethylene for physiologically based pharmacokinetic modeling in rats, mice and humans. Toxicol Appl Pharmacol 136:289–306, 1996.

100. CE Dallas, XM Chen, S Muralidhara, P Varkonyi, RL Tackett, JV Bruckner. Use of tissue disposition data from rats and dogs to determine species differences in input parameters for a physiological model for perchloroethylene. Environ Res 67:54–67, 1994.
101. M Pelekis, P Poulin, K Krishnan. An approach for incorporating tissue composition data into physiological based pharmacokinetic models. Toxicol Ind Health 11:511–522, 1995.
102. CJ Portier, NL Kaplan. Variability of safe dose estimates when using complicated models of the carcinogenic process. A case study: Methylene chloride. Funda Appl Toxicol 13:533–544, 1989.
103. IARC working group on the use of data on mechanisms of carcinogenesis in risk identification. Consensus report. In: H Vainio, P Magee, D McGregor, AJ McMichael, eds. Mechanisms of Carcinogenesis in Risk Indentification. IARC Scientific Publications, No. 116. Lyon: International Agency for Research on Cancer, 1992, pp 28–29.
104. LJ Kinlen. Immunosuppression and cancer. In: H Vainio, P Magee, D McGregor, AJ McMichael, eds. Mechanisms of Carcinogenesis in Risk Indentification. IARC Scientific Publications, No. 116. Lyon: International Agency for Research on Cancer, 1992.
105. BW Stewart. Role of DNA repair in carcinogenesis. In: H Vainio, P Magee, D McGregor, AJ McMichael, eds. Mechanisms of Carcinogenesis in Risk Identification. IARC Scientific Publications, No. 116. Lyon: International Agency for Research on Cancer, 1992.
106. MT Landi, R Sinha, NP Lang, FF Kadlubar. Human cytochrome P4501A2. In: P Vineis, N Malars, M Lang, A d'Errico, N Caporaso, J Cuzick, P Boffetta, eds. Metabolic Polymorphisms and Susceptibility to Cancer. IARC Scientific Publications, No. 148. Lyon: International Agency for Research on Cancer, 1999, pp 173–195.
107. GJ Hammons, Y Yan, NG Lopatina, B Jin, C Wise, EB Blann, LA Poirer, FF Kadlubar, BD Lyn-Cook. Increased expression of hepatic DNA methyltransferase in smokers. Cell Biol Toxicol 15:389–394, 1999.
108. FY Bois, G Krowech, L Zeise. Modeling human interindividual variability in metabolism and risk: The example of 4-aminobiphenyl. Risk Anal 15:205–213, 1995.
109. FY Bois, A Gelman, J Jiang, DR Maszle, L Zeise, G Alexeeff. Population toxicokinetics of tetrachloroethylene. Arch Toxicol 70(6):347–355, 1996.
110. FY Bois. Analysis of PBPK models for risk characterization. Ann NY Acad Sci 895:317–337, 1999.
111. DG McGregor. Chemicals classified by IARC: Their potency in tests for carcinogenicity in rodents and their genotoxicity and acute toxicity. In: H Vainio, P Magee, D McGregor, AJ McMichael, eds. Mechanisms of Carcinogenesis in Risk Identification. IARC Scientific Publications, No. 116. Lyon: International Agency for Research on Cancer, 1992.
112. JK Haseman, JE Huff. Species correlation in long-term carcinogenicity studies. Cancer Lett 37:125–132, 1987.
113. LS Gold, L Bernstein, R Magaw, TH Slone. Interspecies extrapolation in carcinogenesis: Prediction between rats and mice. Environ Health Perspect 81:211–219, 1989.

114. RL Kodell, AP Basu, DW Gaylor. On interspecies correlations of carcinogenic potency. J Toxicol Environ Health 48:231–237, 1996.
115. DA Freedman, LS Gold, TH Lin. Concordance between rats and mice in bioassays for carcinogenesis. Regul Toxicol Pharmacol 23:225–232, 1996.
116. WW Piegorsch, GJ Carr, CJ Portier, DG Hoel. Concordance of carcinogenic response between rodent species: Potency dependence and potential underestimation. Risk Anal 12:115–121, 1992.
117. RL Dedrick, PF Morrison. Carcinogenic potency of alkylating agents in rodents and humans. Cancer Res 52:2464–2467, 1992.
118. JM Kaldor, NE Day, K Hemminki. Quantifying the carcinogenicity of neoplastic drugs. Eur J Cancer 24:703–711, 1988.
119. BC Allen, KS Crump, AM Shipp. Correlation between carcinogenic potency of chemicals in animals and humans. Risk Anal 8:531–544, 1988.
120. K Crump, B Allen, A Shipp. A choice of dose measure for extrapolating carcinogenic risk from animals to humans: An empirical investigation of 23 chemicals. Health Phys 57 (suppl 1):387–393, 1989.
121. EAC Crouch, R Wilson. Interspecies comparisons of carcinogenic potency. J Toxicol Environ Health 5:1095–1118, 1979.
122. DW Gaylor, JJ Chen. Relative potency of chemical carcinogens in rodents. Risk Anal 6:283–290, 1986.
123. B Metzger, E Crouch, R Wilson. On the relationship between carcinogenicity and acute toxicity. Risk Anal 9:169–177, 1989.
124. L Bernstein, LS Gold, BN Ames, MC Pike, DG Hoel. Some tautologous aspects of the comparison of carcinogenic potency in rats and mice. Funda Appl Toxicol 5:79–87, 1985.
125. DA Freedman, LS Gold, TH Sloan. How tautologous are interspecies correlations of carcinogenic potencies? Risk Anal 13:265–272, 1993.
126. EAC Crouch, R Wilson, L Zeise. Tautology or not tautology? J Toxicol Environ Health 20:1–10, 1987.
127. US Environmental Protection Agency. Draft report: A cross-species scaling factor for carcinogenic risk assessment based on equivalence of $mg/kg^{3/4}$/day. Fed Reg 57(109):24152–24173, 1992.
128. EJ Freireich, EA Gehan, DP Rall, LH Schmidt, HE Skipper. Quantitative comparisons of toxicity of anticancer agents in mouse, rat, hamster, dog, monkey, and man. Cancer Chemother Rep 50:219–244, 1966.
129. PS Schein, RD Davis, S Carter, J Newman, DR Schein, DP Rall. The evaluation of anticancer drugs in dogs and monkeys for the prediction of quantitative toxicities in man. Clin Pharmacol Therapeut 11:3–40, 1979.
130. K Watanabe, FY Bois, L Zeise. Interspecies extrapolation: A reexamination of acute toxicity data. Risk Anal 12:301–310, 1992.
131. CC Travis, RK White. Interspecific scaling of toxicity data. Risk Anal 8:119–125, 1988.
132. D Hattis. Pharmacokinetic principles for dose rate extrapolation of carcinogenic risk from genetically active agents. Risk Anal 10:303–316, 1990.
133. R Doll, R Peto. The Causes of Cancer. Quantitative Estimates of Avoidable Risks of Cancer in the United States Today. Oxford: Oxford University Press, 1981.

134. DG Hoel. Incorporation of backgrounds in dose-response models. Fed Proc 39:73–67, 1980.
135. WK Lutz. Endogenous genotoxic agents and processes as a basis of spontaneous carcinogenesis. Mutat Res 238:287–295, 1990.
136. D Krewski, DW Gaylor, WK Lutz. Additivity to background and linear extrapolation. In: S Olin, W Farland, C Park, L Rhomberg, R Scheuplein, T Starr, J Wilson, eds. Low Dose Extrapolation of Cancer Risks: Issues and Perspectives. Washington DC: ILSI Press, 1995, pp 105–121.
137. FS Yeh, MC Yu, CC Mo, S Luo, MJ Tong, BE Henderson. Hepatitis B virus, aflatoxins and hepatocellular carcinoma in Southern Guangxi, China. Cancer Res 49:2506–2509, 1989.
138. CJ Portier, AJ Bailer. Testing for increased carcinogenicity using a survival-adjusted quantile response test. Funda Appl Toxicol 12:731–737, 1989.

15

Biologically Based Cancer Risk Assessment Models

Suresh H. Moolgavkar
Fred Hutchinson Cancer Research Center, Seattle, Washington

I. INTRODUCTION

Compelling evidence from histopathological, epidemiological, and molecular biological studies indicates that malignant transformation is the end result of an accumulation of alterations in the cellular genome (1). Most cancers are clonal, suggesting that a single cell that has acquired these changes is capable of giving rise to a malignant tumor. Although malignant cells from a number of tissues exhibit numerous mutations, recent work suggests that only a few critical alterations are required for malignant transformation (2).

The role of cell proliferation kinetics in malignant transformation is also being increasingly appreciated (3). Increases in division of susceptible cells, either normal stem cells or altered cells on the pathway to malignancy, can increase the probability of cancer in two ways. First, because mutation rates are proportional to cell division rates, an increase in the latter leads to an increase in the former. Second, an increase in cell division rates without a compensatory increase in rates of apoptosis leads to an expansion of populations of susceptible cells, which can increase greatly the probability of cancer. Clonal expansion of intermediate cell populations is the essential feature of promotion. Thus, carcinogenesis involves successive genomic changes, each change resulting in possible further disruption of cellular kinetics, which, in turn, accelerates the acquisition of more mutations. Phenotypically altered cell populations in the intermediate stages on the pathway to malignancy have been observed in a variety of epithelial tissues including liver, pancreas, lung, kidney, and urinary bladder in both experimental animals and humans and are generally believed to represent precancerous lesions (4, 5).

Of particular importance in the risk assessment context are the concepts of initiation and promotion with their implied classification of environmental carcinogens into initiators and promoters. Experimental findings, historically developed from results obtained in the mouse skin carcinogenesis system, led to the operational definitions of the stages of initiation, promotion, and progression. These observations in the mouse skin system have since been extended to other experimental systems, most notably the rodent liver. Initiation is thought to involve (one or more) genomic alterations; promotion is the clonal expansion of initiated cells; and progression is the conversion of one (or a few) of the population of initiated cells, probably by acquisition of one or more genomic alterations, into fully malignant cells.

The observation that both mutation and cell proliferation kinetics play essential roles in carcinogenesis suggests the framework for a mathematical model for carcinogenesis that incorporates the essential biological features of the process. The first multistage models of carcinogenesis were proposed by Nordling (6) and by Armitage and Doll (7). About that time, it had been observed that the age-specific incidence curves of many common human cancers increased roughly with a power of age. This observation could mathematically be derived from a multistage model for carcinogenesis. These early models ignored cell proliferation, although somewhat later Armitage and Doll acknowledged the possibility of clonal expansion of intermediate cells in a two-stage model of carcinogenesis. They showed that a two-stage model with clonal expansion of intermediate cells could explain the age-specific incidence curves of the common human carcinomas (8). These ideas were later extended and embellished by Knudson (9) and by Moolgavkar and Venzon (10) and Moolgavkar and Knudson (11). Similar models were proposed by Kendall (12) and by Neyman and Scott (13).

Ideally, risk assessments should be based on epidemiological studies because they offer two obvious major advantages over experimental studies. First, epidemiological studies are done in the species of ultimate interest, thus finessing the difficult problem of interspecies extrapolation. Second, estimates of risk can be directly obtained for levels of exposure that are close to those typical of "free-living" human populations. Epidemiological studies are often conducted in cohorts with occupational exposures to the agent of interest, which are typically higher than exposures in the general population. Nevertheless, these exposures

Biologically Based Models

are much closer to those in the general population than exposures used in experimental studies. Some of what epidemiological studies gain in the way of relevance over experimental studies is lost in precision, however. It is generally true that both exposures and disease outcomes are measured with less precision in epidemiological studies than in experimental studies, leading, possibly, to bias in the estimate of risk. Another potentially serious problem arises from the fact that human populations, particularly occupational cohorts, are rarely exposed to single agents. When exposure to multiple agents is involved, the effect of the single agent of interest is often difficult to investigate. Finally, lifestyle factors such as diet and, in particular, smoking, have a profound effect on human cancer risks. These potential confounders are often difficult to control in epidemiological studies, leading to possible bias in the estimates of risk associated with exposure to the agent of interest.

When appropriate epidemiological studies are not available, which is unfortunately the case for most agents of interest, the risk assessor has to rely on experimental data. The twin problems of low-dose and interspecies extrapolation then raise their ugly heads. Although purely statistical methods have been used to address these problems, the consensus now appears to be moving toward the use of quantitative methods with strong biological underpinnings.

The main focus of this chapter is the use of biologically based models for cancer risk assessment. Mathematical models of carcinogenesis have a wider role, however. First, they provide a mathematical framework within which the quantitative aspects of cancer in populations (whether populations of humans in epidemiological studies or inbred populations of animals in experimental studies) can be viewed and questions about them asked. For example, models of carcinogenesis have been used for analyses of various epidemiological (14, 15) and experimental (16, 17) data sets, and biological interpretations of the data have been sought. Second, from the statistical point of view, cancer models provide a rich class of hazard functions for analyses of time-to-tumor data. These models provide a very convenient way to incorporate time- and age-dependent exposure patterns into analyses. Moreover, analyses of epidemiological data using conventional statistical methods often require rather artificial assumptions, such as multiplicativity of relative risks or additivity of excess risks. Third, the analyses of intermediate lesions on the pathway to malignancy is greatly facilitated by the use of cancer models (18–20).

In the risk assessment context, a comprehensive model would incorporate not only the fundamental biological principles underlying carcinogenesis in a pharmacodynamic model, but also the principles underlying intake and disposition of the environmental agent under consideration in a pharmacokinetic model. We are far from developing such a comprehensive model, although some elements of each component of the comprehensive model are in place. In this chapter, I cover only pharmacodynamic models. I begin with a brief nontechnical discussion of multistage cancer models and illustrate their use in risk assessment by giving two examples: one of a risk assessment based on epidemiological data and a second of a risk assessment based on experimental data.

Current multistage models make the following fundamental assumptions: (a) cancers are clonal, that is, malignant tumors arise from a single malignant progenitor cell;(b) each susceptible (stem) cell in a tissue is as likely to become malignant as any other;(c) the process of malignant transformation in a cell is independent of that in any other cell; and(d) once a malignant cell is generated, it gives rise to a detectable tumor with probability 1 after a constant lag time. The last two assumptions, which are made for mathematical convenience, are clearly false. Methods for relaxing these assumptions are currently being investigated (21, 22).

For more details of the ideas developed here, I refer the reader to a recent monograph (23).

II. MULTISTAGE MODELS

A. Some Fundamental Concepts

The concept of incidence rate is of fundamental importance in epidemiology. Although the word incidence is often used rather loosely in the toxicological literature, it has a precise meaning in epidemiology and biostatistics (the concept of incidence in epidemiology is identical to that of hazard in biostatistics). A measure of disease frequency, the incidence rate is the rate (per person per unit time) at which *new* cases of a disease appear in the population (whether of humans or experimental animals) under study. For example, because the incidence rates of many chronic diseases vary strongly with age, a commonly used measure of frequency is the age-specific incidence rate. Usually, 5-year age groups are used: the age-specific incidence rate of a cancer in the 5-year age group 45–49, for example, is estimated as the ratio of the number of *new* cases of cancer occurring in that age group in a single year and the number of individuals in that age group who are *cancer-free* at the beginning of the year. Strictly speaking, the denominator should be not the total number of cancer-free individuals at the beginning of the year, but the person-years at risk during the year. This is because some individuals contribute less than a full year of experience to the denominator, either because they enter the relevant population after the year has begun (e.g., an individual may reach the age of 45 sometime during the year) or because they may leave the population before the year is over (e.g., an individual might reach the age of 50 or die during the year). For a more detailed discussion of the concept of person-years, see Rothman and Greenland (24). I note here that the concept of incidence (or hazard) rate is an instantaneous concept and is most precisely de-

Biologically Based Models

fined in terms of the differential calculus. For a discussion see, for example, Kalbfleisch and Prentice (25) or Cox and Oakes (26).

Another commonly used measure of disease frequency is the probability of developing the disease during a specified period of time. In quantitative risk assessment, one is often interested in the lifetime probability (often called lifetime risk) of developing cancer and the impact of environmental agents on this probability. The incidence rate and the probability of disease are completely equivalent in the sense that if either is known as a function of age, the other can be computed. Precisely, one has the relationship: $P(t_1, t_2) = 1 - \exp(-\int_{t_1}^{t_2} h(s)\, ds)$, where $P(t_1, t_2)$ is the probability of developing disease between the ages t_1 and t_2, and $h(s)$ is the incidence (hazard) rate at age s. If the hazard rate is small (as it typically is for cancer in human populations), then the probability is well approximated by the cumulative incidence, $P(t_1, t_2) \sim \int_{t_1}^{t_2} h(s)\, ds$.

B. Multistage Models and Incidence Functions

As I mentioned earlier, multistage models were first proposed (6, 7) to explain the observation that the age-specific incidence curves of many common adult carcinomas increase roughly with a power of age. In the fitting of these early models to epidemiological data, a couple of mathematical approximations, which greatly simplified the analyses, were made. The accuracy of these approximations depends crucially on the fact that cancer in human populations is a rare disease. The approximations are inaccurate when used for analyses of experimental data in which tumors occur with high probability because the experiments are conducted at high exposure levels of the agent under investigation. Even for epidemiological data, use of these approximations could lead to misinterpretations of the data. Yet, the approximate form of these models has been, and continues to be, used for data analyses.

Without going into technical detail, I would like to discuss some of the consequences that follow from the multistage nature of carcinogenesis. These consequences can be derived only by a consideration of the exact solution of the underlying mathematical model. The approximate solutions lead to different conclusions. I should mention here that the linearized multistage (LMS) procedure makes even further simplifications, and therefore should not be considered to be based on biological principles.

The incidence function arising from the concept of multistage carcinogenesis has three important features. These features are not peculiar to a specific multistage model; rather they are true for a broad class of models and may be thought of as consequences of multistage carcinogenesis.

The first consequence following from a consideration of the exact solution of multistage models is that the incidence function cannot rise indefinitely but must approach a finite asymptote. Recall that the age-specific incidence curves for many cancers increase roughly with a power of age. What the asymptotic behavior of the theoretical incidence curve says is that this increase cannot occur indefinitely: with increasing age, the incidence must flatten out. In fact, exactly this behavior had been noted for many of the tumors: the observed incidence at higher ages deviated below the incidence that would have been predicted by a power law relationship. Within the framework of the approximate solution of multistage model, various explanations were given for this observation. For example, it was suggested that cancer cases were being undercounted in the older age groups and that there was something special about the biology of extreme old age that protected against cancer. These explanations may or may not be correct; however, they do not have to be invoked. The exact solution of the multistage model predicts exactly this behavior of the incidence function. Thus, the flattening out of the age-specific incidence curve at the older ages is a mathematical consequence of multistage carcinogenesis and could have been predicted if misleading approximations had not been used (27–29).

The second consequence following from multistage carcinogenesis has to do with the behavior of the age-specific incidence curve after exposure to an environmental carcinogen ceases. Suppose that exposure to an environmental agent increases the rate of one or more of the steps in the carcinogenic pathway, for instance, by increasing the rates of specific mutations or of cell divisions. Suppose also that these rates revert back to background rates after exposure stops. Then the exact (but not the approximate) solution predicts that the incidence function after exposure ends will eventually approach the incidence function in those individuals who were never exposed. This phenomenon has been observed among exsmokers. Incidence rates of lung cancer in exsmokers appear to approach the incidence rates in nonsmokers about 15 years after quitting smoking. A similar phenomenon has been observed among the survivors of the atomic bombings in Hiroshima and Nagasaki. One of the explanations given for this phenomenon is that repair of damaged tissue must have occurred after exposure stopped. This could well be true, but again this explanation need not be invoked. The reversion of the incidence function to background rates is a mathematical consequence of multistage carcinogenesis, providing inappropriate approximations are not used (27–29).

The third consequence has to do with the incidence of second malignant tumors among individuals who have already had one. A computation using the exact solution shows that the age-specific incidence of second malignant tumors is higher than the age-specific incidence of the first malignant tumor (at the same age). Although the incidence of second malignant tumors is difficult to study in human populations because of the treatment intervention after the occurrence of the first tumor, animal experiments appear to show that this is indeed true. The explanation has been advanced that physiological and immunological changes after the first malignancy renders the animal susceptible to a second tumor. This may be true, but the higher incidence of second tumors is a logical consequence of multistage carcinogenesis (30).

C. The Armitage-Doll Model

I gave a precise definition of the concept of incidence (or hazard) because this is the fundamental concept used in fitting multistage models to data. As noted above, the Armitage-Doll model, on which the current LMS procedure for low-dose extrapolation is based, was originally proposed to explain the regularity of the age-specific incidence curves of the common human carcinomas. The age-specific incidence curves of many common adult carcinomas increase roughly with a power of age. Equivalently, a plot of the logarithm of the age-specific incidence curve against the logarithm of age is a straight line, although as pointed out above, the observed age-specific incidence rates fall below this line at the older ages. For these so-called log-log cancers, Armitage and Doll (7) showed that a multistage model (without consideration of cell proliferation kinetics) predicts this behavior approximately, with the slope of the straight line being one less than the number of stages required for malignant transformation.

The LMS procedure, which is a widely used default procedure for low-dose extrapolation of cancer risks, has its origins in the Armitage-Doll multistage model. To understand the properties of this procedure, it is necessary to understand the Armitage-Doll approximation. The multistage model postulates that a malignant tumor arises in a tissue when a single susceptible cell undergoes malignant transformation through a finite sequence of intermediate stages, the waiting time between any stage and the subsequent one being exponentially distributed. Suppose now that malignancy occurs in n stages and that λ_i is the parameter of the waiting time distribution in stage i. That is, the rate of transition from stage i to stage $i+1$ is λ_i, which implies that the mean sojourn time in stage i is $(\lambda_i)^{-1}$. Then, under the assumptions that malignancy is rare and that each of the transition rates is much smaller than the lifespan of the species being studied, the multistage model leads to the following approximation for the hazard function:

$$h(t) \sim \{N\lambda_0\lambda_1\ldots\lambda_{n-1}t^{n-1}\}/(n-1)!,$$

where N is the total number of susceptible cells in the tissue.

That is, as mentioned above, the hazard function increases with a power of age one less than the number of stages required for malignant transformation. This form of the multistage model is the basis of the LMS procedure for low-dose extrapolation, which involves a further approximation. It is clear that, with all the approximations made, the LMS procedure is simply an exercise in curve-fitting; the hazard function used in the LMS procedure can be substantially different from the exact hazard function generated by the multistage model, particularly when experimental data with high tumor incidence is considered. See the recent International Agency For Research on Cancer (IARC) monograph (23) for details.

D. The Two-Stage Clonal Expansion Model

The two-stage clonal expansion model (10, 11, 27) posits that malignant transformation of a susceptible cell is the result of two specific, rate-limiting, hereditary (at the level of the cell), and irreversible events. I emphasize that this model should not be interpreted as positing that carcinogenesis results from two specific locus mutations. Rather, this model is best interpreted within the initiation-promotion-progression paradigm of chemical carcinogenesis. Initiation, which confers a growth advantage on the cell, is a rare event, and can be thought of as the first rate-limiting step. Mathematically, the process of initiation can reasonably be modeled by a time-dependent Poisson process. Thus, the two-stage clonal expansion model posits that the arrival of cells into the initiated compartment follows a Poisson process, and that promotion consists of the clonal expansion of these cells by a stochastic birth and death process. Finally, one of the initiated cells may be converted into a malignant cell, and this conversion may involve one or more mutations. From the mathematical perspective, incorporating more than one mutation in the progression step is equivalent to having a time-dependent second mutation rate in the two-stage model. This possibility can explicitly be tested in the data. When more biological information (on the number of stem cells or on intermediate lesions on the pathway to malignancy, for example) is available, the model can be extended to include this information (31).

Because the two-stage clonal expansion model explicitly considers both genomic events and cell proliferation kinetics, it provides a flexible tool for incorporating both genotoxic and nongenotoxic carcinogens in cancer risk assessment. Within the framework of the model, an environmental agent acts by affecting one or more of its parameters. Thus, an environmental agent may affect the rate of one or both stages (initiation and progression) or of cell proliferation kinetics of normal or initiated cells. A purely genotoxic agent would be expected to increase the rates of initiation or progression or both. A pure promoting agent might increase the cell division rate or decrease the rate of apoptosis of initiated cells (and possibly also of normal cells, but this would be expected to have less of an effect on cancer incidence). Compensatory or regenerative cell proliferation occurs at necrogenic doses of many different agents and may explain their promoting effects.

Because the two-stage clonal expansion model explicitly considers initiation and promotion, it can be used for analyses of the intermediate lesions, such as the enzyme-altered foci in rodent hepatocarcinogenesis experiments or the papillomas in mouse skin painting experiments, that arise in many model systems for the study of chemical carcinogenesis. The requisite mathematical expressions required for the analyses of quantitative information on intermediate lesions have been developed and applied to analyses of both altered hepatic foci (17, 19) and papillomas on the mouse skin (20). I would like to emphasize that explicit consideration of cell division and apoptosis (rather than just the net proliferation) of initiated cells has important implications. If the rate of apoptosis is greater than zero, then an initiated cell may die without giving rise to a detectable clone. Thus, although initiation is believed to be irreversible on the level of the single cell, it may be reversible on the level of the tissue, that is, initiated cells may die without giving rise to initiated clones (Fig. 1). For quantitative risk assessment, analyses of intermediate lesions allows the dose-response curve to be extended to lower doses than would be possible with consideration of malignant lesions alone, because intermediate lesions generally occur at doses that are often too low for the appearance of malignant lesions, with the typical experimental protocol involving at most a few hundred animals.

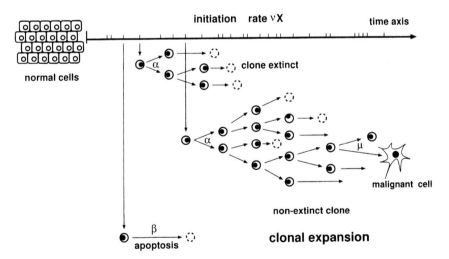

FIG. 1. The two-stage clonal expansion model. Initiation may occur either spontaneously or in response to an initiating agent. Initiated cells may divide or undergo apoptosis. The clonal expansion of initiated cells is promotion. A significant number of initiated cells may never give rise to clones because they become extinct by apoptosis. The probability of extinction of small clones of initiated cells can be high. Eventually, an initiated cell may acquire further genetic changes to become malignant.

1. Examples

I consider two examples of quantitative cancer risk assessment using the two-stage clonal expansion model, one using epidemiological data, and the other based on experimental data. Emissions from coke ovens are complex mixtures of polycyclic aromatic hydrocarbons (PAH), and epidemiological data on exposure and lung cancer are available in cohorts of steel workers. For the man-made mineral fibers, the epidemiological studies have so far shown no evidence of increased risk of lung cancer or mesothelioma. The occupational cohorts in which these studies have been conducted are small, however, and the studies, which are continuing, have not been ongoing for a long enough period for detection of an effect, if there is one.

a. Emissions from Coke Ovens and Lung Cancer. Emissions from coke ovens contain a complex mixture of PAH and have been associated with lung cancer in occupational cohort studies of coke oven workers. Both the United States Environmental Agency (USEPA [32]) and the IARC [33] have determined that coke oven emissions are human carcinogens. In 1984, the USEPA used a simple version of the multistage model to analyze the then available epidemiological data on cohorts of U.S. steel workers and estimated that the unit risk of lung cancer associated with exposure to coke oven emissions was 6.2×10^{-4}. Here unit risk was defined as the excess probability of lung cancer at age 70 associated with continuous exposure to $1\mu g/m^3$ from birth. The EPA risk assessment was based on follow-up of the cohorts through 1966. Moreover, the EPA had exposure information only on broad cumulative exposure categories from unpublished tables compiled by Lamd. Since the EPA risk assessment, follow-up of the cohorts has been completed through 1982. There are now four times as many lung cancers in the cohort as were present in 1966. Additionally, detailed exposure histories, reconstructed from job descriptions, are now available for all workers in the cohort.

This updated cohort was used to conduct another risk assessment in 1998 (34). The two-stage clonal expansion model was used to analyze the data, and the detailed exposure histories for each worker in the cohort were explicitly considered in the analyses. This ability to explicitly use detailed patterns of exposure is one of the strengths of analyses based on multistage models. A deficiency of the coke oven epidemiological data is the absence of any information on cigarette smoking, which is potentially a strong confounder of the association between exposure to emissions and lung cancer. An indirect adjustment for smoking could be made. It is well known that individuals born at around the same time share many lifestyle factors, including smoking. Therefore, an indirect way to adjust, at least partially, for smoking is to introduce parameters for birth cohort effects in the models used for analyses of the data. Indeed, birth cohort effects turned out to be highly significant in these data. With the two-stage model applied to the greatly expanded data, the results of analyses indicated that, as expected, emissions from coke oven batteries had significant effects on both the rate of initiation and that of promotion. The exposure-response curve was nonlinear at the higher exposure levels but was close to linear at low levels of exposure. The estimated unit risk, as defined above, was 1.5×10^{-4} with 95% confidence interval, derived using Markov Chain Monte Carlo (MCMC) methods, equal to $1.2 \times 10^{-4} - 1.8 \times 10^{-4}$. The interested reader is referred to the recent paper (34) for details.

Because the risk assessment for coke oven emissions was based on epidemiological data, the difficult problem of interspecies extrapolation was completely avoided. There are some concerns, however, about the risk assessment. First, as mentioned above, no smoking information was available, and a potentially important confounder could not be adequately adjusted in the analyses. Second, although the exposure information was much better than in the original EPA risk assessment, it was still crude compared with the precision with which exposure can be measured in experimental studies. Indeed, it is likely that there were substantial exposure measurement errors. How and in which direction these measurement errors biased the estimates of risk is impossible to say. Finally, the unit risk was estimated in a cohort of steel workers. To what extent the estimated risk can be extrapolated to other populations is an open question. Certainly, other factors influencing the risk of lung cancer, such as level of smoking, could modify the effect of coke oven emissions. The next example illustrates that the risk associated with an agent of interest will depend on background cancer rates in a population, which are determined by exposure to other factors.

b. Occupational Exposure to Refractory Ceramic Fibers and Lung Cancer. Refractory Ceramic Fibers (RCF) are man-made mineral fibers, typically made of alumina, silica, and other metal oxides. Because of their ability to withstand high temperatures, RCF are used as insulation materials, primarily for lining furnaces and kilns. Because of the known carcinogenicity of asbestos, there is concern that the man-made fibers may present a risk of lung cancer and mesothelioma, particularly in the occupational setting. In 1988, IARC evaluated the then available information on RCF and classified these fibers as possible human carcinogens (35).

I summarize here a recent risk assessment (36) for RCF based on two long-term oncogenicity studies in male Fischer rats conducted to assess the potential pathogenic effects associated with prolonged inhalation of RCF. Because this risk assessment was based on toxicological data, both low-dose and interspecies extrapolations had to be addressed. Serial sacrifices were conducted during the study, and the fiber burden in the lungs of sacrificed animals was determined by direct counting and reported as number of fibers per milligram of dry lung tissue. With this information, it was possible to determine the temporal pattern of fiber burden in the lung for each animal in the study. This temporal pattern was used explicitly in the two-stage clonal expansion model for analyses of the data. The analogy with the coke oven risk assessment, where detailed patterns of exposure were available for each worker in the cohort, is obvious. The data were consistent with the hypothesis that RCF are initiators of carcinogenesis in the rat lung. A promoting effect could not be ruled out; however, the data set was too small to consider initiation and promotion simultaneously. The likelihood for the initiation model was higher than that for the promotion model. The risk assessment was, therefore, based on treating RCF as an initiator of lung carcinogenesis.

The dose-response relationship appeared to be nonlinear, with an exponential model for dose-related initiation describing the data better than a linear model. The risk of occupational exposure can then be estimated in the following steps. This risk assessment is similar to a recent risk assessment for formaldehyde conducted by the Chemical Industry Institute of Toxicology (CIIT), which is under review by the EPA. A schema for the risk assessment is shown in Fig. 2.

1. Estimate the ratio $I(d)/I(0)$ from analyses of the lung burden data in rats, where $I(d)$ is the rate of initiation at dose (fiber burden) d and $I(0)$ is the rate of background (spontaneous) initiation.

2. Use the two-stage clonal expansion model to fit age-specific lung cancer incidence or mortality in the human population to which the extrapolation is to be made and estimate the parameters of the model, including $I(0)$ for that population. Two populations were used for our extrapolation. First, an occupational cohort of steel workers (not exposed to coke oven emissions) was selected because the requisite data on this population was available (this was a subcohort of the cohort used for the coke oven risk assessment) and because this occupational cohort could be expected to be similar in its smoking and other habits to a cohort of RCF1 workers. A complicating factor is that the workers, although not exposed to coke oven emissions, may have been exposed to other initiators in the workplace. This would tend to increase the estimates of risk based on this cohort. The second population considered was the population of nonsmokers in the American Cancer Society cohort. This population could be considered to be representative of a nonsmoking middle class population in the United States.

3. Use the estimated ratio $I(d)/I(0)$ from the lung burden studies and the estimated $I(0)$ from the epidemiological data to estimate $I(d)$ in the human population by making the assumption that the ratio is the same in humans and rats. This is the fundamental assumption in this risk assessment. The same assumption was made in the recent CIIT risk assessment of formaldehyde.

4. Use a deposition clearance model describing the accumulation of RCF1 in the human lung to generate a lung burden profile (fibers per mg of dry tissue) for exposure scenarios of interest. Various exposure scenarios were considered. Specifically, the EPA default for exposure assumes that human exposure in the occupational setting occurs for 8 hours/day, 5 days/week, 52 weeks/year, starting at age 20 and continuing constant until age 50.

Biologically Based Models

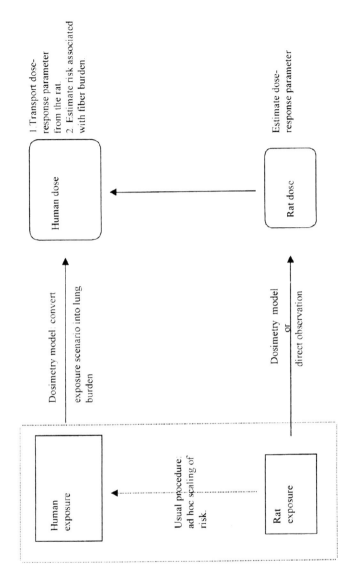

FIG. 2. Schema for transporting risk estimated from animal experiments to human populations. The usual procedure on the left is an empirical scaling procedure. The approach described in the text estimates the dose-dependent rate of initiation in rats and transports it to the human population of interest.

5. Use this lung burden profile together with the background estimated parameters in the human population and the estimated I(d) in that population to generate incidence functions for lung cancer for various lung burden profiles. Quantities of interest in risk assessment, such as the excess probability of tumor at any age, can now be easily estimated.

6. Use MCMC methods to generate distributions (and confidence intervals) for the quantities of interest.

For the EPA exposure scenario, the excess probability at age 70 associated with exposure to 1 fiber/cc in the ACS cohort was estimated to be 3.7×10^{-5}, 95% confidence interval = $2.35 \times 10^{-5} - 5.0 \times 10^{-5}$. For the steel workers cohort, the risk (excess probability) was estimated to be 1.5×10^{-4}, 95% confidence interval = $9 \times 10^{-5} - 2.0 \times 10^{-4}$. This example clearly suggests that risk assessments should, to the extent possible, be population specific. A risk assessment in one population may not be applicable to another. For details, the reader is referred to a forthcoming publication (36).

III. CONCLUDING REMARKS

In this chapter, I have given a brief survey of multistage models of carcinogenesis, with emphasis on their use for quantitative cancer risk assessment. Improvements in risk assessment will depend on the detail and quality of information available from epidemiological and experimental studies, on better understanding of the biological principles underlying carcinogenesis, and on the incorporation of new biological information into cancer models. A good example of a risk assessment based on detailed biological information is provided by a new risk assessment of formaldehyde performed at the CIIT and currently under review at EPA. This risk assessment also uses the two-stage clonal expansion model as the basic pharmacodynamic tool. The two risk assessments considered in this chapter make it clear that quantitative assessments of risk should not be reported as single numbers. Clearly, risk will depend on a number of host factors, such as genetic susceptibility, and modifying factors, such as age at exposure and exposure to other agents. It is also clear that the uncertainty associated with quantitative assessment of risk is much larger than suggested by the confidence intervals derived from statistical procedures. Characterization of this uncertainty is one of the biggest challenges facing risk assessment today.

It is unlikely that in the foreseeable future quantitative cancer risk assessment will be based on a complete understanding of the biological processes underlying carcinogenesis. Use of models based on current understanding of carcinogenesis does, however, identify data gaps that need to be filled. The use of such models can also suggest plausible explanations for observed exposure-response relationships. For example, Lutz and Kopp-Schneider (37) have recently proposed a mechanism for the J-shaped exposure response relationships reported for some carcinogens. Unfortunately, a biologically based approach to quantitative risk assessment requires more data than are usually available for agents of regulatory interest. Thus, a tiered approach to risk assessment, which would use default procedures when the data are sparse but use as much information as possible when data are available, would appear to be the best option. Notwithstanding the fact that such a flexible approach cannot easily be codified into a set of rules, there are encouraging signs that regulatory agencies in the United States are willing to consider it.

REFERENCES

1. H Vainio, PN Magee, DB McGregor, AJ McMichael. Mechanisms of Carcinogenesis in Risk Identification. IARC Scientific Publications 116, 1992.
2. WC Hahn, CM Counter, AS Lundberg, RL Beijersbergen, MW Brooks, RA Weinberg. Creation of human tumour cells with defined genetic elements. Nature 400:464–468, 1999.
3. SM Cohen, LB Ellwein. Cell proliferation in carcinogenesis. Science 249:1007–1011, 1990.
4. P Bannasch. Preneoplastic lesions as end points in carcinogenecity testing. 1. Hepatic preneoplasia. Carcinogenesis 7:849–852, 1986.
5. P Bannasch. Preneoplastic lesions as end points in carcinogenicity testing. II. Preneoplasia in various nonhepatic tissues. Carcinogenesis 7:849–852, 1986.
6. CO Nordling. A new theory of the cancer inducing mechanism. Br J Cancer 7:68–72, 1953.
7. P Armitage, R Doll. The age distribution of cancer and multistage theory of carcinogenesis. Br J Cancer 8:1–2, 1954.
8. P Armitage, R Doll. A two-stage theory of carcinogenesis in relation to the age distribution of human cancer. Br J Cancer 11:161–169, 1957.
9. AG Knudson. Mutation and cancer: Statistical study of retinoblastoma. Proc Natl Acad Sci 68:820–823, 1971.
10. SH Moolgavkar, DJ Venzon. Two-event models for carcinogenesis: Incidence curves for childhood and adult tumors. Math Biosci 47:55–77, 1979.
11. SH Moolgavkar, AG Knudson. Mutation and cancer: A model for human carcinogenesis. J Natl Cancer Inst 66:1037–1052, 1981.
12. DG Kendall. Birth-and-death processes and the theory of carcinogenesis. Biometrika 47:316–330, 1960.
13. J Neyman, E Scott. Statistical aspects of the problem of carcinogenesis. In: Fifth Berkeley Symposium on Mathematical Statistics and Probability. University of California Press: Berkeley, California, 1967, pp 55–77.
14. SH Moolgavkar, EG Luebeck, D Krewski, JM Zielinski. Radon, cigarette smoke and lung cancer: A reanalysis of the Colorado Plateau uranium miners' data. Epidemiology 4:204–217, 1993.

15. M Kai, EG Luebeck, SH Moolgavkar. Analysis of solid cancer incidence among atomic bomb survivors using a two-stage model of carcinogenesis. Radiat Res 148:348–358, 1997.
16. EG Luebeck, SB Curtis, FT Cross, SH Moolgavkar. Two-stage model of radon-induced malignant lung tumors in rats: Effects of cell killing. Radiat Res 145:163–173, 1996.
17. EG Luebeck, SH Moolgavkar, A Buchmann, M Schwarz. Effects of polychlorinated biphenyls in rat liver: Quantitative analysis of enzyme altered foci. Toxicol Appl Pharmacol 111:469–484, 1991.
18. A Dewanji, DJ Venzon, SH Moolgavkar. A stochastic two-stage model for cancer risk assessment II. The number and size of premalignant clones. Risk Anal 9:179–187, 1989.
19. SH Moolgavkar, G Luebeck, M de Gunst, RE Port, M Schwarz. Quantitative analysis of enzyme altered foci in rat hepatocarcinogenesis experiments. Carinogenesis 11:1271–1278, 1990.
20. A Kopp-Schneider, CA Portier. Birth and death differentiation rates of papillomas in mouse skin. Carcinogenesis 13:973–978, 1992.
21. EG Luebeck, SH Moolgavkar. Simulating the process of carcinogenesis. Math Biosci 123:127–146, 1994.
22. GC Yang, CW Chen. A stochastic two-stage carcinogenesis model: A new approach to computing the probability of observing tumor in animal bioassays. Math Biosci 104:247-258, 1991.
23. S Moolgavkar, D Krewski, L Zeise, E Cardis, H. Moller, eds. Quantitative Estimation and Prediction of Human Cancer Risk. IARC Scientific Publications No. 131. Lyon: International Agency for Research on Cancer, 1999.
24. KJ Rothman, S Greenland. Modern Epidemiology. 2nd ed. New York: Lippincott Raven, 1997.
25. JD Kalbfleisch, RL Prentice. The Statistical Analysis of Failure Time Data. New York: Wiley, 1980.
26. DR Cox, D Oakes. Analysis of Survival Data (Monographs on Statistics and Applied Probability 21). New York: Chapman and Hall, 1984.
27. SH Moolgavkar, EG Luebeck. Two-event model for carcinogenesis: Biological, mathematical and statistical considerations. Risk Anal 10:323–341, 1990.
28. MP Little. Generalizations of the two-mutation and classical multistage models of carcinogenesis fitted to the Japanese atomic bomb survivor data. J Radiat Prot 16:7–24, 1996.
29. WY Heidenreich, EG Luebeck, SH Moolgavkar. Some properties of the hazard function of the two-mutation clonal expansion model. Risk Anal 17:391–399, 1997.
30. A Dewanji, SH Moolgavkar, EG Luebeck. Two-mutation model for carcinogenesis: Joint analysis of premalignant and malignant lesions. Math Biosci 104:97–109, 1991.
31. SH Moolgavkar, EG Luebeck. Multistage carcinogenesis: Population-based model for colon cancer. J Natl Cancer Inst 84:610–618, 1992.
32. USEPA, Carcinogen Assessment of Coke Oven Emissions. Washington, DC: United States Environmental Protection Agency. Office of Health and Environmental Assessment, 1984.

Biologically Based Models

33. IARC Monographs on the Evaluation of the Carcinogenic Risk of Chemicals to Humans, vol 34. Polynuclear Aromatic Compounds, Part 3, Industrial Exposures in Aluminium Production, Coal Gasification, Coke Production, and Iron and Steel Founding. Lyon, France: IARC, 1984.
34. SH Moolgavkar, EG Luebeck, EL Anderson. Estimation of unit risk for coke oven emissions. Risk Anal 18:813–825, 1998.
35. IARC Monographs on the Evaluation of the Carcinogenic Risk of Chemicals to Humans, vol 1. Man-made Mineral Fibers and Radon. Lyon, France: IARC, 1988.
36. SH Moolgavkar, EG Luebeck, J Turim, L Hanna. Quantitative assessment of the risk of lung cancer associated with occupational exposure to refractory ceramic fibers. Risk Anal 19:599–611, 1999.
37. WK Lutz, A Kopp-Schneider. Threshold dose response for tumor induction by genotoxic carcinogens modeled via cell-cycle delay. Toxicol Sci 49:110–115, 1999.

Index

2-Acetylaminofluorene (2-AAF), 79, 321–323
Acetylators, rapid and slow 276, 282–283
Acetyltransferases, 276
Acridine orange, 102, 173
Acrylates, 237
Adenomatous polyposis coli (APC), 2
Adenoviruses, 16
   E1A gene, 16
   E1B genes, 16
Administered dose, PBPK, 334
Aflatoxins, 79, 204, 272, 274–275, 284, 323, 328–330
   B_1, 79, 204, 272, 274–275
   cancer risk assessment, 323, 328–330
   DNA adduct, 284
   human hepatocellular carcinomas, 274–275, 328–330
   M_1, 272
Aflatoxin B_1-N^7-guanine DNA adduct, 328–330

Age effect, carcinogenesis 330–331, 358
Alar, 334
Aldehydes, 74
Alkaline elution (AE), 236
Alkylhalide, 278
Alkyl phosphotriesters, 250
Alkyl sulfates, 51
α, β-Unsaturated carbonyls, 74
α, β-Unsaturated carboxylates, 74
$\alpha 2\mu$-Globulin, 54, 302–305, 314
Alveolar rhabdomyosarcomas (A-RMS), 4
   Pax 3 gene, 4
   FKHR gene, 4
Ames test, (see *Salmonella typhimurium* mutation test)
2-Aminobiphenyl (2-ABP), 79
4-Aminobiphenyl (4-ABP), 79, 276, 338
   hemoglobin adduct, 281–282
Aneuploidy, human, 2
   induction, 100–101, 250

Angiotensin II receptor antagonists, 198
Angiotensin-converting enzyme (ACE) inhibitors, 198
Animal bioassay, tumor incidence, 247
Anticarcinogenic effect, 264
Antihistamines, 176
Apoptosis, 13, 361
Apurinic site, 52
Apyrimidinic site, 52
Armitage and Doll multistage model, 325, 332, 354, 359–360
Aromatic amine, 283
Arylamines, 276, 280
Aryl hydrocarbon (Ah) receptor, 56, 307–308
Aryl receptor nuclear translocator (ARNT), 307
Asbestos, 52
Ataxia telangiectasia (AT), 6
5-Aza 2′-deoxycytidine, 53, 57
Aziridinium ions, 74
Azo dyes, 86, 122

Background carcinogen exposure, 341–343
BALB/3T3 cell transformation assay, 209
Barbiturates, 211
Bayesian techniques, 337
BCR gene, 4, 11
Benchmark method, risk analysis 300–301, 314, 325
Benzene, 273, 323
Benzidine, 86
Benzo[a]pyrene, 79, 333
Benzo[a]pyrene-7,8-diol-9,10 epoxide (BPDE), 279
  DNA adduct, 280
Bioavailability, 117

Biologically based cancer risk assessment model (*see* two-stage clonal expansion model)
Biomarkers, 52, 269–286
  application, 280–285
  complex mixture exposure, 280–283
  effect (cytogenetics and mutations), 273–275
  environmental carcinogen exposure, 284–285
  exposure (protein and DNA adducts), 272–273
  occupational carcinogen exposure, 283–284
  susceptibility (metabolism and DNA repair), 276–279
Biophore, SAR, 83
Bisphenol A, 55
Bleomycin, 175
Bloom's syndrome (BS), 6
1,3-Butadiene, 278
Butylated hydroxyanisole, 54

C3H10T1/2 cell transformation assay, 209
Cabonium ions (alkyl-, aryl-, benzylic), 74
Caffeine, metabolism 277
California Proposition 65, 326
Cancer, low dose linearity, 321
  common cancer, 321
  rare cancer, 321
Cancer potency, 40, 60, 321, 325
  from maximum likelihood technique, 325
Cancer risk assessment models, 60–61, 85, 259, 266, 319–343, 359–360
  factors, 330–343
  Armitage and Doll multistage model, 325, 359–360

Index

[Cancer risk assessment models]
  multistage polynomial, 259 323–326, 359
  two-stage clonal expansion model, 266, 325, 354, 360–366
  role of structure-activity relationship (SAR), 85
Carcinogenesis, 47–57, 62–63, 250–255, 266, 297–314, 320, 353–354, 360
  additive effect, 266
  comparison of genotoxic and nongenotoxic mechanisms, 62
  genotoxic mechanisms 48–52, 63, 320
  initiation and promotion, 354, 360
  nongenotoxic (epigenetic) mechanisms, 53–57, 254, 297–314
  mechanisms, genotoxic and nongenotoxic 47–57, 59
  mode of action, 255
  modulation of, 250–254
  probability 353–354
  susceptibility factors, 254
Catalase, 259
Cell cycle, 8–11,
  Checkpoints, 8–9
  DNA synthesis (S phase), 8, 58
  G0 phase, 8
  G1 phase, 8
  G2 phase, 8
  inhibitors, 9–11, 58
    cyclin-dependent kinase inhibitors, 9–11
    INK4 (inhibitors of CDK4), 10, 58
  mitosis (M phase), 8
  restriction point (R point), 8
  transition point, 8

Cell Proliferation, in carcinogenesis, 53–56, 258, 298, 302, 308–312, 314, 326
  clonal expansion, 53–56
  regenerative cell proliferation, 54, 258, 298, 302, 308–312, 314, 326
  growth factor induced proliferation, 54
Centromere specific DNA probes, micronuclei 168
Chemical Right-to-Know (RtK), USEPA High Production Volume (HPV) chemicals 244
Chinese hamster lung cells, (CHL, V79), 92, 100, 240
Chinese hamster ovary cells (CHO/HGPRT) mutagenicity test, 92, 98–100
  OECD guidelines, 233–234, 240
Chinese hamster ovary cell AS52 (CHOAS52/XPRT) mutagenicity test, 92, 98–100, 235, 240
Chloramphenicol acetyltransferase (CAT), 278
Chlordane, 306
Chloroethylene oxide, 51
Chloroform, 55, 314
  cancer risk assessment, 308–311, 326
Chlorothalonil, 54
Chromosome aberrations, 2–3, 11–13, 32, 100, 120, 139, 163, 166, 240, 257, 273–285
  aneuploidy, 3, 100
  biomarkers, 273–285
  break, 100
  deletion, 3, 240
  endoreduplication, 100, 163

[Chromosome aberrations]
  gap, 100
  gene conversion, 240
  homologous mitotic recombination, 240, 257
  in genetic toxicology test, 32
  inversion, 11, 13
  mitotic nondisjunction, 139, 240
  Philadelphia chromosome, 3
  polyploidy, 100
  premature chromosome condensation, 166
  pulverization, 120
  translocation, 2, 3, 11, 12, 13, 100, 240
Chromosome aberration tests (*see* genetic toxicology tests)
Chronic myelogenous leukemia, 4
Ciprofibrate, 306
Clastogens, 100–101
Clean Air Act (USEPA), 86
Clear cell adenocarcinomas, 332
Clofibrate, 306
Coke oven emissions, 362–363
Colorectal adenocarcinomas, 2, 8, 278
  Bax gene, 6
  deleted-in-colon carcinoma gene (DCC), 3
  hereditary nonpolyposis color cancer (HNPCC), 6
Cockayne's syndrome (CS), 6
Comet assay, 127–128, 195, 202
  And micronucleus, 177
  double DNA strand break, 127–128
  single DNA strand break, 127–128
Comparative genomic hybridization (CGH), 4
Computer-assisted structure-activity relationship (SAR), 82–85
Connexin, 57

CREST (calcinosis, Raynaud phenomenon, esophageal dismotility, sclerodactyly and telangiectasis) antikinetochore antibody, 168
Cross-species cancer risk extrapolation, 339–341
  animal size, 341
  average daily dose, 399
  concordance, 339
  cumulative dose, 399
  pharmacokinetics, 340
  pharmacodynamics, 341
  surface area, 339
Cyclin/CDK complexes, 9–10
  cyclin D/CDK4, 11
  cyclin D/CDK6, 11
  dephosphorylation, 10
  phosphorylation, 10
Cyclin-dependent kinases (CDK), 9, 10
  CDK4, 10
  CDK6, 10
Cyclin-dependent kinase inhibitors, 9–11 (*see also* cell cycle inhibitors)
  $p21^{cip1}$, 9, 10
  $p27^{kip1}$, 9, 10
  $p57^{kip2}$, 9, 10
Cyclophosphamide (CP), 143
Cytochalasin B, 164, 170
Cytochrome P450 enzymes, 83, 276–279
Cytosine arabinoside, 175
Cytotoxicity, 257, 298

DDE (1,1-dichloro-2,2-bis(p-chlorophenyl)-ethylene, 272
DDT (2,2-bis[p-chlorophenyl]-1,1,1-trichloroethane), 272
Delaney clause, 250
De minimis risk, 300

Depurination, 250
6'-Diamidino-2-phenylindole (DAPI), 173
Dichloromethyl radicals, 326
1,2,3,4-Diepoxybutane (DEB), 176
Di(2-ethylhexyl)phthalate (DEHP), 306
Di(2-ethylhexyl)adipate (DEHA), 306
Diethylstilbestrol, 55, 332
Diethylnitrosamine, 321–323
7,12-Dimethylbenz(a)anthracene, 331
Dimethylhydrazine, 334
Dimethylnitrosamine, 202
2,4-Dinitrotoluene, 202
1,4-Dioxane, 176
Dioxin (see 2,3,7,8-Tetrachloro-dibenzo-p-dioxin)
DNA, hypomethylation, 2
DNA adducts, 51–52, 124–127, 250, 272–273, 202, 280, 282, 283–285, 328–330,
  alkylated DNA adduct, 51, 250, 272
    AFB_1-N^7-guanine adduct, 284, 328–330
    7-alkyl-2-deoxyguanosine adducts, 282
    BPDE-DNA adduct, 280
    I-compounds, 52, 126
    7-ethyl-deoxyguanosine adduct, 282
    7-methyl-deoxyguanosine adduct, 282
    N^3-alkyladenine DNA adduct, 51
    N^7-alkylguanine DNA adduct, 51, 282
    N-hydroxy-MOCA adduct, 283
    N^2-3-ethenoguanine DNA adduct, 51

[DNA adducts]
  [alkylated DNA adduct]
    O^6-alkylguanine DNA adduct, 51
    O^4-alkylthymidine DNA adduct, 51
    8-oxo-2-deoxyguanosine, 273
    PAH-DNA adduct, 280, 283
  biomarkers, 272–285
  detection, 125
    accelerator mass spectrometry, 127
    immunological (ELISA and RIA) method, 125–126
    ^{32}P-postlabeling method, 125–126, 202, 280
    spectroscopic (HPLC, GC) method, 127
  dosimeter, 125
  in genetic toxicology testing, 124
  macromolecular (bulky) DNA adduct, 51, 52
DNA hybridization, in situ, 101 (see also Fluorescence in situ hybridization (FISH))
DNA polymerase, 6
DNA repair, 6–8, 250
  and age, 279
  genes, uvrB, 95
  mismatch repair, 6
  nucleotide excision repair (NER), 6
DNA repair capacity (DRC), 278
DNA synthesis, 20, 58
  micronucleus, 166
  telomerase, 20, 58
  telomeres, 20
DNA tumor viruses, 16–18
  Simian virus 40 (SV40), 16
    large T antigen, 16
    small T antigen, 16

[DNA tumor viruses]
  human adenoviruses, 16
    E1A gene, 16
    E1B gene, 16
  Human papilloma virus (HPV), 18
    E6 gene, 18
    E7 gene, 18
Dose-response relationships, 60, 63, 255–259, 263–264
  dose response curves, 60, 63
    J shape, 263–264
    linear, 60, 259
    sigmoid, 60
  dose and DNA damage, 255–257, 259
  DNA damage and genetic change, 255, 257–258
Double minutes, 11 (*see also* gene amplification)
Drosophila sex-linked recessive lethal assay, 235

ED10, effect dose 301, 325
Effective dose, PBPK, 334
Electrophiles, electrophilic species 74–76, 124
Embryonal rhabdomyosarcomas (E-RMS), 4
  WT2 gene, 4
  GOK gene, 4
  $p57^{kip2}$ gene, 4
  IG/FII gene, 4
Enzyme altered foci, rodent liver, 360
Enzyme-linked immunosorbent assay (ELISA), 125–126, 272, 285
Epidemiology, 85, 328–330, 355
  advantages, 328
  analysis of data, 328–330
  disadvantages, 355
Episulfonium ions, 74
Epoxide, 74

Epoxide hydrolase (EPHX), 285
Escherichia coli mutagenicity test, 92, 94–98, 233–234
  DNA repair gene, uvrA, 96–97
  OECD guidelines, 233–234
  plasmid, pKM101, 96–97
  test strains, genotypes, 96–97
    WP2uvrA, 96–97
    WP2uvrA (pKM101), 96–97
  tryptophan genes (trp^- genes), 96–97
Estrogen, 55
Ethidium bromide, 127
Ethylene dibromide, 333
Ethylene oxide, 273
Exposure assessment, 270, 272–273, 280–285, 362
  biomarkers of exposure (protein and DNA adducts), 272–273
  complex mixture exposure, 280–283, 362
  environmental carcinogen exposure, 284–285
  occupational carcinogen exposure, 283–284

Fanconi's anemia (FA), 6
Fatty acid β-oxidation enzyme, 56
FIFRA (Federal Insecticide, Fungicide and Rodenticide Act) 221, 239
Flow cytometry, micronucleus, 175
Fluorescence in situ hybridization (FISH), 4, 51, 273
  fluorochrome, 4
  fluorescein isothiocyanate (FITC), 4
  in micronucleus assay, 101
  multifluor in situ hybridization (M-FISH), 6

Index

Formaldehyde, 248, 255, 262, 274, 366
Forestomach tumor, rodent 54
Free radical, DNA damage 52, 74, 250, 259, 326

Gap junctions, 57
Gas chromatography/mass spectrometry (GC/MS), 127, 272
Gastric adenocarcinomas, 278
Gene amplification, 2, 11, 13
   double minutes, 11
   homogeneously staining regions (HSR), 11
Gene expression, 53
   hypermethylation, 53
   hypomethylation, 53
   methylation, 53
Genetic polymorphism, 254, 271, 277–278
   P450 genes, 277–278
Genetic toxicology, germ cell vs. somatic cell, 29
Genetic toxicology, in quantitative risk assessment, 298
Genetic toxicology, test design, 115–118, 186
   dose levels, 116
      MTD (maximum tolerated dose), 116
   duration of exposure, 118
   ICH guidelines, S2A, S2B, 186
   number of animals, 115
   selection of target tissue for analysis, 118
   route of administration, 117
   sex, 115
   species, 115
Genetic toxicology tests, complementary, 113–132
   alkaline elution (AE), 236
   Comet assay, 127–128

[Genetic toxicology tests, complementary]
   DNA adducts, 124–127
   Drosophila sex-linked recessive lethal assay, 235
   sister chromatid exchanges, 236
   transgenic animal mutations assays, 128–131
Genetic toxicology tests, deficiencies, 190
Genetic toxicology tests, regulatory 32–34, 91–104
   bacterial mutagenicity tests, 30, 31, 32–41, 92, 94–98, 188, 233, 237, 240, 242–243
   Escherichia coli mutagenicity test, 92, 94–98
   reasoning, ICH 190–191
   Salmonella typhimurium mutagenicity test (Ames test), 30, 31, 32–41, 92, 94–98
   germ cell mutagenicity tests, 93, 114, 123, 241
   alkaline elution, 93
   mouse dominant lethal, 93, 114, 123, 236
   mouse heritable translocation, 93, 132, 236, 244
   mouse specific locus, 93, 131–132, 236, 244
   rat testicular tissue (spermatogonial) chromosome aberration, 93, 120
   rat testicular tissue sister chromatid exchanges, 93
   rat testicular unscheduled DNA synthesis (UDS), 93
   history, 30
   ICH guidelines, S2B 188
   in vitro chromosome aberration test (ABS), 33, 34–41, 92, 100–102, 188, 235, 244

[Genetic toxicology tests, regulatory]
  [in vitro chromosome aberration test]
    Chinese hamster lung cell (CHL), 92
    Chinese hamster ovary cells (CHO), 92
    human peripheral blood lymphocyte (HPBL), 92, 100, 171
    reasoning, ICH 193–194
  in vitro tests, 29, 33–41
    rat unscheduled DNA synthesis test (UDS), 102, 121–122
  in vivo rodent chromosome aberration test, 33, 93, 100–102, 188, 233–234, 236, 240, 242–245
    rat bone marrow 93, 119
    reasoning, ICH 194–195
  in vivo rodent micronucleus test, 33–41, 100–102, 188, 233–234, 236–237, 240, 242–245
    mouse bone marrow erythrocytes, 33–41, 93, 101–102
    mouse peripheral blood erythrocytes, 93, 102
    reasoning, ICH 194–195
    rat bone marrow erythrocytes, 93
    rat peripheral blood erythrocytes, 93
  in vivo tests, 33–41, 93
    rat unscheduled DNA synthesis test (UDS), 93, 102–103, 122, 202, 236
  mammalian cell gene mutation test, 32, 92, 98–100, 233–235, 240

[Genetic toxicology tests, regulatory]
  [mammalian cell gene mutation test]
    Chinese hamster ovary cells (CHO/HGPRT), 92, 98–100, 240
    Chinese hamster ovary cell AS52 (CHOAS52/XPRT), 92, 98–100, 235, 240
    Chinese hamster lung cell (V79/HGPRT), 240
    mouse lymphoma L5178Y test (MLA), 32, 34–41, 92, 98–100, 235, 240, 242–243
    reasoning, ICH 192–193
  OECD guidelines, 221–245
  prediction of carcinogenicity, 34–41
Genomic imprinting, 57
  loss of imprinting (LOI), 57
Genotoxicity, definition 91
Germ cell mutagenicity tests, (see genetic toxicology tests)
Giemsa, 102, 119, 172
Glass fibers, 176
GLP (Good Laboratory Practice), 94, 222
Glucuronic acid, 276–279
Glutathione, 276–279
Glutathione S-transferases (GST), 277, 285
Glycophorin A (GPA), 274

Hazard identification, 73, 85, 87,104
Hepatitis B virus (HBV), 274–275
  and aflatoxin, 284–285
  surface agent (HbsAg), 284
Heptachlor, 305
Heteroploid conversion, 58, 100
HGPRT (hypoxanthine guanine phosphoribosyltransferase), 98–99

Index

High performance liquid chromatography (HPLC), 127, 272
Histidine biosynthesis genes, Ames test, 96
Hitedness, carcinogenesis, 14, 60
  one-hit theory, 60
  two-hit theory, 14
  multihit theory, 60
Homogeneously staining regions (HSR), 11, 51 (*see also* gene amplification)
Hormesis, 323
Hormones, 55
Human diploid fibroblasts, neoplastic transformation in vitro, 58
Human hepatocellular carcinomas (HCC), 274–275, 285
Human heterogeneity, 262, 270, 276, 279, 299, 337–339
Human peripheral blood lymphocyte (HPBL) chromosome aberration test, 92, 100, 171 (*also see* genetic toxicology tests)
Hyaline droplets, 54, 302–305
Hydrogen peroxide, 52, 259
8-Hydroxyguanine, 52
8-Hydroxydeoxyguanosine, 52
1-Hydroxyprene, 283
Hypermethylation, 53
Hypochlodrite, 52
Hypolipidemic drugs, 56, 306
Hypomethylation, 53
Hypoxanthine phosphoribosyltransferase (HGPRT), 92, 98–100, 240, 274

IARC (International Agency for Research on Cancer), human carcinogens 91

ICH (International Conference on Harmonisation) guidance on genotoxicity and carcinogenicity, 94, 185–212
  background, 187
  battery of genotoxicity tests, 188
  bacterial reverse mutation assay, reasoning, 190–191
  future, genotoxicity testing, 201–204
    hazard identification, 201
    site of contact, 202
    tumor-related genes, 203–204
  guidelines, development process, 186
  ICH, S2A and S2B 186
  in vitro chromosome aberration test, reasoning, 193–194
  in vivo test, rodent hematopoietic cells, reasoning, 194–195
  mouse lymphoma tk assay, reasoning, 192–193
  mutation signature (mutation fingerprint), 203
  other acceptable tests, 195
  result interpretation, 196–201
    cytotoxicity, 198
    threshold, 199–201
ICH (International Conference on Harmonisation) guidance on carcinogenicity, 204–212
  dose selection, 204–205
    maximum tolerated dose (MTD), 205
    rodent to human plasma AUC ratio, 205
    rodent to human surface area ratio, 205
  future, carcinogenicity, 210–212
    integration, genotoxicity and carcinogenicity, 210–211

[ICH (International Conference on Harmonisation) guidance on carcinogenicity]
[future, carcinogenicity]
testing of nongenotoxic compounds, 211–212
ICH, S1A, S1B, S1C, 204
in vitro models, 209–210
BALB/3T3 cell transformation assay, 209
C3H10T1/2 cell transformation assay, 209
Syrian hamster embryo cell transformation assay (SHE assay), 209
rodent models, 205–209
newborn mouse model, 209
traditional, 205–206
transgenic models, 206–208
p53$^{+/-}$ knockout mice, 206–208
v-Ha-ras transgenic mice (TG.AC mice), 206–208
XPA$^{-/-}$ deficient mice, 207–208
Human c-Ha-ras transgenic mice (TgHras2), 207–208
I-compounds, 52, 126
ILSI (International Life Science Institute), 208–209, 309–310
Image analysis, micronuclei, 175
Immunoglobulin gene, 12
regulatory elements, 11
Immunohistochemical staining assay (IHC), 125–126, 272
Imprinting, 57
loss of imprinting (LOI), 57
Incidence rate, cancer 356
Inflammation, 298
INK4 (Inhibitors of CDK4), 10 (*see also* cell cycle inhibitors)
p16^{INK4a}, 10, 58

[INK4 (Inhibitors of CDK4)]
p15^{INK4b}, 10, 58
p18^{INK4c}, 10, 58
p19^{ARF} (p19^{INK4}), 10
Insulin-like growth factor 2 gene (IGF2), 57
Integrated Risk Information System (IRIS), 326
Interactive factors, carcinogenesis 254
aflatoxin and liver caner, 271
aflatoxin and hepatitis B virus (HBV), 328–330
smoking and alcohol, 254
smoking and radon, 254
smoking and lung cancer, 271, 358
smoking and p53 expression in bladder cancer, 281
Intercellular communication, 57
Interspecies scaling factor, 311
Inverse exposure-rate effect, 343

Kinetochore, 101

LED10, 10% increase of cancer risk, 304
Lifetime probability (lifetime risk), 357
Li-Fraumeni syndrome (LFS), 18
d-Limonene, 54
Lindane, 305
Linearized multistage procedure, 60, 63, 298, 357, 359
Lipid peroxidation, 127,
Liquid chromatography/mass spectrometry (LC/MS), 272
Lithium, 55
Liver angiosarcomas, 331
Liver tumor, rodent 55, 305–306
LOH (loss of heterozygosity), 49, 275

Index

Lowest observed effect level
(LOEL), 300
Lung cancer, human, 362–366
   coke oven emissions, 362–363
   refractory ceramic fibers (RCF),
       363–366
Lupiditine, 55

Manifestation time, transgenic mice
       130
Markov Chain Monte Carlo
       (MCMC) method, 363
Maximally tolerated dose (MTD),
       ICH carcinogenicity, 205
Metabolic activation, phase I 34, 97,
       276–279
   Aroclor 1254-induced rat liver S9,
       97, 100, 142–143
   cecal microflora, 122
Metabolic detoxification, phase II
       276–279
Methylation, DNA 53, 57, 253
4,4-methylene-bis(2-chloroaniline)
       (MOCA), 283
4,4'-methylenebis(N,N-dimethyl-
       benzenamine), 315
Methylmethansulfonate (MMS),
       143
4-(methylnitrosamino)-1-(3-pyridyl)-
       1-butanone (NNK), 272,
       280
Methyl tertiary butyl ether (MTBE),
       303–305
Methyltransferase, 53
   methyltransferase deficient mice,
       57
Micronucleus test, 34–41, 93,
       101–102, 161–179,
       242–245 (*see also* genetic
       toxicology tests)
   biological consequence, 166

[Micronucleus test]
   biomarker, 178, 273–275
   CREST (calcinosis, Raynaud
       phenomenon, esophageal
       dismotility, sclerodactyly
       and telangiectasis) anti-
       kinetochore antibody,
       168
   detection of aneuploidy, 101
   DNA hybridization, in situ, 101
   kinetochore, 101
   formation of micronuclei,
       162–163
   in vivo rodent micronucleus test,
       33–41, 93, 101–102, 240
       mouse bone marrow erythro-
           cytes, 33–41, 93, 101–102,
           240
       mouse peripheral blood erythro-
           cytes, 93, 102, 234
           advantages, 194
           incorporation to general toxi-
               cology studies, 195
       PCE/NCE (and PCE/RBC)
           ratio, 102
       rat bone marrow erythrocytes,
           93
       rat peripheral blood erythro-
           cytes, 93
       reasoning, ICH 194–195
   in vitro micronucleus test,
       161–179, 195
       alternative methodologies,
           174–175
           automated analysis, 175
           bleomycin amplification
               assay, 175
           cytosine arabinoside assay,
               175
           mitotic shake-off method, 174
           microwell method, 174

[Micronucleus test]
  [in vitro micronucleus test]
    analysis, 173
      mononucleated cells, 173
      binucleated cells, 173
      multinucleated cells, 173
      proliferation index (PI) 174
    Big Blue rat 2 λ/lacI transgenic fibroblast, 176
    cytokinesis block, 164
    data interpretation, 176–179
    method, 168–173
    structural activity relationship, 176–177
  low volume microwell method, 167
  p53 in micronuclei, 167
Microsatellites, 6
Mitotic nondisjunction, 139
Molecular dosimetry, 271
Molecular epidemiology, 269–286
Monte Carlo simulation, 338
Mouse lymphoma test (MLA) mutagenicity assay, 32–41, 92, 98–100 (*see also* genetic toxicology tests)
  correlation with micronuclei, 177
  large colonies, 99, 147, 155
  L5178 TK$^{+/-}$, 98–99
  microwell method, 139–155
    calculations, 147–149
    criteria for data evaluation, 149–150
    experimental procedures, 143–147
    ICH guidelines, microwell, 193
    modified procedure, 154
    mutation frequency, 148
    relative suspension growth (RSG), 143
    relative total growth (RTG), 148
    statistics, 149

[Mouse lymphoma test (MLA) mutagenicity assay]
  mutation at p53 gene, 154
  OECD/FDA guidelines, 233–234, 237, 240, 242–243
  reasoning, ICH 192–193
  small colonies, 99, 147, 155
  percentage, 148
  soft agar procedure, 141
  THMG medium (thymidine, hypoxanthine, methotrexate, glycine), 142
  threshold, 178
Mouse peripheral blood erythrocytes micronucleus assay, 93, 102 (*see also* genetic toxicology tests)
Multifluor in situ hybridization (M-FISH), 6
Multihit carcinogenesis, (*see* multistage carcinogenesis)
Multistage carcinogenesis, 59, 249
  hits, 14
  one-hit theory, 60
  two-hit theory, 14
  multihit theory, 3, 60
Multistage polynomial, 323–326, 356–359
  assumptions, 356
  concepts, 356
Mutagenic potency, 40
Mutations, 2, 11, 13, 32, 51, 95, 273–275
  amplification, 2, 11, 13, 51
  biomarker, 273–275
  deletion, 2, 51
  in genetic toxicology test, 32
  insertion, 2, 13, 51
  inversion, 2, 11, 51
  point, 2, 51
  reverse mutation (reversion), 95
  translocation, 2, 11, 13, 51

Index

Mutation signature (mutation fingerprint), 203
Mutational "hits", 14

Nafenopin, 306
N^3-Alkyladenine DNA adduct, 51
N^7-Alkylguanine DNA adduct, 51
2-Naphthylamine, 276
Neoplastic transformation, 58
  in vitro human, 58
  in vitro rodent, 58
N^2-3-Ethenoguanine DNA adduct, 51
Nitrenium, 74
Nitroaromatics, 122
Nitrogen mustards, 51
Nitrogen oxide, 52
N-nitrosamines, 51. 280
N-nitrosoureas, 51
Nondisjunction, 101
Nonylphenol, 55
Nongenotoxic (epigenetic) mechanisms, 53–57, 63,
  Definition, 298
  cancer risk assessment, 297–314
No observed effect level (NOEL), 300
Normochromatic erythrocyte (NCE), 34, 101–102
  in micronucleus test, 34, 101–102

O^6-Alkylguanine DNA adduct, 51
O^6-Methylguanine DNA adduct, 51
O^4-Alkylthymidine DNA adduct, 51
OECD (Organisation for Economic Co-operation and Development), 94, 221–245
  background, 222, 229–230
  guidelines, 223–227
  USEPA/OPPTS (Office of Pollution Prevention and Toxics) guidelines, 228–239
    existing chemicals, 233–235

[OECD (Organisation for Economic Co-operation and Development)]
  [USEPA/OPPTS (Office of Pollution Prevention and Toxics) guidelines]
    new chemicals, 236–239
    testing scheme, 231–239
    three-tier system, 233, 235
    trigger carcinogenicity assays, 235
    trigger, second tier assays, 235–236
    trigger, third tier assay, 236
    TSCA Section 5, 236, 239
    Two-test battery, 237–239
  USEPA/OPP (Office of Pesticide Program) guidelines, 239–242
    Data Call-In (DCI), 241
    FIFRA Section 3, 239
    initial battery, 240
    confirmatory testing, 241
Oncogenes, 11–13
  activation, 11, 12
  c-abl, 4, 11
  deregulation, 13
  c-myc (cellular Myc), 11, 12
  l-myc, 13
  Max (Myc-associated factor X), 13
  n-myc, 13
  polymorphisms, 254
  protooncogenes, 11
  ras, 11, 203, 274
    mutation, 11
    expression, 280
  ras (Kirsten), 2
  v-mic, (viral-Myc), 12
Oxidative lesions, 250, 257, 273, 310
  (see also free radical)
Oxonium ions, 74

Oxygen, reactive, 52 (*see also* free radical)
   superoxide, 52, 259
   hydryoxyl radical, 52, 306
   peroxyl radical, 52
   alkoxyl radical, 52
   singlet oxygen, 52
Ozone, 52

p53 related proteins, 19
p53 tumor suppressor gene, 3, 18–20, 58, 167, 203, 274–285, 358
   biomarker, 274–285
   hotspot, 274–275
   in apoptosis, 18
   in cell cycle arrest, 18
   immortality, 58
   in Li-Fraumeni syndrome (LFS), 18
   in micronuclei, 167
   loss of heterozygosity, 275
   mutation, 18, 204
   nuclear overexpression in smokers, 281, 358
   regulation protein, mdm-2, 19
   viral (SV40) complex, 19
Papilloma virus (HPV), 18
   E6 gene, 18
   E7 gene, 18
Partition coefficient, 337
Perchlorate, 55
Performance of genetic toxicology tests 30, 31, 34–41, 74
   ABS and MLA combination 37
   ABS and MN combination 38
   concordance, 31, 34–41
   positive predictivity, 31, 34–41
   prevalence, 31, 34–41
   SAL, ABS and MN combination, 39

[Performance of genetic toxicology tests]
   SAL and ABS combination, 36
   SAL and MLA combination, 36
   SAL and MN combination, 38
   SAL and rodent carcinogenicity, 41
   sensitivity, 31, 34–41
   single assay, 35
   prediction, structural activity relationships (SAR), 74
Peroxidases, 259
Peroxides, 74
Peroxisome proliferation, 52, 56, 211, 306–307, 314, 306–307
   mechanism of action, 306–307
   receptors (peroxisome proliferator-activated receptors (PPAR)), 56,
     in humans, 56
     in rodents, 56
   receptor subtypes (α,β,γ), 56
Peroxynitrile, 52
Pharmacokinetics, 334–337
Phenacetin, 205
Phenobarbital, 313
Phenotypic expression, 99
Phenoxy acid herbicides, 56
Phenylalanine hydroxylase, 1
Phenylketonuria (PKU), 1
Philadelphia chromosome, 3, 11
Phosgene, 310
Phosphorylation, 10
Phthalates, 307
Phthalate esters, 56
Physiologically based pharmacokinetic (PBPK) model, 63, 300, 334
   Advantages, 335–336
Plasticiers, industrial, 306
Poisson distribution, 360

Index

Polychlorinated biphenyls, 55
Polychromatic erythrocyte (PCE), 33, 101–102
   in micronucleus test, 33, 101–102
Polycyclic aromatic hydrocarbon (PAH), 77–79, 280, 362
Polymerase chain reaction (PCR), 273
Potassium thiocyanate, 313
Predictive-Toxicology Evaluation Project (USNIEHS), 84
Premanufacturing notification (PMN, USEPA), 86, 236
Prolatin, 211
Proliferating cell nuclear antigen (PCNA), 10
   cyclin-CDK-PCNA, 10
Proliferative senescence, 20
Prostate cancer, 259
Protein adduct, 272–285
   hemoglobin (Hb) adduct, 272, 281–282
   albumin adduct , 272
   aflatoxin-albumin adduct, 285
Protooncogenes, 11
Public health goal (PHG), drinking water standard 303

Quantitative structure-activity relationships (QSAR), 82
Quinone, and quinoid intermediates 74, 177

Radioimmunoassay (RIA), 125–126, 272
Radiation, ionizing 250, 273–274
Radon, 323
Ras oncogene, 11, 203, 274 (*see also* oncogenes)
   mutation, 11
   expression, 280

Refractory ceramic fibers (RCF), 363–366
Regulatory agencies, 221–222, 242–243
   USEPA, 221–222, 242
   USFDA, 242–243
Regulatory testing guidelines, 93, 94, 114
   ICH (International Conference on Harmonization) guidelines, 94
   OECD (Organisation for Economic Co-operation and Development), 94
   OPP (Office of Pesticide Program, USEPA), tier I and tier II, 114
Renal tumor, male rats, 54, 302–305
Renal tubule cell tumor, 54
Resource Conservation and Recovery Act (RCRA, USEPA), 86
Retinoblastoma (Rb), 10, 14
   Knudson's two hit theory, 14–15
Retinoblastoma gene (Rb gene), 10, 14, 18
   In breast cancer, 14
   in mortality, 58
   In osteosarcomas, 14
Retrovirus, 11
Reverse mutation, revertant 95 (*see also Salmonella typhimurium* mutation test)
Rhabdomyosarcoma (RMS), 4
   Alveolar (A-RMS), 4
   embroynal (E-RMS), 4
   FKHR (fork head in rhabdomyosarcoma) gene, 4
Risk characterization, 60–61, 63 (*see also* cancer risk assessment models)
Rodent bioassay, concordance between rat and mouse 40

Rodent fibroblasts, neoplastic transformation in vitro 58

S9 metabolic activation (homogenate of liver fraction), 34
Saccharin, 54
Salmonella typhimurium mutation test, 30–41, 92, 94–98, 305
  deficiencies, 190
  DNA repair gene, uvrB, 95–96
  histidine genes (his⁻ genes), 95–96
  methods, 97
    plate incorporation assay, 97
    preincubatoin assay, 97
  OECD/FDA guidelines, 233–234, 237, 240, 242–243
  permeability gene, rfa, 95–96
  plasmid, pKM101, 95–96
  plasmid, pAQ1, 95–96
  prediction of carcinogenicity, 31, 34–41
  reasoning, ICH 190–191
  test strains, genotypes, 95–96
    TA97, 95–96
    TA97a, 95–96
    TA98, 95–96
    TA100, 95–96
    TA102, 95–96
    TA1535, 95–96
    TA1537, 95–96
Senescence, in vitro 58
Simian virus 40 (SV40), 16
Sister chromatid exchanges, 236
  biomarker, 273–275
Skin basal cell carcinomas (BCC), 279
Skin painting experiment, 360
Sodium ascorbate, 54
Somatic cell mutation theory of carcinogenesis, 48, 91

Spectral karyotyping (SKY), 6
Structural alerts, 76, 94
Structure-activity relationship (SAR), 73, 77–78, 235
  Chemical structure of genotoxic carcinogens, 78
  correlation with micronuclei, 176
  criteria structural for mutagens and carcinogens, 77, 80–81
  mechanism based SAR, 74
  molecular flexibility, 80
  molecular polyfunctionality, 80
  molecular size and shape, 79
  resonance stabilization, 81
  substituent effects, 79
  spacing/distance between reactive groups, 80
Structure-activity relationship (SAR), computer programs, 82–85, 222, 241
  biophore, 83
  COMPACT (Computer-Optimized Molecular Parametric Analysis of Chemical Toxicity), 83
  DEREK (Deductive Estimation of Risk from Existing Knowledge), 84
  Ke (electrophilicity) carcinogenicity model, 83
  MULTICASE/CASE (computer automated structure evaluator), 82–83
  OncoLogic Cancer Expert System, 84
  quantitative structure-activity relationships (QSAR), 82
  TOKAT (Toxicity Prediction by Komputer Assisted Technology, 82–83
  toxicophore, 84

Index

Styrene, 273, 335
Superoxide dismutase, 259
Susceptibility, 262, 270, 276, 279, 299
Syrian hamster embryo (SHE) cells, 177, 209
  micronuclei 177
  cell transformation, 209

TD25, dose causing 25% excess cancer risk, 340
TD50, dose causing 50% excess cancer risk, 340
Telomerase, 20–21
  in DNA synthesis, 20
  in immortality, 58
Telomeres, 20–21
2,3,7,8-Tetrachlorodibenzo-p-dioxin (TCDD), 56, 79, 248, 264–265, 307–308
Tetrachloroethylene (perchloroethylene), 54, 335, 339
6-Thioguanine (6-TG), 98
Thioureas, 55, 313
Threshold, 60–61, 199–201, 254–259, 302, 320–323
  carcinogenic 60, 63, 263–264, 297
    low dose linearity, 320
  genotoxicity, 199–201, 259
  micronuclei induction, 178
  tumor promoting agents, 258
Thymidine kinase (tk), 32, 98–99, 139
Thyroid-pituitary disruption, 312
Thyroid stimulating hormone (TSH), 55, 312–313
Thyroid tumor, 312–313, 331–332
  rodent, 312–313
  human, 331–332
Thyroxine, 66
Time-to tumor, 259–262

Tobacco smoke, 273, 280 (*see also* interactive factors, carcinogenesis)
  Environmental tobacco smoke (ETS), 282
Toxicophore, 84
Toxic Substance Control Act (TSCA, USEPA), significant new use rule (SNUR), 86
Transcription factor, E2F 11
Transgenic models, mutagenicity 128–131, 202
  Big blue, mouse128–131
    β-galactosidase, 129
    cII, 129
    lac, 129
    lac repressor genes, 129
    λ phage shuttle vector (λLIZ), 129
    X-gal, 129
  Big Blue Rat 2 λ/lacI transgenic fibroblast, and micronuclei, 176
  MutaMouse, 130–131
    LacZ, 130
    phenylgalactose (P-gal) 130
  manifestation time, 130
  germ cell mutation, 131
Transgenic models, carcinogenicity 212, 206–208
  $p53^{+/-}$ knockout mice, 206–208
  v-Ha-ras transgenic mice (TG.AC mice), 206–208
  $XPA^{-/-}$ deficient mice, 207–208
  Human c-Ha-ras transgenic mice (TgHras2), 207–208
2,4,6-Trichlorophenol, 86
Trifluorothymidine (TFT), 32, 98, 142
Trihalomethanes (THM), 311
Tryptophan biosynthesis, 97

TSCA (Toxic Substance Control
	Act), 221, 236
Tumor-specific glycoprotein (TSG),
	14
Tumor suppressor genes (TSG),
	14–17, 254
	p53 gene, 3
	polymorphisms, 254
	Rb gene, 14
Two-stage clonal expansion model,
	61, 266, 325, 354 (*see also*
	cancer risk assessment
	models)
	Initiation-promotion-progression,
	360

Ulcerative colitis, 278
Ultraviolet radiation, UVB 204
Uncertainty factor (UF), 300–301
Unit risk, 321
Unsaturated aledhydes, 250
Unscheduled DNA synthesis test
	(UDS), rat, 93, 94,
	102–103, 121–122, 202,
	236 (*see also* genetic
	toxicology tests)
	biomarker, 278

[Unscheduled DNA synthesis test
	(UDS), rat]
	rat unscheduled DNA synthesis
	test (UDS), 93, 102–103,
	121–122
	in vitro UDS assay, 102, 121–122
	in vivo-in vitro UDS assay, 103,
	122
Urethane, 334
Urinary bladder tumor, male rats, 54
USEPA Gene-Tox Program, 234
USNTP database of carcinogenicity,
	34

Vinyl chloride, 51–52, 248, 255, 331,
	334
Vinyl sulfones, 237

Weight-of-evidence, risk assessment,
	87, 240, 243, 305, 309
Wilms' tumor, 14, 16

Xeroderma pigmentosa (XP), 6, 8
	complementation groups, 8
XPRT (xanthine-guanine phosphori-
	bosyltransferase), 98–99